Geographies of Globalization

Second

D1245209

Geographies of Globalization second edition offers an animated and fully updated exposition of the geographical impacts of globalization and the contribution of human geography to studies and debates in this area. Energetic and engaging, this book:

- critically appraises the concept and processes of globalization from a geographical perspective;
- debates the historical evolution of globalized society;
- illustrates how the core principles of human geography – such as space and scale – lead to a better understanding of the phenomenon;
- analyses the interconnected economic, political and cultural geographies of globalization;
- examines the impact of global transformations 'on the ground' using examples from six continents;
- discusses the three global crises currently facing the world – inequality, the environment and unstable capitalism most recently manifested in the Great Recession;
- articulates a human geographical framework for progressive globalization and approaching solutions to the problems we face.

Boxed sections highlight key concepts and innovative work by geographers as well as topical and lively debates concerning current global trends. The book is also generously illustrated with a wide range of figures, photographs and maps. *Geographies of Globalization* is essential for geography undergraduates and postgraduates studying the phenomenon in both dedicated and related courses and highly recommended for those in allied disciplines such as Sociology, International Relations, Development Studies and Anthropology.

Warwick E. Murray is Professor of Human Geography and Development Studies at Victoria University of Wellington, New Zealand. He has held university positions in the UK and Fiji and has been a visiting professor at universities in Europe and South America. He is President of the Australasian Iberian and Latin American Studies Association. He has served as editor on a number of journals including *Asia Pacific Viewpoint* and *Journal of Rural Studies.*

John Overton is Professor of Development Studies and Human Geography at Victoria University of Wellington, New Zealand. He has held university positions at four other institutions including the Australian National University. He is past President of the New Zealand Geographical Society and former Director of the Commonwealth Geographical Bureau. He has served as an editor for a range of journals including *Asia Pacific Viewpoint.*

Routledge Contemporary Human Geography Series

Series Editors:

David Bell, Manchester Metropolitan University

Stephen Wynn Williams, Staffordshire University.

This series of texts offers stimulating introductions to the core subdisciplines of human geography. Building between 'traditional' approaches to subdisciplinary studies and contemporary treatments of these same issues, these concise introductions respond particularly to the new demands of modular courses. Uniformly designed, with a focus on student-friendly features, these books will form a coherent series which is up-to-date and reliable.

Existing Titles:

Cultural Geography
Mike Crang

Development Geography
Rupert Hodder

Political Geography
Mark Blacksell

Tourism Geography, 2nd edition
Stephen Williams

Urban Geography, 4th Edition
Tim Hall and Heather Barrett

Forthcoming:

Geographies of Globalization, 2nd Edition
Warwick Murray and John Overton

Praise for the Second edition

'Genuinely global, unlike much of its competition, the new edition of *Geographies of Globalization* continues to provide an extremely well-written, enjoyable, thoughtful and occasionally provocative analysis of globalization. Already highly successful, the book is well on the way to becoming a classic.'

Professor John Connell, School of Geosciences,
University of Sydney, Australia

'A first rate introduction to globalization encompassing economic, cultural, political and environmental processes and perspectives. This accessible and well-structured textbook combines a critical discussion of various theories of globalization with wide-ranging and up-to-date examples, presenting globalization as a dynamic and geographically unequal phenomenon that is central to understanding the modern world.'

Professor Michael Woods, Department of Geography and
Earth Sciences, Aberystwyth University, UK

'A comprehensive, up-to-date and eminently readable critical exploration of the idea that as globalization marches on, geography and its core principles matter more than ever for understanding the process, its challenges, and its impacts on places from the local to the global scale.'

Emeritus Professor Peter Daniels, Department of Geography,
University of Birmingham, UK

'Globalization became a buzzword in the 1990s. Two decades on and writing from the southwestern Pacific, Warwick Murray and John Overton provide an excellent review of the debates: looking backwards, forwards and beyond the polemics.'

Professor James D. Sidaway, Department of Geography,
National University of Singapore, Singapore

Reviews for first edition

'I am certain that *Geographies of Globalization* will make an excellent text for many geography courses that focus on globalization.'

Annals of the AAG

'The book is very well written, carrying the reader along with all the zest and enthusiasm that characterise a winner of one of the 2006 national tertiary teaching awards. Reading it often seems like being in a high-energy classroom.'

New Zealand Journal of Social Sciences

'A valid contribution to the globalisation literature as an introductory level or foundation text, combining key themes and empirical case studies with some key theoretical ideas. As part of the Routledge Contemporary Human Geography Series the book serves its role as a teaching aid, providing a concise introduction to the subject while is also amenable to delivery as, or as an accompaniment to, an undergraduate lecture course.'

Tim Vorley, University of Leicester

'*Geographies of Globalization* is written in a very clear, accessible and concise manner and is a book that offers students something of a route-map through the uncertainty, confusion and misunderstandings that surround this now widely debated phenomenon.'

New Zealand Geographer 2006

'This book will travel well beyond the discipline of geography and will be equally useful for students of a range of other social science disciplines.'

Marcus Power, Department of Geography, University of Durham

'*Geographies of Globalization* is a must-read. It offers reasons why geographers have been marginal to the wider globalization debates, an agenda for rectifying this issue, and a call to arms to geographers on how to make their discipline distinct and valuable to the debates.'

Canadian Geographer Review

Routledge Contemporary Human Geography

Geographies of Globalization

Second edition

Warwick E. Murray and
John Overton

Routledge
Taylor & Francis Group

LONDON AND NEW YORK

First edition published 2006
Second edition published 2015
by Routledge
2 Park Square, Milton Park, Abingdon, Oxon OX14 4RN

and by Routledge
711 Third Avenue, New York, NY 10017

*Routledge is an imprint of the Taylor & Francis Group, an
informa business*

© 2015 Warwick E. Murray and John Overton

British Library Cataloguing in Publication Data
A catalogue record for this book is available from the British
Library

Library of Congress Cataloging in Publication Data
Murray, Warwick E.
  Geographies of globalization / Warwick Murray, John
  Overton.—Second edition.
    pages cm.—(Routledge contemporary human
    geography series)
  1. Human geography—Economic aspects.   2. Human
  geography—Political aspects.   3. Globalization.
  I. Overton, John.   II. Title.
  GF50.M87 2014
  303.48'2—dc23   2014012613

ISBN: 978-0-415-56761-9 (hbk)
ISBN: 978-0-415-56762-6 (pbk)
ISBN: 978-0-203-86019-9 (ebk)

Typeset in Franklin Gothic
by Keystroke, Station Road, Codsall, Wolverhampton

MIX
Paper from
responsible sources
FSC
www.fsc.org   FSC® C013604

Printed and bound by CPI Group (UK) Ltd, Croydon, CR0 4YY

Contents

Plates

Figures

Tables

Cartoons

Maps

Boxes

Acknowledgements

Warwick thanked a lot of important people in the first edition, and all of those acknowledgments stand for this, the second edition. He would like to reiterate his enormous gratitude to his wife and daughters, Ximena, Gabriela and Francisca, the latter two of whom have recently begun to ask rhetorical questions such as: "Daddy, what is globalization? Please don't answer that it will take too long!" They are a reminder every day of the wonderful things that globalization can and does produce.

John wishes to thank Jane Overton for her endless patience in enduring the chaos that writing days at the Institute have wrought upon the household.

Both John and Warwick would like to say a special thank you to Andrew Mould of Routledge for his extra-special patience. They would also like to add a number of postgraduate students to the list for their stimulating presence and in many cases work on this edition: Colin Kennedy, Catherine Jones, Karly Christ, Monica Evans, Lorena de la Torre and Laura Barrett have been especially helpful. Jo Heitger produced the maps and helped with the illustrations and the word of the day. Many thanks to the friendly academic and administrative staff at SGEES, VUW, Wellington who create such a stimulating and pleasant place to work. Thanks to the Paekakariki Institute for Social Sciences for hospitality and inspiration, and providing a perfect place to study the drafts.

We would like to end with a song which, in the spirit of globalization, you can watch performed by Warwick on YouTube if you search for it. When the first edition was submitted, all the cases of 'globalization' were spelt 'globalisation' with an s – here in New Zealand we use the letter s not z for '-ise'. Warwick asked if the editors could change them back to include s not z and was told no as it might damage sales in the USA. He thought it captured some of the story of globalization nicely and wrote a satirical ditty reflecting on it!

Globalization Is Not Spelt with a Zed

Globalization is not spelt with a zed
I implore you to realise
We must defend our s's
Lest we Americanise

It's all part of a conspiratorial plan
To dictate everything we do
Not content with legal, military, and economic monopolies
They are controlling our language too

Come along and resist with me
Use Queen's English and decolonise your head
And join the anti-globalisation movement
But not the one not spelt with a zed

Warwick Murray and John Overton, Paekakariki Institute for
Social Sciences and Victoria University of Wellington, New Zealand

Abbreviations

AFTA	ASEAN Free Trade Area
AGM	Anti-globalization movement
ANCOM	Andean Community
APEC	Asia-Pacific Economic Cooperation
ARENA	Action, Research, Education Network, Aotearoa
ASEAN	Association of Southeast Asian Nations
AUSAID	Australian Agency for International Development
BMG	Bertelsmann Music Group
BRIC	Brazil, Russia, India and China
CAMRA	Campaign for Real Ale
CARCOM	Caribbean Common Market
CENTO	Central Treaty Organization
CET	Common external tariff
CJD	Creutzfeldt-Jakob disease
COMECON	Council for Mutual Economic Assistance
DIRECON	Office of Regional Trade Agreements, Chile
DMZ	Demilitarized Zone
EC	European Community
ECLA	Economic Commission for Latin America
ECSC	European Coal and Steel Community
EFTA	European Free Trade Area
EMI	Electric and Musical Industries Ltd
EPZ	Export processing zones
ETA	Euskadi Ta Askatasuna (Basque Fatherland and Liberty) – Basque Separatists
EU	European Union
EURATOM	European Atomic Energy Commission
EZLN	Ejercito Zapatista Liberacíon Nacional
FAO	Food and Agriculture Organization

FDI	Foreign direct investment
FLO	Fairtrade Labelling Organisations International
FTAA	Free Trade Area of the Americas
G3	USA, Japan and the EU
G7	Canada, France, Germany, Italy, Japan, United Kingdom, USA
G8	G7 + Russia
GATT	General Agreement on Tariffs and Trade
GBS	General Budget Support
GCC	Global commodity chain
GDP	Gross domestic product
GFC	Global Financial Crisis
GM	Genetic modification
GNP	Gross national product
GUUAM	Georgia, Ukraine, Uzbekistan, Azerbaijan and Moldova
HDI	Human Development Index
HIPC	Heavily indebted poor countries
HYV	High-yielding variety
IBRD	International Bank for Reconstruction and Development (World Bank)
IEA	International Energy Agency
ILO	International Labour Office
IMF	International Monetary Fund
INGO	International Non-governmental Organization
IPCC	International Panel on Climate Change
ISI	Import substitution industrialization
ITS	International Trade Secretariats
KFC	Kentucky Fried Chicken
LDC	Less developed country
LETS	Local Economic Trading Schemes
MAD	Mutually assured destruction
MDG	Millennium Development Goal
MDRI	Multilateral debt reduction strategy
MERCOSUR	Mercado Común del Sur (Common Market of the Southern Cone)
MIRAB	Migration, remittances, aid and bureaucracy
MNC	Multinational company
MTV	Music television
NAFTA	North American Free Trade Agreement
NATO	North Atlantic Treaty Organization
NGO	Non-governmental organization

NIC	Newly industrializing countries
NIDL	New international division of labour
NSM	New social movement
NTAX	Non-traditional agricultural exports
NZAID	New Zealand Agency for International Development
OECD	Organization for Economic Cooperation and Development
OPEC	Organization of Petroleum Exporting Countries
PGA	People's Global Action
PIC	Pacific Island country
PRSP	Poverty Reduction Strategy Papers
R&D	Research and development
RAMSI	Regional Assistance Mission to the Solomon Islands
REDD	Reducing emissions from deforestation and degradation
RIAA	Recording Industry Association of America
RTA	Regional Trade Agreement
SAP	Structural Adjustment Programme
SEATO	Southeast Asia Treaty Organization
SWAps	Sector Wide Approaches
TNC	Transnational corporation/company
TNI	Transnationality Index
TPP	Trans-Pacific Partnership
TRIAD	USA, Japan, EU
UN	United Nations
UNCTAD	United Nations Conference on Trade and Development
UNDP	United National Development Programme
UNEP	United Nations Environment Programme
UNESCO	United Nations Educational, Scientific and Cultural Organization
UNFCCC	United Nations Framework Convention on Climate Change
UNIFEM	UN Development Fund for Women
USSR	United Soviet Socialist Republics
WB	World Bank (International Bank for Reconstruction and Development)
WEF	World Economic Forum
WHO	World Health Organization
WMD	Weapons of mass destruction
WST	World Systems Theory
WTO	World Trade Organization

Part I

Globalization in three dimensions

Globalization and place – the death of geography?

- Globalization and the death of geography
- Protest against globalization – a global justice movement?
- Competing discourses of globalization
- Defining globalization
- Geography and the study of globalization
- New globalizing spaces?
- This book – 16 geographical questions about globalization

Globalization and the death of geography

As Warwick was sprinting to his inaugural lecture in 2012, entitled *Globalization and Geography in Crisis*, the Deputy Vice Chancellor of the university saw him racing to the theatre with an inflatable globe in his hand. The Deputy Vice Chancellor said, "you're right, it *is* a small world!" We live, if the hyperbole is to be believed, in a 'global village', or at least some of us do. The term that is most commonly used to refer to this apparent shrinking is 'globalization'.

In January 2009, Warwick sat in Chonchi on the island of Chiloe, a small isolated rural locality in the south of Chile, watching BBC *Mundo* whilst simultaneously listening to the internet broadcast on CNN of Barack Obama's inauguration. This was the installation of the first black President of the USA who had dual African and European heritage and was born in Polynesia; he couldn't help but feel in awe of the reach and intricate complexity of globalization. John felt the same in September 2010 in Cape Town, when he learned of the Canterbury earthquake in New Zealand, being able to watch reports on BBC television and listen to internet radio from Christchurch talkback stations. He was actually better informed about the damage than some relatives without electricity in Christchurch. These moments of a globalized imagination are surely becoming more and more common for many people. Many reading this book now would have personal examples with accompanying mental images that speak of the vastly interconnected and complex world in which we live (see Plate 1.1).

We live in a world that our grandparents, and perhaps even our parents, could never have imagined in their wildest dreams. For most, what has changed most dramatically in the world has been technology: cell phones, computers, digital cameras and the like. Yet perhaps equally startling have been changes in the way the world is structured and connected. For example, it is remarkable to note how communication has globalized over the last 25 years. This introductory chapter was written in part whilst sitting in the Duke Humphrey's library which is the oldest part (approaching 500 years) of the Bodleian Library at Oxford University – the oldest library in Europe and undisputedly a centre of learning for many centuries. When we compare the fixed nature of the books and artefacts in this library to the accessibility that is granted through search engines such as Google, online encyclopaedias such as Wikipedia and video-sharing technologies such as YouTube, we can only be astounded by how things have changed.

As such, globalization is tied up with the twentieth-century revolution in transport and communications technology that has captured the popular imagination. It is possible today, for example, for a member of the general public to circumnavigate the world in just over a day on commercial airlines. Sixty years ago, the air journey from England to Australia took about a week. In 1870, it would take 70 days for surface mail to travel from London to New Zealand. With the advent of the telephone and faxes, e-mail, and more recently internet-based social media such as Twitter, communication across the world, and even beyond, can be instant.

Increasingly used to rationalize a wide range of economic and political policies, and to explain a plethora of cultural, social and economic processes and outcomes, the concept of 'globalization' has assumed enormous power. Despite this, it is not always well defined or critically appraised in either popular or even academic usage. A common image of globalization is one of a process that unfolds like a blanket across the globe, homogenizing the world's economies, societies and cultures as it falls. Everywhere becomes the same, boundaries don't matter and distance disappears, so the argument goes. If communication is now instant, if where we live no longer matters, and if things are the same wherever we are, does globalization spell the end of geography?

Plate 1.1 *Dubai Mall – the epitome of high neoliberal globalization?*
Shopping malls are ubiquitous. Wherever they are in the world, they often contain similar chains of shops, they sell global brands and they are arguably the most common sites where people from all over the world connect with the global economy.

Source: Photo by authors

The end of geography?

People have been predicting the death of geography for nearly four decades on the basis of the trends discussed above. Toffler (1970) led this perspective in *Future Shock*, arguing that the evolution of transport and communications technologies and the intensified flows of people that have resulted imply that the principal source of diversity is no longer space. Twenty years ago, the subtitle of O'Brien's *Global Financial Integration* proclaimed 'the end of geography' – a situation involving 'a state of economic development where geographical location no longer matters' (O'Brien, 1992, p. 1) (see Cartoon 1.1) (see Dymski, 2009 for a strong rebuttal of this argument).

Cartoon 1.1   *Homogenized destinations*

Source: Louis Hellman

There can be little doubt that, relative to the past, people and processes in 'far-off' places can have instant 'local' impacts. Geographers and other social scientists have used a number of other terms for this process including 'annihilation of space by time', 'time–space convergence' and 'time–space compression' (see Map 1.1) (Harvey, 1989, 2007). Economic *contagion* became a new buzzword in financial sectors, following the Asian 'crisis' of 1997 which quickly spread around the Pacific Asia region from Thailand to South Korea and on to Japan. This term was used again to describe the alarming velocity at which the credit crunch in the USA in 2007, and the eventual Global Financial Crisis of 2008/9 which it precipitated, raced across the globe – the ramifications of the collapse had a rapid and tangible impact on economies all across the world to varying degrees.

But this 'shrinking' is by no means just economic in nature – it appears to be permeating all spheres of human activity and experience. Twenty-four-hour news providers and other broadcasters, such as CNN, Al Jazeera and the BBC, report unfolding political, social and cultural events – such as the death of Hugo Chávez in Venezuela in 2013, the London Olympics opening ceremony in 2012 or the devastating Japanese tsunami of 2011 – in real time to people's homes and offices across the planet. Also the news providers increasingly rely on text and footage in the reporting of such phenomena from participants on the ground during such events, as was witnessed with the capture and gruesome killing of former Libyan leader Muammar Gaddafi in 2011.

In the political and diplomatic sphere the impact of WikiLeaks, the website established by Julian Assange in 2006 to broadcast whistle-blower-generated leaks of sensitive government communications across the world, will resonate for many years to come and continues to reveal state secrets and classified interactions. Depending on which view one takes, this example of the globalization in the political sphere can be seen as highly liberating (some credit the spark for the Arab Spring as the revelation of corrupt activities through WikiLeaks, for example) or as tantamount to informational terrorism that has cost lives, as some members of the US government have argued in making the case for the arrest and prosecution of Assange and the whistle-blower Edward Snowden in 2013.

In the environmental sphere, carbon dioxide emissions in the USA or China have a direct impact on small island states as sea level rises and cyclone events become more common in localities such as Grenada in the

Caribbean or Niue in the Pacific. Feasibly, then, we could talk of increased cultural, social, environmental and political *contagion* given the new fluidity of global flows. It is clear that the way that many of us experience the world is shifting in sometimes dizzying ways and that a revolution in technologies of interaction has played a central role in this (see Box 1.1 and Plate 1.2).

The core aim of this book is to illustrate that as the processes of globalization fundamentally alter the way people, commodities and information flow and interact, this creates new and complex geographies. Popular notions of what globalization is often misunderstand what geography is, both as an entity and an academic discipline, and fail to

---

Box 1.1

Transportation and communication milestones

Date	Event
c. 60,000 BC	*Homo sapiens'* colonization of Australia implies the construction of seaworthy boats
c. 2900 BC	Egyptian hieroglyphic script fully developed
c. 2500 BC	Chariots in Mesopotamia, skis in Scandinavia
c. 1200 BC	Phoenician alphabet of 22 letters invented
c. 220 BC	Three-masted ships in Greece
c. 270	Magnetic compass invented in China
c. 650	Chinese scholars develop technique for printing texts from engraved wooden blocks
1403	Moveable metal type first used for printing in Korea
1522	Spanish ships returning from Magellan's voyage complete the first circumnavigation of the world
1807	First steamboat service begins operation on Hudson river, USA
1825	Stockton–Darlington railway, Britain, opens with locomotive designed by George Stephenson
1843	SS Great Britain is first propeller-driven ship to cross the Atlantic
1866	First transatlantic cable is successfully laid
1876	Telephone is invented by British scientist Alexander Graham Bell
1878	First electric railway is demonstrated in Germany by Werner von Siemens
1886	Four-wheeled petrol-powered automobile is built by German engineer Gottlieb Daimler

Box 1.1
continued

Date	Event
1901	Marconi transmits radio signals across the Atlantic
1903	Wright brothers make first sustained flight of powered aircraft
1908	Mass production of cars begins with Model T Ford in USA
1927	First public broadcast of television is transmitted in Britain
1941	First flight of jet-powered plane, with engine designed by British engineer Frank Whittle, takes place
1951	US engineers build UNIVAC I, the first commercial computer
1957	Soviet Union launches an artificial satellite – Sputnik I
1962	US communications satellite Telstar is launched
1964	Japanese railways begin running high-speed 'bullet' trains
1970	Boeing 747 enters service
1975	A personal computer (PC) in kit form goes on sale in the USA
1976	First supersonic flight – Concorde is launched
1981	The Space Shuttle comes into service
1984	A cell phone network is launched in Chicago, USA
1994	World Wide Web is created
1997	Mobile internet technologies emerge (WAP)
1998	Founding of Google – leading internet search site
2003	Founding of MySpace followed by other internet social media sites including Facebook (2004) and Twitter (2005) as well audio- and video-sharing internet websites led by YouTube (2005)
2004	The first private space flight on SpaceShipOne is launched
2007	Airbus A380 – the world's largest passenger aeroplane begins commercial service
2007	Google Street View is introduced
2012	Twitter's registered users globally passes 500 million

Source: Compiled by authors

appreciate how contemporary geographers define central components of their analysis such as space, place, scale and location. There are two notes of caution that need to be sounded about the 'shrinking world' concept immediately. First, *the relative distance between some places and people has become greater*. For example, the income gap between the poorest and richest countries and peoples has increased over the past 50 years (Potter *et al.*, 2008). Those who are hooked up to the internet may enjoy rapid communications with distant friends who can appear to be just around the corner, but the majority who do not have access to such technologies have become *relatively* more isolated. The fact that

Plate 1.2 *Different modes of transport in Tahiti*
Although transport methods have evolved over time, as illustrated in Box 1.1, it is not uncommon to see different modes from various ages existing side by side.
Source: Photo by the authors

'shrinking' technology can drive places apart is illustrated in Map 1.1, which shows the simultaneous time–space convergence and divergence of the world measured in terms of the cost of a minute-long phone call from the USA in 2000. In short, rather than smothering the Earth like a blanket, globalization has cast a net across it, and this has increased spatial differentiation accordingly.

A second point is that it is tempting to run away with the idea that technology itself is *driving* the processes of spatial shrinking and that therefore globalization is technologically determined. Underpinning technological change are social, cultural, economic and political processes and human agendas. Globalization processes are *facilitated* by technological change but are driven by much more fundamental forces. The study of globalization has political overtones precisely because it widens gaps, creates unevenness and creates winners and losers. This is why globalization can inspire fear and loathing among some while being

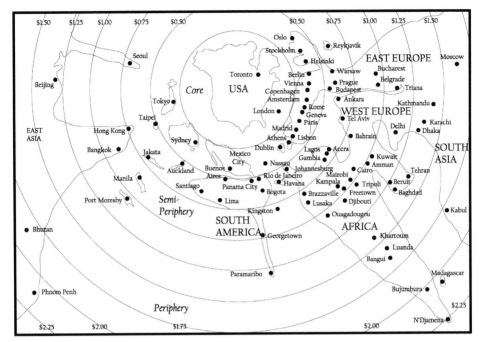

Map 1.1 *Time–space convergence and divergence*

Source: Adapted from Gwynne *et al.* (2003)

seen as a saviour by others. In short, globalization is about both processes and political-economic agendas.

These initial qualifications to the 'shrinking world' concept expose some of the weaknesses of 'geography is dead' arguments. Contemporary research in human geography and some other social sciences points towards the differentiating impacts of 'global' processes as they interact with 'local' places, institutions and people. It is indeed true that the world's economies and cultures are increasingly interconnected, and that 'global' forces are penetrating even the most remote and peripheral regions and localities on Earth. However, as such processes are articulated and resisted in *specific* places, complete with particular histories, societies and environments, they create more uneven geographies. These geographies of globalization, although increasingly dynamic, sometimes transient and therefore harder to grasp, are none the less significant and have real social, economic and political implications. Underpinning this book is the proposition that understanding globalization, and attempting to regulate and reform it, requires that geography be taken more, not less, seriously.

Protest against globalization – a global justice movement?

Globalization has been a site of enormous struggle over recent years. The latest wave of protests was manifested in the Occupy movement of 2011 which began in New York, USA on Wall Street and led to significant events in over 40 countries. The Occupy movement was precipitated in part by the consequences of and, in particular, the responses to the Global Financial Crisis. In particular it reacted to the perceived injustice of the various bailouts and fiscal stimulus packages which favoured those that had benefitted from the model of globalization that had crashed from 2007 onwards, namely large corporations and especially the financial capital sector. The movement was a large conglomerate of people from all walks of life including students, organized labour environmentalists, feminists and many other groups. In many ways, the relatively short-lived but spectacular Occupy movement can be seen as the latest incarnation of the anti- and alter-globalization movement (see extensive discussion in Chapter 5), targeted most specifically against neoliberalism. The Occupy movement itself came at the end of a particularly turbulent set of years following the financial crisis which was to varying degrees and in different ways responsible for the precipitation of major global events worldwide as diverse as the Iranian protests (2009), the Arab Spring (2010), the United Kingdom riots (2011), the Brazilian World Cup marches (2013) and the Chilean Winter (2010/11) (see Box 1.2 and Plate 1.3). It seemed like popular protest peaked in around 2011, and much of the anger vented in different ways and for different reasons was targeted broadly at globalization as currently practised – seen by many as responsible for the highly regressive impacts of the financial crisis. However, the crisis should not be seen as the initiation of discontent; prior to this, the food crisis and a third oil price crisis among other things provoked vast outpourings of dissatisfaction. Indeed, if these protests are broadly grouped together as reactions to the model of globalization as currently practised, their roots are intertwined inextricably and go back a long way.

Images of 'anti-globalization' protests are not new and have been flashed across our television screens for many years, beginning with Seattle in 1999 (Cumbers *et al.*, 2008). Predictably, mainstream media have often portrayed such protests as violent and anarchistic. There can be no doubt that huge passions have been stirred, but in general we have witnessed the evolution of a peaceful global protest movement (see Plate 1.3). Although this is complex, diverse and incoherent in many respects (see Chapter 5), there can be no doubt that the movement has fulfilled a major purpose in bringing the issue of globalization and its perceived regressive impacts to the attention of the global public (McFarlane, 2009). One of the great ironies, as is explored in Chapter 5, is that the movement has used the technologies of globalization, notably the internet, to spread its message. Based on this, and given that some quarters of the movement are not arguing for a rejection of globalization *per se*, it has been more common over the recent past to refer to it as the 'alter-globalization' movement. The movement has attracted some heavyweight intellectual supporters, such as Noam Chomsky (2012) and David Harvey (2011) who are both vocal supporters of the Occupy movement and its ancestors. But it has also attracted some less likely ones like former World Bank chief economist and economics Nobel Prize winner Joseph Stiglitz, who wrote prophetically in 2002:

> [f]ormerly uneventful meetings of obscure technocrats discussing mundane subjects such as concessional loans and trade quotas have now become the scene of raging street battles and huge demonstrations. The protests at the Seattle meeting of the World Trade Organisation in 1999 were a shock. Since then, the movement has grown stronger and the fury has spread. Virtually every major meeting of the international Monetary Fund, the World Bank, and the World Trade Organisation is now the scene of conflict and turmoil. The death of a protestor in Genoa in 2001 was just the beginning of what may be many more casualties in the war against globalization. . . . It is clear to almost everybody that something has gone horribly wrong.
>
> (Stiglitz, 2002, pp. 3–4)

Yet 'anti-globalization' did not start in Seattle 1999, as the mainstream media might have us believe. There have been similarly motivated protests and battles across the world in response to corporate globalization's greatest facilitator, neoliberalism (see Chapter 4), for decades. More generally, resistance to capitalism has a long and turbulent history over two centuries long. Neither are protests always as spectacular as those that have made the recent headlines; daily resistance to globalization and neoliberalism continues perpetually in the lives of many marginalized in both the rich and poor worlds. There can be little doubt, however, that the movement is unprecedented in terms of its breadth and reach. To what extent globalization was responsible for the ongoing financial crisis is a matter of debate. Some might argue we are entering a new wave of globalization based on the outcomes and responses. Others might suggest that we are entering the final phase of capitalism which is in its death throes. There is no agreement on these issues as we will see later in this book, but we must agree that globalization has become the most contentious concept of our time.

What has all of this got to do with time–space compression and the popular notions associated with globalization as a process? In general, the 'anti-globalization movement' has a particular standpoint in terms of what it believes globalization is – what might be termed 'corporate/ neoliberal globalization'. The movement's supporters, to the extent that it is possible to conflate their opinions into one idea, believe that there is a powerful state-sponsored corporate agenda at work which seeks to spread free-market capitalism and that 'shrinking' processes provide a 'spatial fix' which helps achieve that goal (Harvey, 1989, 2011). But this is only one way of conceiving of the phenomenon – the globalization debate is one of competing discourses about the nature of geographical interactions.

Competing discourses of globalization

The reaction of the so-called anti-globalization movement illustrates the contested nature of globalization – it is perceived and represented in many ways, none of which are necessarily 'correct' (Jones, 2010; Rupert, 2000; Schirato and Webb, 2003; Ritzer, 2011; Scholte, 2000). Before attempting more formal definitions it is useful to try to boil down some of the broad contours of the debate, which are then taken up in depth through the remainder of this book. This debate is not just abstract and has real political ramifications. Although discourses overlap (see Chapter 2), it is useful to divide normative perspectives into three camps.

Pro-globalization

This perspective holds that the spread of market economics, competition, free trade and Western democracy – that is to say globalization as currently practised – are important *progressive* trends. Capitalism is seen as a moral good which promotes economic growth and efficiency and leads to global welfare gains overall. This is the type of view that is

Box 1.2

Embedded in the Chilean Winter

The Chilean Winter is the term used to refer to the wave of very large protests led by Chilean university students in 2010/11. The target of the protests was the perceived injustice of the contemporary educational system in Chile which, whilst available to all, was of variable quality, with a particular gulf in quality between state and private institutions at all levels. In particular the rise in tuition fees, perceived falling quality and lack of investment in state-run universities were the main targets of protest. From the 1980s onwards, although the number of places available for tertiary education in Chile had increased dramatically, these were almost exclusively in private institutions that had been established under the neoliberal dictatorship of Pinochet (1973–1989). At the core of the complaints was the idea that education had remained available to those who were able to afford it and this perpetuated the already deep inequalities in Chilean society. Anecdotal evidence suggests that those participating saw their protest as part of a broader global movement opposed to social injustice and inequality. In an immediate sense however the uprising was rooted in the protests of secondary school students beginning in 2006 against the perceived slowness of reform under the centre-left President Michelle Bachelet. It rapidly evolved into a much broader popular movement against inequality which the movement argued was inherent to the neoliberal Chilean political economy. This manifested itself in geographical, gender-based and ethnic-based differentiation despite decades of rapid economic growth which had failed to trickle down. It is no coincidence that the movement peaked just following the election of the centre-right President Piñera who was set to continue the neoliberal policies established in the 1970s. The movement also rose at the peak of the negative flow-on effects of the recession precipitated by the financial crisis, which were less obvious in Chile due to vast copper wealth but nevertheless impacted the poor more profoundly than any other group.

Led by a new generation of charismatic student leaders – most notably Camila Vallejo who, for her role, was voted *The Guardian*'s as well as *Time Magazine*'s person of the year in 2011 – the movement became very popular, and at times approval ratings among the public were above 90 per cent. Protests, including traditional marches together with less orthodox 'happenings' such as the world's largest 'kiss-in' for peace, became national events and often attracted hundreds of thousands of participants. In August 2011, a two-day general strike was called in solidarity with the students. With renewed growth in Chile and falling unemployment in 2013, the movement is less widely supported; but the legacy of this latest wave of anti-neoliberal protests in Chile will be felt in national politics – formal and informal – for many years to come as the issues raised and new

Box 1.2
continued

Plate 1.3 *The Chilean Winter*

Source: Photo by Colin Kennedy

leaders and groupings precipitated are now firmly implanted in the public imagination. Colin Kennedy is a geographer who investigated the movement for a PhD in 2011, and he spent some time embedded there. He concluded, on the basis of interviews with leaders and other interested parties, that the Chilean Winter was the outcome of a much longer system of educational inequality in Latin America perpetuated by neoliberalism and solvable only by a partial return to Latin American theories of development that seek to reform globalization from within (see Box 8.3 in Chapter 8). (See Kennedy, 2013; Kennedy and Murray, 2012.)

associated with the IMF and its sister institutions, the World Bank and the WTO. In general, most national governments are of this opinion at the present time, especially those that are most powerful in the global capitalist economy such as the USA and the UK. The right-wing media also supports and disseminates this view (Herman and Chomsky, 1988; Chomsky, 2012). Pro-globalists often conceive of globalization as a continuation of the logic of modernization and enlightment progress, and as something that is an inevitable part of human evolution. Underlying this view is usually the idea that technology is the driving force of progress. The political ramifications are that 'there is no alternative' and citizens and nation-states will have to learn to participate for better or for worse. This view is closely associated with the *hyperglobalist* take on globalization (see Chapter 2).

Anti-globalization

As explored briefly above, this perspective holds that globalization is a threat to local society and environment in a way that echoes colonialism of the past. Some argue that those who bear the heaviest load are already marginalized groups, especially in the poor world. Under this view, globalization is seen as increasing – not decreasing – the unevenness of development and perpetuating inequality in ways that are not necessarily reversible. Anti-globalists argue that pro-globalists make globalization *seem* inevitable because it is in the interests of those who promote it – the business class, industrialists and rich countries. The political ramifications tend towards the revolutionary and against the institutions and corporations of global capitalism (Korten, 1995; Ritzer, 2011).

Alter-globalization

This view posits that the nature of globalization is not predetermined and does not follow a given evolutionary path. Rather, it is a consequence of human actions and particular political choices. The political implications of this view are reformist: citizens and nation-states have a role to play in resisting and regulating it, and alternative and progressive globalizations are possible (see Chapter 9). As such, globalization can yield both positive and negative outcomes depending on how it is constructed. The major task then is changing the *nature* of globalization through human action – but not destroying it. This view is associated to a large degree with the *transformationalist* school of thought (see Chapter 2), and the majority of the work undertaken in human geography to date falls into this camp – although it has tended to be a little more radical than the weak or passive transformationalist view, as we will see.

Defining globalization

Given these multiple and overlapping discourses, how might globalization be defined? Not surprisingly, no single definition exists (see Bisley, 2007 for a critical review). Daniels *et al.* (2012, p. 512) highlight the controversy over use of the term in 'defining' it in the following way:

> A contested term relating to the transformation of spatial relations that involves a change in the relationship between space, economy and society.

Sparke (2009b, p. 308) echoes the breadth of perspectives from a range of viewpoints arguing that it is a:

> big buzzword in political speech and ubiquitous analytical category in academic debate, globalization today operates rather like modernisation did in the mid-twentieth century as the key term of a master discourse about the general state of the world.

Illustrating the competing views, the following two early definitions represent a hyperglobalist view and a sceptical view respectively (see Chapter 2 for a full discussion of these terms):

> Information, capital, and innovation flow all over the world at top speed, enabled by technology and fuelled by consumers' desires for access to the best and least expensive products.
>
> (Ohmae, 1995, in Dicken, 1998, p. 4)

> Globalization seems to be as much an overstatement as it is an ideology and an analytical concept.
>
> (Ruigrok and van Tulder, 1995, p. 22)

In a leading early book on globalization, *Global Transformations*, Held *et al.* (1999, p. 16) emphasize a transformationalist argument:

> [Globalization is] a process (or set of processes) which embodies a transformation in the spatial organisation of social relations and transactions – assessed in terms of their extensity, intensity, velocity and impact – generating transcontinental or interregional flows and networks of activity, interaction and the exercise of power.

In seeking a definition, it is useful to consider what a completely globalized world would look like: a single society, a homogenized global culture, a single global economy and no nation-states? That is clearly not what we have. We are therefore living in a *globalizing* world – where processes do appear to be taking us closer to the above situation. Building on the work of Robertson which highlights the development of a 'global consciousness' (1992, p. 7; see also 2003), Waters (2001, p. 5) offers a sociological definition:

> A process in which the constraints of geography on social and cultural arrangements recede, in which people become increasingly aware that they are receding.

Human geographical work has tended to build upon transformationalist and sociological work. A definition reflecting this perspective is found in Cloke *et al.*'s *Introducing Human Geographies* (2005, p. 36; see also 2013):

> The economic, political, social and cultural processes whereby: a) places across the globe are increasingly interconnected; b) social relations and economic transactions increasingly occur at the intercontinental scale; and c) the globe itself comes to be a recognizable geographic entity. As such, globalization does not mean that everywhere in the world becomes the same. Nor is it an entirely even process; different places are differently connected into the world and view that world from different perspectives. Globalization has been occurring for several hundred years, but in the contemporary world that scale and extent of social, political and economic interpenetration appears to be qualitatively different to international networks in the past.

There are, of course, many more definitions, and we will come across some of them later in the book, one of the most important being Harvey's (1989, 2011) concept of time–space compression (see Chapter 2).

We have attempted to summarize our opinion on what globalization is below and to catch the spirit of a geographical perspective. There are two versions – a long one and a short one. The short one reads:

> Globalization is a collection of dialectical human-agency-driven processes which create local–local and person–person networks of inclusion/ compression that increasingly transcend territorial/national borders and stretch to become global in proportion.

For the long version, take the short one and add the following:

> The processes are dialectical in that the relative social distance of those not on the net from those on the net widens as the intensity

and extensity of the processes increase. Thus globalization simultaneously creates spaces of exclusion/marginalization leading to an increase in social, economic, political and cultural unevenness across space. For those on the net however, globalizing technologies – in transport and communications – give the impression that the world is one system that is becoming smaller. The rise of such processes is intimately tied up with the rise and expansion of capitalism and thus goes back to the first 'global' empires. Given the endless capitalist imperative of increasing profits, new technologies have been developed which have greatly increased the velocity of capital, information and ideas, especially over the past three decades. Individuals, social groups, governments and transnational bodies both respond to and create the processes of globalization in their struggles to create and protect viable and sustainable livelihoods and identities, and supranational organizations have been developed to attempt to regulate and direct such flows. From a political viewpoint, globalization has often been recast as an agenda or a discourse where the self-interest of certain groups leads them to accord normative or moral status to the processes.

You must arrive at your own definition – and students are consistently creative on the subject of what globalization means (see Box 1.3 and Map 1.2). The above definition, however, is the one that is assumed for the remainder of this book, and many of the terms within it will be elaborated upon. In short, this text takes the explicit view that globalization is real, inherently geographical in its nature, and creates unevenness and inequality.

Geography and the study of globalization

The term 'globalization' was first used in the mid-1960s and entered into popular usage in the mid-1990s. Few deny there is something going on 'out there' which is radically restructuring the world. There is, as already noted, disagreement on what drives it, how it should be theorized and what impacts it is having. Theoretical and empirical research on these issues has generated a vast and growing literature in the social sciences. The International Bibliography of the Social Sciences (IBSS) records articles published in the most prestigious academic journals. The first article with the term 'globali*ation' (that is, spelled with either an s or a z) in either the title or the keywords appeared in 1983. In 1990 only 15 articles included the term. By 1998 this figure had risen to over 1,000, in 2003 it stood at 2,909, and it peaked in 2006 at 3,634. In 2012 there

Box 1.3

Student definitions of globalization

These definitions are reproduced from Victoria University, New Zealand, from class competitions which ran through the 2000s for third-year students on a human geography course – before any substantive lectures took place. Students were given ten minutes to write their definitions. Note how many students tend to define globalization in terms of 'loss of diversity', often with a political motive behind this:

'Globalization is seen, both positively and negatively, as the increased connectiveness of peoples/cultures. The primary issue is who is globalization for? And, who has the power at present?'

'The spread of ideas, theories and practices through the realms of commerce, culture, society. The ideas that are spread are those of the dominant cultures and societies and often result in the loss of diversity and imposition of ideas, values, practices, etc. upon others, not always best suited to those they are imposed upon.'

'The process of making any given part of the world indistinguishable from any other part.'

'Countries linked through world systems (e.g. economically, politically, etc.) yet incorporating these into their pre-existing cultures.'

'Globalization is the increased communication, transportation of goods, services, people and ideas around the world.'

'People need people, and people need adventure. These two ideas have led us to where we are in the world. This is the phenomenon known as globalization.'

'Globalization is everything interconnected working to make the rich richer and move the poor further into the depths of despair and deprivation. Globalization is a frenzy of mass media and booming industry. It is labour at its cheapest and fashion at a high price. It is stripping the Earth of its natural resources for every last drop of Mother Nature's juices.'

'The sharing of new technologies with the aim of making life easy and "better".'

'Is the gradual disappearance of the illusion of space created by physical distance, through various technologies created by the human race.'

'Globalization is the fact that words such as "globalization" with a z are used in New Zealand when the correct spelling is globalisation with an "s".'

'McDonald's and Starbucks.'

'I believe that globalization is the next stage in the growth of humanity that has been brought about through both modernization and curiosity.'

'Globalization is the gradual awareness of the human race that its existence is intrinsically interconnected to even the tiniest particle in the universe and our lives

Box 1.3
continued

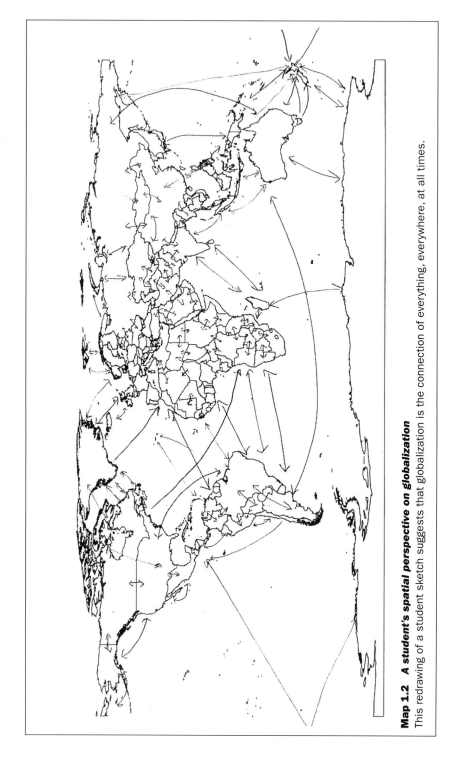

Map 1.2 A student's spatial perspective on globalization

This redrawing of a student sketch suggests that globalization is the connection of everything, everywhere, at all times.

Box 1.3
continued

and actions are completely dependent on everything that surrounds us.'

'I think globalization is a very complex idea that affects people differently, which creates different perceptions of globalization and its

advantages and disadvantages – this makes it hard to define.'

'A constant and unstoppable process of gradual change which we must own and try to influence, not just be influenced by.'

were 1,621 articles with this as a keyword listed in the IBSS. In the 20 years from 1983, 12,859 articles dealing substantively with globalization issues were listed in the IBSS; and, to date, there have been 37,505 overall. Since the middle of the 2000s this number has dropped and, by 2012, stood at just over 1,500. This halving in six years most probably represents dissatisfaction in academia with the term which had become a catch-all category, rather like 'sustainability' had in the 1990s. There was a rapid rise in articles using 'global*ation' and 'geography' as key words together in the second half of the 1990s, rising from 1 in 1993 to 72 in 2000. In 2002, 152 articles of this nature were listed and the number peaked in 2006 at 161. There has been a drop over recent years, however, falling to 64 in 2012. This measure gives us only a feel for the changing importance of the topic as many articles deal with issues surrounding globalization less explicitly. Increasingly, since the middle of the 2000s, articles that in the past may have been considered as being about globalization have not explicitly used this term in either the title or the keyword. Arguably the move away from use of globalization as a term especially is marked in geography, with its unique and frustrating habit of moving debates beyond terms that are too quickly perceived to be passé (Dicken, 2004; Murray, 2009).

Although much of human geography is intimately tied to globalization debates and the contribution of geographers to those debates rose in absolute terms until the mid-2000s, geography books and articles on the subject are not referred to a great deal by academics in other disciplines. Some authors – such as Harvey (1989, 1995, 2000, 2003, 2011); Sassen (2000, 2001, 2011); Amin and Thrift (2004); Knox and Taylor (1995); Taylor and Flint (2000); Jones (2010); Johnston *et al.* (2002); Massey (1984, 1995); LeHeron (1993); Cox (1997); Corbridge *et al.* (1994); Knox and Agnew (1998); Peet (1991); Smith (1984, 2003); Daniels and Lever (1996); Dicken (2003, 2011); Gwynne *et al.* (2003); Jones (2010); Sparke (2013) and Swyngedouw (1997) to mention a few – have made significant interventions, some of which will be explored later in this book. In general, however, geographers' work on globalization is influenced by research from beyond the discipline far more than the other way around (see Dicken, 2004).

So what does geography have to offer? What is intriguing for those who study globalization, and possibly confusing for those who are new to it, is that there appear to be 'globalizing' and 'localizing' forces simultaneously at work in politics, culture and the economy. For example, in Europe we have witnessed, over the past 50 years, the evolution of an integration movement culminating in a political and economic union of 27 countries as of mid 2014. In 2002, 12 of these countries adopted a common currency, the Euro. And yet, at the same time, Tony Blair's Labour government devolved power to the Scottish Parliament, the Welsh Assembly and, with less success, the Northern Irish Assembly. In 2014 there will be a referendum on the issue of Scottish independence as a result of this. Similar political centralizing and decentralizing forces are at work in Spain where Catalonian and, to a lesser extent, Basque independence are actively discussed and quite feasible in the near future. In the economic sphere, we have witnessed the evolution of a globalized financial sector, coordinated through a number of world cities across the planet such as London, New York, Tokyo, Singapore (see Plate 1.4) and Dubai (see Plate 4.2). Through the operation of this sector, economic change in one region can be instantly transmitted to another through currency and stock markets; for example, as was devastatingly illustrated through the workings of the Global Financial Crisis (GFC) (see Chapter 7 for a critical analysis of this and the terminology used to describe it). Simultaneously, in countries such as Australia and the UK, Local Economic Trading Schemes (LETS) have evolved which bypass these circuits of capital entirely and establish non-monetary exchange. Finally, in the cultural sphere, wherever you go in the world and tune into commercial radio, it is likely at some point that you will hear a song by One Direction, Lady Gaga, Coldplay or Adele. However, as global music 'brands' have risen to prominence on the airwaves, a 'world' music scene has evolved which markets 'local' non-Western musical styles to global markets through retail outlets and concerts such as the successful WOMAD series in Western countries (see the global music industry case study in Chapter 6). As subsequent chapters will show, it is too simplistic talk of global and local processes in all spheres. Using the music example above, there has been an active attempt by many groups to incorporate elements of the global and local in their music: witness the flamenco, funk and hip hop fusion of *Ojos de Brujo* from Barcelona, for example.

Plate 1.4 *Singapore – a unique place in the financial space of flows*
Source: Photo by the authors

As we will argue, the global and the local exist on a continuum of scale and their interaction is constant, dynamic and produces hybridized outcomes that are neither global nor local in essence – they are 'glocalized' (this is explored in greater detail in this book). What is crucial to understand from a geographical point of view, however, is which of these is dominant – the global tendency or the local tendency – in any particular political, cultural or economic pattern or phenomenon.

How do we make sense of these apparently oppositional processes at different scales? Human geography offers a distinct framework for understanding the complexities of globalization in at least four ways:

1 *The discipline is inherently concerned with space and, in particular, the interaction of processes, structures and agents at different scales of analysis.* The debate surrounding the relative importance of structure and agency in human geography has raged for many years. Most geographers now concede that – although the relative power of each varies across time and space, and this power is contingent on a range of political, economic and environmental factors – structure and agency play a mutually determining role in the local spatial outcomes of broad processes. Furthermore, as already noted, the 'global' is constituted by 'local' – everything, ultimately, has a local expression, even if it is stretched to become 'global'. For example, in the cultural example above, it is wrong to assume that Coldplay produces 'global' music. Their music combines particular elements emanating from specific places (psychedelic rock, jangle rock, folk and elements of

mainstream pop principally), and this hybrid is marketed through networks stretching across the globe that link together particular sites of consumption. This conception avoids the determinism of globalist economic discourses and escapes the idealized localism of some anthropological viewpoints.

2 Linked to the first point, *human geography 'peoples' discourses of globalization*. Globalization is constructed from neither 'above' nor 'below'; rather it is constantly refashioned by interaction between people and institutions on these various scales. Although there is no doubt that some are more powerful than others in shaping the processes, it is nevertheless driven by humans. For example, the financial sector is not some kind of floating global force; it comprises individual decision-makers located at particular nodes in the network. Through the problematization of scale as an objective entity, some human geographies have warned of the futility of attempting to separate the 'global' and the 'local'. These forces are not oppositional; they are part of the same *dialectical process*.

3 *Human geography is inherently eclectic*. Geographers deal with economic, cultural, political and environmental processes and the spatial impacts of these for, and due to, the actions of people. Globalization involves change in each of these spheres, which is one reason why it has proven so difficult for academics in their traditional disciplinary roles to understand it. For example, the evolution of the European Union is much more than an economic union. Its roots are political and cultural and its impacts wide-ranging. Human geography is in a strong position to reflect upon such change.

4 More recently human geography, in some sub-disciplinary areas at least, has stepped away from goals of objective scientific reasoning towards partial accounts. *The rise of critical and moral geographies* are important as they provide academic ground for the discussion of the 'rights' and 'wrongs' of globalization, which are crucial given the important trends that are at stake in the debate.

New globalizing spaces?

To conclude this introductory chapter, we return to the death knells sounded for geography in some quarters. The two views on this important debate are outlined below. The essence of the argument in this book is that new globalized spaces are being forged which are changing the nature of geography without eroding its relevance.

Geography is dead

The argument that geography is dead, or at least less relevant today, put forward by some hyperglobalist economists (see Chapter 2) and other non-geographers may be characterized as follows. Time–space compression renders distance unimportant. Locality has less meaning since we live in a 'global village'. Difference is declining and culture homogenizing. Thus, for example, under this argument transnational corporations (TNCs) are completely 'footloose' and spread identical products across the planet. In the cultural sphere, global brands such as Nike or Taylor Swift or the BBC send out identical messages which are interpreted in the same way across world society. This idea sees globalization as deterministic and inevitable, subject to its own inexorable logic. Territories are made less salient by the transcendence of process across and above them.

Geography is reborn

In contrast, the more realistic argument that geography has new life through globalization might be as follows. Distance, as measured *in an absolute sense*, is indeed less important; but place, space, locality and the *relative* distance between these things are not. 'Global' processes are actually stretched 'local to local' processes, and they unfold in localities that have a unique history and character. In the economic sphere, TNCs choose to locate in particular places due to a complex mixture of local characteristics in the destination area and factors specific to the firm that are influenced by the nature of their source location. In culture, hybridity becomes the new normality as 'global' trends mix with local ones to create new cultures; for example, Nuyorican salsa or Polynesian rock. Geography is indeed transforming, both as a concrete reality and as a discipline that seeks to analyse those changes, but it is not irrelevant. This new, highly volatile unevenness makes geography – 'the study of spatial differentiation' – more important than ever as the world fragments and recrystallizes along new political, economic, social and cultural lines. It is especially important to understand this as the inequalities that result from the juxtaposition of globalizing process and black holes in the net are increasingly obvious (see Plate 1.5).

Plate 1.5 *Black holes in the net*
A highway passes over a deprived community in Manila, Philippines
Source: Photo by Donovan Storey

Altered spaces?

It is in the area of technological change, especially the rise of the internet, where the death of geography is often most readily proclaimed. *The Economist* (15 March 2003) ran an article entitled the 'Revenge of geography' where it was argued that: 'it was naïve to imagine that the global reach of the internet would make geography irrelevant. Wireline and wireless technologies have bound the virtual and physical worlds closer than ever' (p. 19). The article goes on to explain that the process of using such technology is inherently bound by place and 'real' or territorial space. In particular, it is claimed:

> Actually, geography is far from dead. Although it is often helpful to think of the internet as a parallel universe, or an omnipresent cloud, its users live in the real world where limitations of geography still apply. And these limitations extend online. Finding information relevant to a

particular place, or the location associated with a specific piece of information, is not always easy. This has caused a surge of innovation, as new technologies have developed to link places on the internet with places in the real world – stitching together the supposedly separate virtual and physical worlds.

This argument is echoed in the groundbreaking work of geographers such as Kitchin and Dodge (2002) in their chapter 'The emerging geographies of cyberspace', which argues that cyberspace exhibits uneven and real geographies where 'online interactions are not divorced from those offline, but rather are contexted by them' (p. 353). It is important to bear such arguments in mind when the globalization of social media, such as Facebook and Twitter, is referred to, for example. The input into what are local noticeboards with global distribution networks are created and consumed in specific locations and are thus linked reflexively with local geographies. This was painfully revealed during the financial crisis as the notion that the financial sector 'floats' above real places was laid bare as thousands of localities, especially already relatively impoverished ones, battled unemployment, increasing poverty and political instability between 2007 and 2012.

The concept of space is important here; those who believe that geography is rendered less important have misinterpreted the discipline and some of its core elements. Crucially confusion exists concerning the difference between *geographic* space and *absolute* space. Geography is seen as something innate, usually represented on a topographic map upon which processes unfold. This is, of course, a central component of geographical analysis, but it is not the whole story. A wide body of literature in geography has argued that space has many meanings that are transformed by contemporary globalization (see Chapter 2). Geographers need to work extra hard to keep up with the challenges of these new spaces and to convince others of their importance.

This book – 16 geographical questions about globalization

Essentially this book is about exploring the various concepts, theories, processes and impacts of globalization from a geographical perspective. It seeks to reflect critically on the current and potential role of geography for understanding globalization more effectively. We are going to have to carefully define concepts such as 'global' and 'local' and to think deeply about the implications of increased interaction within and between these geographic scales. We must make two important points here. First, the study of globalization is inherently multidisciplinary; there is geography in the work of non-geographic specialists and vice versa. Second, the literature that has been generated concerning globalization is overwhelming; it would take a life's work to summarize the main works and their implications, and we have only ten brief chapters to survey some of the main contours here. Some important issues are left out, although readers are pointed to where they can find further information. The purpose of this book is to highlight the most important ideas concerning the processes and agendas of globalization and to provide signposts to further study.

In order to achieve the above objectives, the book is separated into three sections that flow from the conceptual to the concrete, the general to the specific and the global to the local. Each of these sections contains individual chapters that expand upon specific aspects of the organizing themes. *Part I: Globalization in Three Dimensions*, which includes this chapter, introduces and critically appraises the concept of globalization. It then goes on to discuss various 'theories' and perspectives (Chapter 2); drawing out the spatial implications of the various viewpoints and placing them in historical context (Chapter 3). *Part II: Globalization in Three Spheres* breaks down and analyses the process and outcomes of globalization in overlapping spheres in three chapters – economic (Chapter 4), political (Chapter 5) and cultural (Chapter 6), emphasizing the links and breaks between change in these various areas. Case studies and examples from a wide range of countries are inserted throughout the text, sometimes in boxed form. In setting out these examples, an explicit attempt has been made to draw on materials and research from across the globe – something to which a coherent geography of globalization should aspire and something which many human geography books in the UK or the USA do not always achieve. *Part III: Globalization and Three Crises* deals with three crises of special concern to human geographers with regard to studies of globalization – the unstable and uneven nature of neoliberal globalization (Chapter 7); development and inequality (Chapter 8) and the environment (Chapter 9). The argument is made that, in spite of the overwhelming public concern with the latter in recent years, each are equally important and interwoven in ways that makes solving one contingent upon solving the other two. Ultimately these three crises are interlinked because they are derived from and manifest the model of globalization that has been practised and diffused across the world.

In the final chapter (10) the theoretical, historical and empirical analyses are brought together in order to reflect on the changing nature of globalization and the implications of this for both geographies on the ground and the way they are studied and researched. Particular attention is paid to the possibility of transformative *progressive* globalization (as opposed to 'globalization as currently practised') and the role geography might play in helping to forge this.

There are 16 questions on globalization relevant to geographers which the book sets out to address, and these are returned to throughout and in the concluding chapter explicitly:

1 How can globalization be defined?
2 Is globalization new?
3 What drives globalization, now and in the past?
4 Is contemporary globalization different from the globalization of the past?
5 Is it homogenizing global society?
6 Is globalization a process or an agenda?
7 How does globalization alter our concepts of space, place and scale?
8 Is globalization the same thing as internationalization?
9 Is globalization eroding the power of the nation-state?
10 Is globalization bad for the environment?
11 Is globalization bad for the 'poor world'?
12 Does globalization create more winners than losers?
13 What has been the impact of the Global Financial Crisis on globalization?
14 Can globalization be reformed?
15 If so, how should it be reformed?
16 What role can geography and geographers play in forging an alternative progressive globalization?

Running throughout the book, then, is an explicit engagement with the often-assumed deterministic power of globalization. Centrally, it is argued that human geographies of difference remain as important as they have ever been. This general conclusion, illustrated through appropriate examples, is likely to be of relevance to human geography students who never cease to be amazed by the increasing complexity and heterogeneity of their own localities despite dominant discourses of homogenization.

Further reading

Dicken (2004): This article suggests that geographers are missing the boat in terms of studying globalization and sets a convincing agenda for how geographers might make a difference.

Held *et al.* (2007): There is no better introduction to the globalization and anti-globalization perspectives than is found in this book.

Jens-Uwe and Warrier (2009): This dictionary should be a useful starting point for many students of globalization.

Jones (2010): This collection brings together some excellent readings from thinkers on globalization and pays particular attention to the implications for and role of geography.

Perrons (2009): This collection on globalizing failures is a challenging but valuable read in terms of its critical stance on the impacts of globalization from a geographical point of view.

Ritzer (2011): This book provides a clear and concise overview of the main debates in globalization studies from a sociological perspective. See also Ritzer, 2008 and 2010.

Sidaway (2012): Political and development geographer Sidaway reflects on geographies of globalization to date and the relationship with area studies.

Sparke (2009b): This entry in *The Dictionary of Human Geography* is a lively introduction to the many strands of geographical thinking on globalization. See also Sparke (2013).

Taylor *et al.* (2002): They introduce the collection *Geographies of Global Change* and discuss the relationship between geography and globalization.

Globalizations and *Global Networks* are two journals that often have excellent cutting-edge articles on global issues, sometimes related specifically to geography and often from allied fields.

② Globalization and space – contesting theories

A world of theory – key ideas and themes

How can we approach thinking about globalization? What concepts exist that have tried to make sense of this often bewildering topic? How long has such thinking existed? What are the roots and drivers of globalization and how have these changed over time? How are these ideas informed by and how do they inform concepts from human geography? The following two chapters investigate theories and histories of globalization, respectively, and introduce a range of ideas and concepts that seek to begin to answer such questions. Two central themes run through both chapters: first, globalization does not homogenize space and place, rather it leads to new and ever more complex geographies; second, globalization is not new and, although thinkers are divided as to when it began, most are agreed that it has a heritage that is considerable. These two points are important as they remind us that globalization as a process, agenda and concept is complex and multifaceted, and that understanding it requires looking back and looking broadly.

A number of other key points evolve from these core ideas. For example, we will see in our theoretical discussion that, although work in this area blossomed through the 2000s, thinking about globalization – even though it was not called that – goes back hundreds of years. Another key point is that, given the inherently uneven nature over time of that which we term globalization, we must also bear in mind that ideas concerning its nature and operation are highly contested; there is no single definition or theory and certainly no consensus on the nature of its impacts. Furthermore, theories are highly dynamic and respond to the increasingly rapidly evolving politics, economics and culture of the world and the localities which comprise it. We will see that one's opinion on globalization turns on politics, socio-economic context, culture and personal positionality. This is not to say that all ideas regarding globalization are equal; some are more worthy of attention than others, and here we attempt to deal with the major ones in our estimation. However it is impossible to either a) cover all aspects of thinking in this regard or b) overgeneralize with regard to opinions concerning the driving forces, process, and outcomes of globalization. As such, the following two chapters should be seen as an entry point into the broad debates that concern globalization, intended to stimulate further interest and study rather than offer objective truths. Notwithstanding this, we do offer our thoughts on the nature, impacts and chronology of globalization and hope that these are useful in framing thought and provoking critique in the mind of the reader.

Early theories – Marx to the global village

In order to illustrate the long heritage of the idea, Waters (2001) reminds us that 'curiously, globalization, or a concept very much like it, put in an early appearance in the development of social science' (p. 7). He traces the links between some of the early classical accounts of the 'internationalism' of Saint-Simon, Comte and Durkheim among others. Durkheim, for example, spoke of the implications of industrialization in terms of weakening state structures and nation-state culture, leading to the 'dismantling of boundaries between societies' (Waters, 2001, p. 8). In another example, Weber talked of the spread of Calvinism and rational thinking in Europe and eventual dominance of Western culture. Marx had a globalization theory of sorts too, which was more formally elaborated by Lenin. He argued that the bourgeoisie (industrial class), in their search for profits and extraction of surplus from the proletariat (working class), would expand their geographical horizons drawing peripheral countries into the system as providers of cheap labour and raw materials. His conclusion was that this would eventually lead to revolution – echoing the utopian internationalism of the earlier ideas of others including Saint-Simon. Two interesting points stand in Marx's account: first, he is explicit that the underlying logic for the expansion of capitalism is cultural as well as economic; second, Marx views this process as one of 'the interdependence of nations and recognizes the continuing existence of the nation-state' (Waters, 2001, p. 9). This latter point in particular went on to influence Marxist dependency theory, world systems theory and modernization theory.

Writing in the 1970s, Wallerstein (1980) proposed World Systems Theory (WST), arguing that the capitalist mode of production would supersede others as it expanded across the globe. For Wallerstein, and for dependency theorists who employ similar perspectives, this would not necessarily lead to an eventual utopian outcome as Marx predicted, and it would continue indefinitely. Central to Wallerstein's ideas is the centrality of the nation-state and coexistence of *different* world systems and cultures of which capitalism is the dominant one. Unifying all of these ideas considered thus far was the central idea that the rise and spread of capitalism was linked to the diffusion and spreading of economic relations across space – an idea later referred to as globalization.

A global village?

Among the most influential ideas of the second half of the twentieth century with respect to globalization was that of Canadian literary critic Marshall McLuhan, namely his concept of the 'global village' (1962, 1964). This concept has entered the popular imagination in ways that were probably not intended by the author and has been central in the evolution of a popular global consciousness (Robertson, 2003; Ritzer, 2011). After the Second World War, critics noted that social relations were extending greatly over space due to the evolution of communications technology, some of which had been advanced during the war effort. In the 1960s, McLuhan captured the essence of this and coined the term, arguing that 'the world has become compressed and electrically contracted, so that the globe is no more than one village' (McLuhan, 1964, p. 4). Although often employed in a simplistic manner subsequently, McLuhan's original concept was complex. It referred principally to the evolution of social relations and the cultural infrastructures used to communicate these over time. Human history was divided into three phases based on the dominant mode of communication:

- oral;
- writing/painting;
- electronic.

Over time, argued McLuhan, each phase had been superseded by another mode and this had spatial ramifications. Oral communication, by definition, restricted social interaction to relatively short distances; for example, within villages. The advent of writing and painting made the transport of ideas and cultural markers across space a lot easier leading to an increase in the 'extensibility' of human relations. The arrival of the electronic age and the even greater stretching of social relations had the paradoxical effect of once again making oral communication more common. This mode echoed the first period, but this time the 'village' was stretched globally. In the introduction to *Understanding Media*, McLuhan sums this up in a way that predicts contemporary social media: '(t)oday, after more than a century of electric technology, we have extended our central nervous system in a global embrace, abolishing both space and time as far as our planet is concerned' (1964, p. 3).

The construct is often misunderstood as a utopian reflection of reality and one which promotes the idea of homogenization. The author himself was often ambivalent about the outcomes of 'electric contraction'. For example, he writes: 'the aspiration of our time for wholeness, empathy and depth of awareness is a natural adjunct of electric technology . . . There is a deep faith to be found in this attitude – a faith that concerns the ultimate harmony of all being' (1964, p. 5). In other writings, however, McLuhan refers to the global village as a place of 'terror', 'uncertainty' and 'tribalism'. The 'global village' idea has been co-opted by politicians, business persons and academics of all shades to justify and illustrate their particular world views. Employed uncritically, it fails to capture the unevenness of access to electronic technology that characterizes the world – the oft-cited 'digital divide', for example. There is no doubt, however, that by injecting notions of cultural interaction and 'stretched' space into the debate the idea influenced a number of subsequent theorizations, including those of Robertson (1992) and Giddens (1990), as we explore below. McLuhan was also influential in critical media studies and media ecology where his notions of stretched networks that transmit cultural markers are analysed. McLuhan's ideas were influential but did not consider the driving force of what became known as globalization and it is to those debates that we now turn.

Contemporary theses – hyperglobalists, sceptics and transformationalists

From around the mid-1980s explicit theorization about globalization appeared as the term entered into academic and eventually popular usage. Waters (2001) cites figures such as Beck (1992), Rosenau (1990), Lash and Urry (1994) and geographer Harvey (1989) as important early contributors in this endeavour. It is Robertston (1992) and Giddens (1991) who are credited with the major contributions, however, with their ideas concerning the 'relativisation of the individual with respect to higher scales of analysis' and 'time–space distanciation', respectively. It is argued that:

> [t]he significant features of each of their [Robertson's and Gidden's] proposals are first that they are multi-causal or multi-dimensional in their approach, and second, that they emphasise subjectivity and culture as central factors in the current acceleration of the globalization process.
>
> (Waters, 2001, p. 14)

A useful way of reviewing the diversity of ideas is provided by Held *et al.* (1999) in their three 'theses' of globalization schema. These authors do not see an evolving consensus, or even paradigm, arguing that:

> beyond a general acknowledgement of a real or perceived intensification of global interconnectedness there is substantial disagreement as to how globalization is best conceptualised, how one should think about its causal dynamics, and how one should characterise its structural consequences, if any.
>
> (Held *et al.*, 1999, p. 2)

What is particularly noticeable about Held *et al.*'s three-part division is that traditional ideological lines are blurred. In the hyperglobalist camp, for example, we find neo-Marxists such as Greider (1997), together with neoliberals such as Ohmae (1990, 1995). Similarly, in the sceptical group, both conservative and radical views exist. Nevertheless, as we see below, each school conforms around a central theme in terms of the nature and consequences of globalization (see also Chapter 3).

The hyperglobalists

According to hyperglobalists a new age of history is upon us, one that is entirely unprecedented. Ohmae, for example, argues that 'traditional nation-states have become unnatural, even impossible, units in the global economy' (Ohmae, 1995, p. 5). In this argument, the world economy is borderless and characterized by a single global market which functions through transnational networks of production, trade and finance (Khan, 1996). Therefore, emergent forms of governance, which sit above and below the scale of the nation-state, become increasingly important. The divergence of ideologies within this group is highlighted by the fact that both neoliberals and some neo-Marxists form part of the school. The former, epitomized by Ohmae (1995), view the erosion of state regulatory power as a positive factor which signals the victory of capitalism over socialism. It is notable that such rhetoric was written and taken very seriously, in business schools in particular, immediately following the collapse of the USSR in 1989. The second group, the radicals, also argue that global capitalism has triumphed. In contrast, they view global capitalism as both oppressive and regressive (Peet and Watts, 1993; Petras, 1999). Both perspectives posit that economic forces are dominant, that a truly integrated global economy actually exists, and that governments are wedged uncomfortably in between local, regional and supranational bodies of governance, and are thus increasingly irrelevant.

The hyperglobalist thesis argues that new transnational class allegiances have evolved, creating a new global elite that shares an ideological attachment to neoliberalism and consumerism. Some see this as evidence of the first truly 'global civilization', facilitating the spread not only of neoliberal consumerism but also of liberal democracy and new forms of global governance (Strange, 1996; McGrew 2008). Radical hyperglobalists do not celebrate this but do acknowledge the arrival of the 'global market civilisation' (Perlmutter, 1991; Greider, 1997; Friedman, 2007). In sum, '[w]hether issuing from a liberal or radical/ socialist perspective, the hyperglobalist thesis represents globalization as embodying nothing less than a fundamental reconfiguration of the "framework of human action"' (Held *et al.*, 1999, p. 5). Globalization then is reified as an unstoppable and all-determining juggernaut (Klak, 1998; Wolf, 2004).

The sceptics

At the other end of the spectrum are the sceptics who question the very existence of 'globalization' (Held and McGrew, 2007b). Although less vocal, they provide an important counterpoint to the accounts that claim globalization to be deterministic. The leading proponents of this view are

Plate 2.1 *Hyperglobalization: new highway construction in Kenya*
New highways are being constructed at a rapid pace throughout much of Africa and the rest of
the developing world. They are part of a development strategy which seeks to integrate rural
and remote areas with urban centres and into the global economy.
Source: Photo by the authors

Hirst and Thompson (1999; see also Hirst *et al.*, 2009) who use a broad
range of empirical data to argue that the world economy is no more
interconnected today than in the past. Indeed, they argue that during the
era of the Gold Standard, at the end of the nineteenth century, the world
economy was *more* integrated (see Chapter 3 for a fuller analysis of this
issue). Flows of labour, investment and trade would – if the
hyperglobalist thesis were true – be unrestricted and free. This, they
argue, is very far from reality, and labour remains particularly immobile.
For sceptics, national governments remain the central actors in
constructing and regulating the global economy. What we have then is an
internationalized political economy which is fragmenting into powerful
economic blocs, the most powerful being Europe, East Asia and the USA.
The process of regionalism, according to the sceptics, runs in opposition
to globalization, whereas, for other schools, it is seen as a precursor to it
(see Chapter 4 on regionalism).

Sceptics charge that hyperglobalist accounts of globalization are smokescreens set up to obfuscate the real intentions of the powerful capitalist elite. Capitalist national governments, principally those of the G8, are the primary designers of the new global economy. This select group of core countries insists upon free-market reform in poorer countries – claiming that this is necessary for survival in the context of inevitable globalization. Simultaneously, these vociferous proponents of free-market reform are themselves the most protectionist. In this way, sceptics view globalization as an 'agenda' (Firth, 2000) or a 'project' (McMichael, 2004, 2013) which seeks actively to wedge open spaces of opportunity for capital emanating from the advanced capitalist economies (Hirst, 1997). This agenda came about as a result of US-led reconstruction of the global economic order following the Second World War, which created a new agenda of liberalization – designed largely in the interests of the dominant powers at the time (Gilpin, 2001) (see Chapter 4). Neo-Marxists have seen recent 'globalization' as another phase in the expansion of Western imperialism where national governments act as the agents and conduits for monopoly capital (Firth, 2000; McMichael, 2013).

Given the above, the sceptics question the rise of a global 'civilization' or a new global 'civil society'. Rather, deepening inequality heightens neocolonial tensions between civilizational blocs and ethic groupings (Huntington, 1996). The discourse of the 'homogenization' of global culture is seen as part of the West's historic project of domination initiated under colonialism. There are links here, which are just beginning to be explored, with postcolonial scholarship which tends to view globalization discourse as an agenda designed to extend the subordination of postcolonial subjects (see Blunt and McEwan, 2003; Gregory, 2004; McEwan and Daya, 2013; Sidaway, 2000; Sidaway *et al.*, 2003). Writers such as Chomsky (2001) see the recent 'war on terror' as part of the same project, complementing discourses of development and globalization as a strategy for continued dominance by the West. The principal outcome of the sceptical thesis is the exploding of the myth of the erosion of state power. Nation-states are seen as the principal disseminators of the agenda as well as, potentially at least, the principal sites of resistance to it.

The transformationalists

Lying at various points on the continuum between extremes is the transformationalist thesis (see Table 2.1). Scholars in this group argue,

Table 2.1 *Three theses of globalization – a schema*

	Hyperglobalist	Sceptical	Transformationalist
What is happening?	The global era	Increased regionalism	Unprecedented interconnectedness
Central features	Global civilization based on global capitalism and governance	Core-led regionalism makes globe less interconnected than in late nineteenth century	'Thick globalization'. High intensity, extensity and velocity of globalization
Driving processes	Technology, capitalism and human ingenuity	Nation-states and the market	'Modern' forces in unison
Patterns of differentiation	Collapsing of welfare differentials over time as market equalizes	Core–periphery structure reinforced leading to greater global inequality	New networks of inclusion/exclusion that are more complex than old patterns
Conceptualization of globalization	Borderless world and perfect markets	Regionalization, internationalization and imperfect markets	Time–space compression and distanciation which rescale interaction
Implications for the nation-state	Eroded or made irrelevant	Strengthened and made more relevant	Transformed governance patterns and new state imperatives
Historical path	Global civilization based on new transnational elite and cross-class groups	Neo-imperialism and civilizational clashes through actions of regional blocs and neoliberal agenda	Indeterminate – depends on construction and action of nation-states and civil society
Core position	Triumph of capitalism and the market over nation-states	Powerful states create globalization agenda to perpetuate their dominant position	Transformation of governance at all scales and new networks of power

Source: Adapted from Held *et al.* (1999)

often in very different ways, that globalization is real and is restructuring society profoundly (Castells, 1996; Giddens, 1990, 2002). This social change has evolved from historical transitions, a viewpoint that contrasts with the hyperglobalist claim that globalization is an entirely new condition. Tranformationalists argue that globalization is historically and geographically contingent, constructed by human action and, therefore, has no predetermined outcome (Peet, 1991). The role of nation-states is altered, though not necessarily eroded. Novel policy responses are required given that traditional binaries such as international/domestic, internal/external and global/local are collapsed (Rosenau, 1990; Sassen, 2000, 2007). In this context, Rosenau (1990) talks of the rise of 'intermestic' affairs.

Although building explicitly on historical processes of change (see Chapter 3), contemporary globalization is nonetheless different to that of the past. In the economic sphere, for example, we see trade, financial and productive organizations transcending national borders so that the fate of distant groups is interlinked in ways that are qualitatively different from the past (Dicken, 2012) (and see Chapter 4). Cultural exchange is more extensive, leading to the formation of new hybrids. In the political sphere, the links between the sovereign power of nation-states and the territory of those states is disrupted. National polities are no longer governed solely by national governments but also by supranational and international institutions. The formation of transnational social movements, based on new identities that link sub-national groups together across boundaries, challenges the state from below (Routledge, 2013). In this respect, 'sovereignty today is less a territorially defined barrier than a bargaining resource for a politics characterised by complex transnational networks' (Keohane, 1995, cited in Held *et al.*, 1999, p. 9). Economic, cultural and political governance is diffusing away from nation-states to both higher and lower geographic scales simultaneously. According to Rosenau (1990) this has made states more 'activist', as they reach out to manage the transnational affairs that so intimately affect them. Much more will be said concerning the transformationalist theses in what remains of this book, given that it is the perspective which marries best with a human geographical approach.

Geographical implications of the three discourses of globalization

From a geographical viewpoint the hyperglobalist position has important ramifications. The traditional 'core–periphery' model, together with variants on this theme such as 'North/South division' and 'developed/developing world', are seen as outdated given the decline in the importance of the nation-state as the unit of analysis. Rather, new hierarchies of power are evolving, particularly surrounding the *newer international division of labour* (see Chapter 4) and through networks. States still attempt to manage the new divisions that arise between social groups, localities and sub-national regions using redistributive and welfare measures. These, however, are becoming increasingly difficult to sustain and are being reformed. There are thus winners and losers in the new global economy. Echoing neoclassical trade theory, neoliberal hyperglobalists argue that the rise of the global free market, while creating losers, makes everybody better off in the long run and will eventually create a homogeneous affluent and modern society. In contrast, radical hyperglobalists argue that the new patterns of inequality are likely to be permanent features. This harks back to neo-Marxist inspired structuralist and dependency theories developed in Latin America in the 1960s.

For the sceptics, as the vast majority of world trade and investment flows take place within and between the major blocs, the principal geographic outcome of this process is to further marginalize peripheral economies and to make it more difficult for semi-peripheral economies to compete. Significant barriers are often placed around the major blocs: the EU's Common Agricultural Policy, for example, discriminates markedly against agricultural exports from poorer countries, Japan protects its rice sector from imports and the USA controls agricultural imports in general. This means that the geographical embrace of the world economy is actually diminished from what it once was (Gordon, 1988; Weiss, 1998). It also implies an intensification of the structural inequalities that persist in the global economy (Gwynne *et al.*, 2003; Harvey, 2011). Sceptics question the rise of the new and newer international divisions of labour and the existence of truly 'global' corporations (Krugman, 1996). Such companies, it is argued, remain deeply embedded in national territories (Thompson and Allen, 1997), which continue to be the principal 'containers' of economic activity (Daniels and Lever, 1996; Jones 2010). Culture becomes increasingly differentiated along ethnic and racial lines and reflects the unequal distribution of modernization.

The most important spatial implication of the transformationalist perspective is that the global capitalist system forges increased unevenness. This does not replicate traditional core–periphery patterns and is not territorial in essence, but rather social in nature and articulated through networks. However, these divisions are not necessarily locked in for ever, as radical hyperglobalists argue. Within particular places, irrespective of where they sit in the world system as measured by reference to old divisions such as First and Third World, we see 'new hierarchies which cut across and penetrate all societies and regions of the world' (Held *et al.*, 1999, p. 8; Dicken, 2012). Overall, except in the case of neoliberal hyperglobalist accounts, each points towards the increased differentiation of global society, polity and economy as globalization unfolds. The concluding chapter of this book elaborates this discussion and draws up a framework for the spatial implications of globalization in the different spheres of concern (see Table 10.1).

Four outstanding controversies

There exist significant sources of contention in the debate between the three globalization theses. Four of the most important ones are discussed below and responses contrasted.

- *Conceptualization*: What is the nature of globalization?
- *Causality*: What is driving it?
- *Periodization*: Is it new?
- *Impacts*: What are its outcomes?

Hyperglobalists and sceptics alike tend to *conceptualize* globalization as comprising a supposed end-state. This is generally argued to be a global market fulfilling the conditions of perfect competition, perfect integration and factor market equalization. Sceptics such as Hirst and Thompson (see Chapter 3) test the evidence for globalization by comparing empirical trends to this ideal type. Transformationalists, on the other hand, conceive of globalization as a historical process which is 'more contingent and open ended which does not fit with orthodox linear models of social change' (Held *et al.*, 1999, p. 11), and such changes cannot always be fully reflected or captured in terms of empirical data alone. Furthermore, transformationalists, unlike hyperglobalists, argue that the process of globalization is a highly differentiated one in which neither culture nor economics plays the sole determining role.

A second area of contention is *causation*. The central issue here is whether the process is mono- or multi-causal. Many accounts have reduced the phenomenon to the expansionist and, sometimes, imperial tendencies of capitalism (see Harvey, 1989); others, for example, view technology as the driving force (see Ohmae, 1995). More recent literature has argued that a multiplicity of interlocking factors in the realms of culture, politics, economics and technology are relevant (Held and McGrew, 2002, 2007; Robertson, 2003; Ritzer, 2011; Sparke, 2013). As the collection of readings assembled by Roberts and Hite (2000) attests, this debate relates to an older one concerning the diffusion of modernization. Should globalization be seen simply as the expansion of Western modernization, as world systems theory might argue, or is it something more complex that has different root causes and expression across the world? In brief, is globalization more than simply westernization?

A third controversy is whether globalization should be conceived of as something new, and this debate forms the focus of Chapter 3. Writers are divided on the *periodization* of the process. Some see it as a post-1970s phenomenon, some as a twentieth-century process, and others as beginning with the earliest human migrations. Held *et al*. (1999) argue that growing evidence of global trade and cultural links in the pre-modern age force us to be more sensitive to history in our explanations and characterizations of the process. Defining the historical shape of globalization, of course, is as much an exercise in definitions as it is in teasing out complex historical geographies. Ultimately, as with many of the other controversies, it boils down to what we mean by globalization.

Inevitably, there is a great deal of contention over the *impacts* of globalization. At the broadest level there are two schools of thought.

The first sees globalization as eroding the power of the nation-state, leading to an end of the welfare state and the imposition of neoliberal policies across the world (S. Amin, 1997; Cox, 1997). The second is much more critical of the 'deterministic' argument and argues that national state power is adapting and transforming in response to the imperatives of globalization (Dicken, 2012; Ruigrok and van Tulder, 1995). This argument is consistent with a human geographical approach based on the co-determination of the global and the local, building on Giddens' ideas of *structuration* (see 'structuration' entry in *The Dictionary of Human Geography* (Gregory *et al*., 2009)).

Analysing globalization – four important measures

How can we measure and interpret globalization and make coherent judgements with respect to the controversies outlined above? Four measures are critical:

- the extensity of global networks;
- the intensity of global interconnectedness;
- the velocity of global flows;
- the impact propensity of global interconnectedness.

The first is the *extensity* of social relations, which relates to the concept of time–space distanciation and the way that change in one part of the world can have increasingly important ramifications for people at a distance. Together with the stretching implied above, globalization implies an increased *intensity* of transborder and 'distanciated' flows so that the volume of interaction increases. Further, the increased intensity and extensity of such processes implies an increase in the *velocity* of social, economic and political flows. These three strands together imply a greater *enmeshment* of the global and local as the pace, intensity and impact of the linkages rise and interpenetration increases.

Geography and space

Much social theory has concerned itself with time (Gregory, 1994). There is now more attention to space – and this has arisen due partly to the difficult spatial questions raised by the globalization debate. Human geography is inherently concerned with space. Almost all definitions of the discipline incorporate the term – though in competing and variable ways (see Johnston *et al.*, 2009). Along with human–environment interaction, *spatial differentiation* is often the defining core. But the way human geography conceptualizes space has varied across time and place, and geography remains characterized by many different national and regional traditions. This has led to shifting concepts of scale also. This evolution in thinking has mirrored broader paradigmatic shifts in human geography especially in the Anglo-American world. We consider space below and scale in the subsequent section.

Three views of space

There are at least three ways of conceptualizing space – each of which has important implications for the way we understand the world.

- *Absolute space* – Units of territory which can be numerically measured and are ontologically given (i.e. they exist independently of the way they are perceived – they are real). Space is ready to be filled by the features that make up geographies. Sometimes this is referred to as cartographic space – measured in miles, kilometres and so on. *Absolute space is an external given which has neutral discursive meaning.*
- *Relative space* – Space is 'perceived' by humans and this perception may vary according to their culture, available technology and resources. Relative space does not correspond to a fixed unit – it is the 'sense of space' which becomes important. For example, a person in New Zealand may feel much 'closer' to his or her grandmother in the UK with whom he or she communicates via e-mail than with his or her neighbour to whom he or she has never spoken. A settlement that is on a railway line appears much closer than one that is not for those who use rail travel, even though the latter may be closer in an absolute sense. *Relative space is perceived. It constructs and is constructed by human activity and experience.*
- *Metaphorical space* – Space that does not refer to any territorial unit at all. The 'space of views' – thinkers in one paradigm might be spread out but don't share an absolute space. The internet and the advent of cyberspace provide new metaphorical spaces for interaction. Virtual reality is a metaphorical space. *Metaphorical space does not exist 'on the ground' – but it can have real effects.*

Society and space – changing concepts in human geography

The way human geography conceptualizes space has played an important role in the evolution of the subject. In the 1960s and early 1970s a dominant approach in geography was that of *spatial science*. Growing out of spatial sociology and urban ecology, this was a largely quantitative endeavour based on the idea that the approach of the natural sciences (positivism) could be applied to the social sciences. By using sophisticated statistical measures of the distribution of phenomena, it was believed that we could derive spatial laws that would explain and predict human activity. Space, then, could be read as a map of society. This approach gave precedence to the concept of absolute space. The paradigm was largely unsuccessful, partly because the techniques employed failed to capture the complexity of society and the importance of alternative concepts of space; although it is making something of a comeback now as new GIS technologies allow more sophisticated quantitative spatial analysis.

Following the critique of spatial science in some places (the UK in particular), ascendant humanist approaches began to inform a more sensitive reading of space. Relative space in particular was emphasized as the human perception of place was explored. Instead of considering how space could be measured and how society could be read off from this, it asked: How does space *construct* society? This is to say that the relationship between society and space was conceptualized as a two-way flow or co-determinist. Approaches in the 1980s included work on *gendered spaces*, *racialized spaces* and *spaces of poverty*, for example. In the 1990s postmodern approaches informed a more explicit engagement with metaphorical space, especially in the form of *Third Space* (Soja, 1996). In this construct, identities which are increasingly hybrid adopt marginal spaces as locations from which to challenge the dominant groups in society – cyberspaces being one such 'liberating' space (see Little, 2013).

All three concepts of space are important in shaping the way humans experience and interact with the world, and each plays a role in human geography. A major misconception that has caused some to doubt the continued relevance of geography is that it is concerned with absolute space only. Globalization is altering the way time–space operates and is perceived. Time–space geography has formed an important sub-discipline since the groundbreaking work of Hägerstrand (1968, 1975). Building on this foundation, there are three interlinked ways of thinking geographically about the relationship between space and time:

● *time–space convergence;*
● *time–space distanciation;*
● *time–space compression.*

Time–space convergence refers to the *decrease in the friction of distance between places*. Coined by Donald Janelle in the late 1960s (1968, 1969, 1973), it referred to the apparent convergence of settlements linked by transport technology: as transport evolved, travel time would be reduced between them, giving the sensation that they had moved closer together. The velocity at which settlements are moving together may be called the time–space convergence rate. Using Edinburgh and London, he noted that they had converged at a rate of 30 minutes per year over a 200-year period. This measure emphasizes the importance of relative distance and how this is shifted by technology (see Map 1.1 for an example of this approach).

Time–space distanciation refers to the *stretching of social systems across space and time*. The term was coined by sociologist Anthony Giddens (1990) and refers to the interpenetration of people and places over increasingly large distances. Echoing McLuhan's work (see the global village section above) Giddens argues that people interact in two ways: face to face, and remotely through transport and communications technologies. The second has become increasingly important, 'distanciating' the relations between people. Thus people who are not actually physically present in absolute space can be important social actors. Giddens does not argue that this process leads to homogenization; indeed greater distanciation increases the potential for humans to restructure global scale systems. Massey (1991) builds on this concept when she discusses the 'global sense of the local' which pervades our everyday experiences.

Time–space compression refers to the *annihilation of space through time that lies at the core of capitalism* (Harvey, 1989). Concepts of convergence and distanciation do not offer an explanation for why social relations are stretched across space. Geographer David Harvey, in *The Condition of Postmodernity*, provided an argument which has been of central influence in the way geographers think about the relationship between time–space and globalization. He suggests:

> that we have been experiencing, these last two decades, an intense phase of time–space compression that has had a disorienting and disruptive impact upon political-economic practices, the balance of class power, as well as upon social and cultural life.
>
> (1989, p. 284)

He goes on to argue that time–space convergence and distanciation are results of the expansion of capitalist relations of production across the globe. Given that 'time is money', capitalists are constantly seeking ways to speed up circuits of capital to reduce the 'turnover time of capital' – the amount of time it takes to convert investment into a profit. This

Plate 2.2 *Time–space compression: a billboard in Samoa advertises airfares*
Cheaper airfares and improved air services allow Pacific Island people to both migrate and maintain contact with families overseas. Time and space have been compressed for many people as the time and cost of travelling over long distances has fallen dramatically.
Source: Photo by the authors

search for reduced turnover time has led to the development of technologies and policies which facilitate time–space compression. Barriers have been removed through neoliberal discourse and technology has advanced to make capital, goods and people as rapidly transportable as possible (see Plate 2.2).

Harvey links this increased time–space compression to a crisis in accumulation of capitalism as we move from the modern to the postmodern era (see Chapter 4). He is careful to draw out the cultural roots and implications of this. Capitalism itself is a culture, and the impacts of time–space compression in the cultural sphere are important as identity 'melts into air' and people search for 'secure moorings and longer-lasting values in a shifting world' (1989, p. 293). Ultimately the argument places an economic rationale at the core and this has been criticized by new cultural geographers especially. The links with his ideas and WST are apparent when he says: 'We have, in short, witnessed another fierce round in that process of annihilation of space through time that has always lain at the centre of capitalism's dynamic' (1989, p. 293).

Geography and scale – local and global

Scale is a central concept in geography and yet it is something which is often used uncritically (see Manson, 2008). What do we actually mean by 'global'? What do we actually mean by 'local'? If we are to attempt to understand globalization, we must appreciate the roots and implications of these terms. These are not esoteric questions – how we interpret scale affects how we construct the world and, therefore, what we do in terms of policy. In this sense there is a politics of scale – both personal and institutional. For example, what do members of the anti-globalization movement mean when they say they are fighting a global force? What does the environmental movement mean by 'think global, act local'? What do transnational fruit companies mean when they advertise that they are bringing 'local' produce to the 'globe'? What does a 'global' war on terror mean when these actions are concentrated in very specific locations such as Libya, Afghanistan or Iraq?

The production of geographic scale

In an exposition on the role of scale in human geography, Herod notes that 'within human geography scale is usually seen in one of two ways: either as a real material thing which actually exists and is the result of political struggle and/or social processes, or as a way of framing our understanding of the world' (2003, p. 229). This corresponds to the broad division in human geography, which has seen the eclipse of positivist philosophy with a miasma of approaches often referred to as post-positivism. The idea that scale can be thought through critically is something that is really quite new in human geography. The result of this debate in human geography is that scale may be viewed:

> in terms of a process rather than in terms of a fixed entity. In other words the global and the local are not static arenas within which social life plays out but are constantly remade by social actions. This allows us to consider not only how a firm or a political organisation might 'go global' to engage with actors or opportunities not present within its own local spaces of dependence, but also how a particular social actor such as a TNC may

> attempt to 'go local' through tailoring . . . to appeal to consumer tastes in
> different places or to reflect particular communities cultural values.
>
> (Herod, 2003, pp. 233–234)

It is especially important to understand the relationship between
the global and the local. Free-market reform and neoliberalism (see
Chapter 4) has either been strongly recommended or imposed, by
international institutions such as the IMF and the World Bank, on the
basis that it is inevitable that 'global' forces will increasingly determine
economic change in any given place. This conception of the global as
all-powerful and inevitable has had incredible political force over the
past 20 or so years. One of the major pitfalls of many non-geographic
interpretations of globalization then has been a somewhat naïve
concept of scale and, in particular, a 'disabling dichotomy' of
the global and the local. Dicken argues (2004, p. 9) that
'[f]ortunately geographers' conception of scale is considerably
more sophisticated than this'.

Herod (2003, 2009), drawing on the work of numerous geographers
before him such as Peter Taylor and Neil Smith among others,
argues that it is possible to view the concept of scale through five
metaphors: *ladder*, *concentric circles*, *Russian dolls*, *earthworm burrows*,
tree roots (see Figure 2.1). In the case of the first, scale is seen
like rungs on a ladder, moving from the top (global) to the bottom
(local). In this metaphor, scales are seen as separate and arranged in a
strictly hierarchical way. In the case of the second metaphor the
local is conceptualized as a small circle encompassed by ever-larger
circles. Although scales are still distinct, in contrast to the ladder
metaphor the global is not seen as 'above' the local. The Russian dolls
metaphor is similar to the concentric circles in that scales are seen as
distinct and of different 'size'. However, this concept gives the sense that
the various scales fit together to make a complex whole given that they
nest inside each other and the doll is not complete without fitting all of
the pieces together. The metaphors of earthworm burrows and tree roots
provide two radically different perspectives on scale compared to the
other three concepts. They move us away from 'layered' and 'territorial'
notions of scale and suggest the importance of networks. Thus all scales
are interconnected through the burrows or roots which penetrate to
different strata of the soil. In the latter two metaphors we move away
from the idea of bounded scales. When we conceptualize them at
opposite ends of networks, this alters the way we think about the global
and the local.

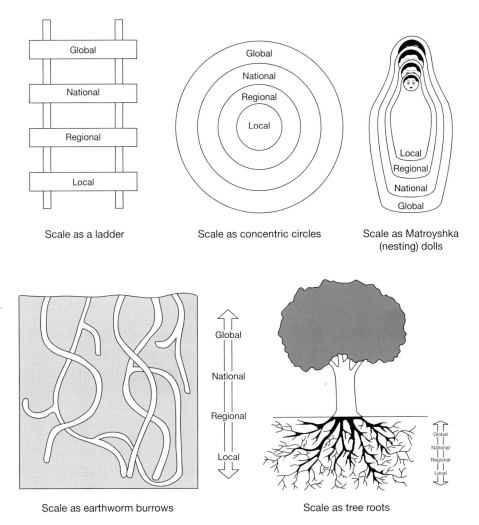

Figure 2.1 *Five concepts of scale*

Source: Adapted from Herod (2009)

Global–local interaction

In many ways, human geography is made distinctive by its simultaneous concern for both the global and the local. As geographers, we want to know more about particular places 'in depth' at the same time as 'broadening our horizons' and learning about the wider world. Concern with the 'global' has peaked at certain times in the discipline. Thus we have seen the rise of geographies associated with imperialism and exploration in the early twentieth century through to environmental and developmental concerns in the 1960s and 1970s, and more recently we have sought to tease out geographies of 'compression'. Concern with the local has permeated this transition and sometimes eclipsed globalism – especially in the 1980s with the rise of new cultural geography.

What has become clear, however, is that we cannot explain one without the other. The global and the local are two sides of the same coin. The call for focusing on big issues is welcome but it will not be successful without simultaneous concern for the local. *The Dictionary of Human Geography* points out that many locality studies have been influenced by the 'upsurge of interest in globalization during the 1990s [which] has involved a recasting of the issue of local specificity in terms of global–local relations' (Painter, 2009, p. 425). How, then, have geographers interpreted the local?

The importance of locality studies is emphasized when we consider what we mean by 'global'. The global is constructed from local action – and when we refer to the global process we are usually describing local-to-local flows that have become stretched across space to become global in extent. This is the process that is referred to in the definition of globalization provided in Chapter 1. In short, to be of use geographies need to be multi-scalar.

Crang (2005, 2013) provides a useful three-part framework in terms of how different thinkers have conceptualized global and local interaction (Figure 2.2; see also Gibson-Graham, 2002):

- world as mosaic;
- world as system;
- world as network.

World as mosaic. This concept sees the world as resembling a jigsaw puzzle with different localities juxtaposed but independent. This might refer to different neighbourhoods or countries, for example. This type of concept is promoted by tourism and other simplifying ways of experiencing and representing the world. There are three features:

- It emphasizes borders.
- Each locality has a unique 'personality' and unified geographical identity.
- Intrusion is seen as a threat to authenticity and/or tradition.

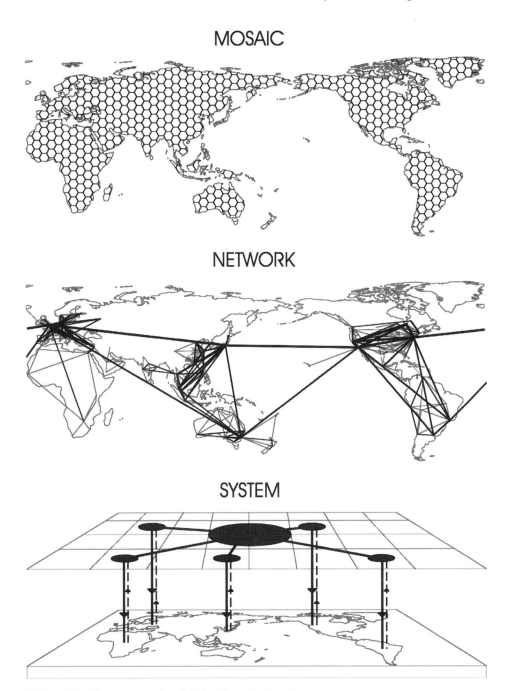

Figure 2.2 *Three concepts of global–local interaction*

Source: Adapted from Crang (2005)

Arguably this concept is no longer relevant as there are no totally isolated places left – it is a useful exercise to try and think of one. Looking closely at different types of borders (Box 2.1) we can see how mosaics may exist but their character and the permeability of their boundaries may vary greatly. Crang argues that mosaics tend to 'fossilize' difference and may be used as part of a defensive 'localism' which attempts to exclude 'outsiders'. Racism, for example, emanates from this world view. Further, by fossilizing difference, places can actually be shaped in the image that the viewers desire. In the case of tourism, it is often 'authenticity' and 'tradition' which tourists wish to see – and this is invariably 'constructed' and 'frozen' to meet this demand.

Box 2.1

A world of borders

The concept of globalization is sometimes associated with the phrase 'a world without borders', referring to the way nation-states have become weakened, restrictions on the movement of people, goods and ideas between states have been eased, and the social and cultural have become more homogenized. Yet, across the globe, borders remain as indicators of the mosaics we live in and are experienced every day by millions of people who travel from one country to another. Borders vary greatly. Some are barely noticeable and people pass easily; others require long delays as documents, vehicles and luggage are checked; and yet others present a barrier to most forms of movement. The examples below illustrate how variable and complex borders have become. (See also Diener and Hagan, 2009.)

North and South Korea

The border between the Democratic People's Republic of Korea (North Korea) and the Republic of Korea (South Korea) is perhaps the world's most militarized and impermeable border. It is a remnant of the type of border that existed during the Cold War, effectively preventing the movement of goods and people from one country to another. The Korean border remains a dangerous place to visit. Since 1953 when an armistice stopped hostilities in the Korean War (a state of war still technically exists), a 4km wide 'demilitarized zone' (DMZ) has separated two heavily armed sides. Incursions, exchanges of fire and tensions have marked this border since its inception. At Panmunjeom on the DMZ there is a joint centre where periodic negotiations take place, and for many years

Box 2.1
continued

this was the only crossing place between the two sides.

Crossing the border is almost impossible for ordinary citizens or foreigners, and any unauthorized attempts would be met by gunfire as this border remains heavily monitored throughout its length (see Plate 2.3). There is some provision for goods and people to cross the border, and a rail link has existed since 2007. However such interaction is tightly monitored by the two sides and is frequently subject to closure. Periodic attempts have been made to lessen tension and encourage some limited movement (such as joint industrial developments and family reunions), but these are liable to interruption.

Whilst this border forms a strong barrier to the movement of people and goods – even information – it is notable that not only is it one not recognized by either side as anything other than a temporary expedient (both lay claim to legitimacy over the whole of Korea), but also it is purely an arbitrary line (the position of both armies when hostilities stopped). There is a high degree of cultural homogeneity across the Korean peninsula and kinship connections exist across the border, despite 50 years of closure. It remains a highly visible manifestation of differences in political ideology and economic organization, yet it has little or no meaning in a cultural or historic sense.

USA and Mexico

Crossing overland between San Diego in USA and Tijuana in Mexico involves long delays at a border crossing when, on the USA side in particular, there are strenuous and thorough efforts to prevent people entering without a correct visa. Every year, thousands of Mexican and other residents attempt to enter the USA in order to find work or reunite with their families. Although not as fortified or closely monitored as the Korean border, the USA–Mexico border remains a place of wire fences, armed border guards and rigorous checking of documentation. Illegal incursions only rarely result in fatal gunfire – instead more passive 'capture and return' methods are used – but there is a high degree of political backing in the USA to keep this a 'hard' border (see Plate 2.4). Although travelling into the USA through this border involves long delays and checks, when moving the other way and entering Mexico most people move easily and quickly usually without even passport checks being made.

Yet this border has a much different character away from the issue of preventing south to north movement of people. Increasingly, it has little meaning as a cultural divide. Hispanic missions, settlement and control existed across a wide area in California, Texas and Arizona before the United States took over the territory largely as a result of war. Also, in the past 50 years there has been a very large and steady stream of migrants, both legal and illegal, moving to the USA from the south; and both the Spanish language and Latino culture are now very much a largely accepted element of the American way of life. Spanish is the second most widely spoken language in the USA, being the primary language of some 37 million people.

Box 2.1
continued

Plate 2.3 *Panmunjeom, Korea*

Source: Photo by the authors

Furthermore, the border is an increasing irrelevance in economic terms. When the North American Free Trade Agreement (NAFTA) was signed in 1994, many tariffs were eliminated or reduced overnight, and commitments were made to phase out remaining barriers. This has greatly promoted cross-border movement of goods and services and eased restrictions on such movement. Some obstacles remain, as in monitoring of traffic for illegal drugs, but generally this border effectively exists to restrict the movement of people, not goods, ideas or information.

The European Union

Europe is a region of diverse cultures and societies, and the historic landscape is marked by many examples of past barriers, differences and restrictions, as it is by interactions and trade. Borders frequently were fought over and redrawn. In many

Box 2.1
continued

Plate 2.4   The Mexico–USA border at Tijuana

Source: Photo by the authors

respects, Europe is the hearth of the nation-state – especially following 1918 and the break-up of the Austro-Hungarian, Ottoman and Russian empires, when the map of Europe was redrawn in the Treaty of Versailles in 1919. This led to the creation of many new countries, most defined by a national identity shaped by language, history and culture. These countries embarked on their own development paths and in many cases they were keen to regulate the movement of people across their borders. Such restrictions were reshaped after 1945. Following the Second World War, some nation-states disappeared, swallowed up by multinational states such as Yugoslavia or the Soviet Union. The Cold War meant that the different political ideologies of East and West were separated by hard boundaries: an 'Iron Curtain' as described by Winston Churchill. Furthermore, even within the Western bloc, barriers existed to trade (as between France and the UK) and cultural divides and national rivalries remained.

However, the evolution of the European Union (see Box 5.2) and later the end of the Cold War have greatly changed the nature of borders in Europe. Although it can be difficult for non-European citizens to enter some countries of the EU, movement across internal borders is extremely easy. In many cases (see Plate 2.5) there is little more than a couple of road signs to mark the border, and passport checks and vehicle inspections are rare. Similarly, trucks, aircraft and railways transport goods easily and quickly within the EU with minimal restrictions.

Box 2.1
continued

Plate 2.5 *The Luxembourg–Germany border*

Source: Photo by Katherine Mihm

European borders (at least within the EU) thus have 'softened' considerably, to the point of becoming irrelevant on a daily basis. Movement is free and economic integration encouraged. Yet, unlike the Korean border or even elements of the USA–Mexico border, these lines retain meaning as cultural divides. Players and fans may be able to move easily between countries to attend matches but when the Netherlands plays Germany at football, or Wales play England at rugby, the rivalries are fierce and the sense of history is deep. Borders, then, are as much cultural imaginaries as they are physical barriers.

World as system. Many social scientists have looked at the interaction of global and local through a global system perspective (WST and dependency, for example). In this case, it is argued that local outcomes are produced through the particular location of the place within the broader global system at that point in time. This type of argument counters explanations which seek to explain local differences solely in terms of 'internal' characteristics (for example, modernization accounts of development theory). This kind of cultural determinism leads to

problematic and sometimes racist explanations. For example, why is
Europe rich and Africa poor? Because there is something inherent in
European culture which makes it more productive and efficient?
A systems-based explanation argues that it is that the history
of the two regions in the context of the global system and the way this
conditions their interactions – Europe as colonizer, Africa as
colonized – that explains this inequality. Systems approaches do not
always suggest that the 'global' conditions the 'local', however.
The way the global unfolds is also influenced by pre-existing
conditions. It is this process of mutual differentiation, which
works both up and down the scale, that produces difference
and diversity.

World as network. More recently, geographers have focused on
the concept of networks to explain links between the 'global' and the
'local' and, in doing so, have reflected critically on the nature of the
former. This focuses on the connections between different people and
institutions located on specific nodes across the world. In this sense, the
local becomes the global and the global the local – something which is
sometimes called 'glocalization' (see discussion later in this chapter).
For example, Massey (1991) talks of the 'global sense of place' she
experiences walking down her home high street in Kilburn. She
emphasizes the links between that particular place and other parts of the
world, which have been transmitted down particular networks. Her
conclusion is that locality is constructed from its interaction with
the global and that the two are inextricably intertwined in character through
such networks. As such, 'localities are always provisional, always in the
process of being made, always contested' (Massey, 1991, p. 29).
This argument supports the idea that what we are really referring to when
we talk about the global is localized links over a large distance.
Think of global money flows, international migration, world music, and
even the World Wide Web – they are all constituted of human action taking
place in specific locations.

In sum, although Crang's (2005, 2013) arguments concerning the
importance of networks as opposed to systems and mosaic interpretations
of global society are compelling, it could be argued that the reality is
even more complicated than the startlingly complex 'architecture' he
proposes (see Box 2.2). Networks are indeed important and becoming
increasingly more so, of that there can be little doubt. However, is it not

possible that networks might more appropriately be characterized as functioning within systems – principally the global capitalist system but also the socialist system, where this still exists? Mosaics are still existent; in world cities, for example, one of the outcomes of globalization has been the creation of areas increasingly differentiated along socio-economic and sometimes ethnic lines (Knox, 2002; Sassen, 2011). What we have is *a system of multi-fibred networks, characterized by multidirectional flows, nestled within a larger system (on the whole, capitalism), which creates at its nodes a mosaicking of space.*

Box 2.2

Global–local interaction as seen in the wine section of a supermarket

We can see examples of Crang's (2013) mosaic/system/network framework for global–local interaction in everyday situations, and we can interpret phenomena simultaneously using each approach. For example, in many parts of the world, supermarkets have a section devoted to selling wine. If we look closely at the wine labels, we can read their geographical origins and character in different ways.

Wine mosaics

If we look at the shelves of expensive wine we may find wines from France and from Bordeaux or Burgundy in particular. Most of the wines from these regions are classified not by the grape variety they were made out of but by their locality of origin, and this is

regulated, codified and protected by laws such as the French *Appellation d'origine Contrôlée* system. This system regulates the style of wine produced from certain specified areas and protects the place name as a brand (such as Champagne) that cannot be used for wine from other regions. Furthermore, the system attempts to designate quality levels for wines from different places. As a result, maps of the premier French wine regions are tightly defined and criss-crossed by lines which separate one wine appellation from another. Sitting on the right side of such a line (in a top-quality designated area) can mean that producers can charge much more for their wine than those from the other side of the line, and their land is worth a great deal more as a result. Map 2.1 shows a portion of a map from Burgundy that demonstrates

Box 2.2
continued

Map 2.1 *The Vosne-Romanée district of Burgundy, France*

Box 2.2
continued

how a relatively small district (Vosne-Romanée) is strictly divided between the highest-quality land and wine (*Grand Cru*), the second tier land (*Premier Cru*), and lower-grade land and wine (*Commune Appellation* and non-appellation 'other' vineyards). This is a mosaic in action. It is based on historical and subjective perceptions of quality; it is subject to political processes of designation; it is mapped and governed by laws; it deeply affects the market for land and wine; and it is manifested in the prices we pay for these wines.

Wine mosaics are not just restricted to France. Seeing the success of long-established French brands and marketing strategies, some producers in New World wine regions (California, Australia, New Zealand, South Africa, Argentina and Chile) at first tried to co-opt French place-based brand names (such as Champagne) but were prevented by law from doing this. Instead they have developed their own appellation systems, dividing their countries into designated wine regions and districts and similarly delimiting and protecting these places. Now some new wine place names (such as Mendoza, Barossa, Stellenbosch, Marlborough or Napa Valley) have become established in the global market, and we can see these manifestations of mosaic building on our supermarket wine shelves.

Wine systems

Elsewhere on the wine shelves, we may see wines from many different countries; but, if we closely examine who makes these wines

(not always apparent from the labels), we might discover that, despite the plethora of brands on display, a relatively small number of global beverage companies dominate the world market for wine. Companies such as the American giant Constellation Brands, Pernod Ricard from France, or Diageo from the UK own wine brands from around the world. Furthermore, these brands span a very wide range from the very expensive brands (some of which may be French appellation-branded wines) to the cheapest bulk wines, often packaged in cardboard casks rather than bottles and with little indication of their specific place of origin or even year of production (indeed they may blend wines from different years and countries!). These global beverage companies may acquire very diverse portfolios of wine brands from across the globe (see Table 2.2).

These wines are linked through systems not just because they are owned by the same companies but also because, as types of wine, they are dependent on one another and the companies need these different wines so they can maximize profits. Although the high-priced luxury wines may have high profit margins for their producers, they are sold in relatively small volumes; and wine companies need the very high turnover of medium- and low-priced wines to provide them with cash flow and higher total profits, even if the profit margin per bottle may be low. Yet in some ways these lower-segment wines need the expensive wines. Consumers are attracted to wine not because it is just a cheap form of alcohol but because its purchase and consumption is often associated with prestige and status,

Box 2.2
continued

Table 2.2   Constellation Brands – brand portfolio

Sector	Place of origin (# of brands)	Main Brands
Wine	Australia (20)	Banrock Station, Hardys, Houghton, Leasingham
	New Zealand (5)	Nobilo, Selaks, Kim Crawford
	South Africa (3)	Flagstone
	South America (3)	Marcus James
	Europe (6)	Ruffino, Mouton Cadet
	Canada (14)	Inniskillin, Jackson-Triggs
	USA (28)	Paul Masson, Robert Mondavi, Estancia, Vendage
Beer	(7)	Corona, Tsingtao
Spirits	North America (6)	Black Velvet Whisky, Paul Masson Brandy
Other liquor	(12)	Stones Ginger Wine

Source: www.cbrands.com/CBI/constellationbrands/OurBusiness/GlobalBrandPortfolio/BrandPortfolio.pdf (accessed March 2010)

compared to beer or many spirits. This prestige and status is 'earned' by the luxury brands – from Champagne and Bordeaux and Burgundy for example – and lower-priced wine can leverage off this perception by consumers to demand a higher price. Thus medium-segment wines will mimic expensive wines by using place of origin and vintage as prominent parts of their label and they will invoke calls to prestige by using brand names which may include words such as 'chateau'. Mutual dependence in the system also works the other way in that cheaper wines help establish wine as a mass product consumed by a growing proportion of the population (wine consumption has increased very rapidly in China in the past two decades, for example). Once wine is established in this way, consumers may in time move up the quality scale and seek out more expensive wines to buy. Wine companies may link their products in quite overt ways, especially when selling through large retail outlets such as supermarket

chains. Thus, for example, if a wine retailer wants to secure a small quantity of a very expensive luxury brand of wine, they may have to agree to order as well a large volume of a cheaper brand from the same company. Therefore, these wine systems are seen in the investment and marketing strategies of global beverage companies. They acquire vineyards around the world, they seek to balance low-cost mass production of cheaper wine with high-priced luxury wine brands, and they develop marketing strategies in ways which see different wine brands, countries of origin and price segments linked together.

Wine networks

The third type of global–local interaction we can see on our wine shelves illustrates how certain networks can affect the wine we buy and consume. If our supermarket is in the United Kingdom, we are likely to be able to

Box 2.2
continued

Plate 2.6 *South African Fair Trade wine in a British supermarket*

Source: Photo by the authors

find bottles of wine with a Fair Trade label, particularly certain wines from South Africa (see Plate 2.6). Looking closely on other shelves, we may also see that some wines are labelled with a certified organic label (such as BioGro). Many consumers will seek out such wines because they may believe in the principles of ethical consumption: buying products which they can be assured of as being produced not using bad labour or trading practices or by not doing harm to the environment.

The marketing of wines in these ways represents how networks might operate and not only cut across national boundaries but also penetrate into the production and marketing decisions of wine companies. The Fair Trade marketing of South African wine presents an interesting case. Prior to 1994, when apartheid was abolished, sales of South African wine on several international markets were boycotted due to the oppressive race-based laws of the country. After this, large quantities of South African

Box 2.2
continued

wine began to reappear on global markets, especially in the United Kingdom. However, a coalition of NGOs brought attention to the persistence of bad labour practices in South African vineyards, such as low wages and the continuance of the practice of paying part of the wages of African workers in the form of cheap brandy. The public campaign that followed called for a boycott of South African wines and brought a swift response from winemakers. Reforms were made by many and some sought Fair Trade certification – the code for Fair Trade in the wine industry was actually developed mainly in South Africa. Now for those South African winemakers that conform to ethical labour standards, Fair Trade certification and labelling of their wines has turned a marketing disaster into a distinct advantage.

The promotion of organic wines has seen a similar interaction between networks of civil society, consumer groups and wine producers. This has effectively transmitted consumer wishes and campaigns into changes in the way wine is produced (by some companies at least). So, as with mosaics and systems, networks have helped shape the nature of the world's wine industry and this is seen in the bottles that line our supermarkets and wine shops.

Glocalization?

There can be little doubt that localities are increasingly impacted by flows coming from 'higher' scales. A major error however is that this process has often been seen as deterministic. Geographer Swyngedouw (1997) coined the phrase 'glocalization' to refer to a two-way relationship between the global and the local (see Sparke, 2009b). As Dicken says, '[s]uch a term helps us to appreciate the interrelatedness of the geographical scales and, in particular, the idea that while the "local" exists within the "global" the "global" also exists within the "local"' (2000, p. 459). Through this concept, geographers have been quite active in countering the 'pure' spaces of flows proposition. Dicken argues that many 'deterritorialization' arguments 'are based on a deep misconception of the nature of spatial processes which are deeply embedded in place' (2000, p. 458). Everything that takes place in the cultural, economic and political spheres is 'grounded' and localized, and requires what Dicken (2012) calls a 'spatial fixedness' to operate.

The concept of glocalization reminds us that the outcomes of globalization are path dependent. Flows of information across the internet move from hub to hub in particular places, and, within networks of supposedly hyperglobalized sectors, such as finance, significant agglomerations of activity crystallize. The City of London, for example, by virtue of its history at the centre of the largest global empire yet known, functions as a major financial centre. Inertia of place is important and 'sunk capital', both in a sociocultural and an economic sense, plays an important role in producing localities with specialized functions and character. At the supply end of the global economy, for example, the non-homogeneous distribution of natural resources such as oil, good land, forests, fishing grounds and so on continues to determine the comparative advantage of regions and localities and the way, therefore, that globalization influences 'development'. But historical patterns also play an important role as global flows move across them. Ask yourself why, for example, Latin America's economy largely comprises earnings from natural resource exports to the core economies of the world. Is it simply because Latin America has many natural resources, such as forestry, metals and minerals, or is to do with the pattern of trade that was established when the regional economy was brought into the orbit of the global capitalist system? (See Murray and Silva, 2004; Barton and Murray, 2009.)

Underlying geographies of agglomeration are thus physical, historical, sociocultural and economic factors and processes. In the case of the latter, for example, economies of agglomeration flow from the economies of scale that are gained by certain functions clustering together – sometimes referred to as 'traded interdependencies'. There are also 'untraded interdependencies'; that is to say, benefits (positive externalities to use economics terminology) which geographic proximity brings that cannot always be strictly measured. Amin and Thrift (1994) identify three sources of such interdependencies: (1) face-to-face contact; (2) social and cultural interaction which establish networks of trust; and (3) enhancement of knowledge and innovation. Underlying such ideas is the notion that human beings are essentially social animals who depend on the benefits that interaction with others yields. There is no direct evidence to suggest that this human social and organizational trait is diminishing in the age of the current communications revolution.

The above discussion leads to the conclusion that spaces of flows and spaces of places need not be mutually exclusive and both are simultaneously in action – spaces of flows operate within and between spaces of places. In this sense, Dicken (2004, p. 9) notes:

> *Bounded political spaces matter.* Some, like the nation-state, matter more than others. In this sense, therefore, we have a very complex situation in which topologically defined networks (for example, of TNCs) both 'interrupt' – and are interrupted by – political-territorial boundaries.

Summarizing the distinct viewpoints on the interaction of the global and the local, Thrift (2000, p. 456; see also Gregory *et al.*, 2009) suggests:

> There are three possible accounts. On one, global processes leave their footprints on places, allowing this little choice but either fall into line or be stamped out. On another, local places 'turn' global processes . . . global processes can only obtain a purchase by fitting in with local cultures. Finally, between these two views, is the one which argues for a process of glocalisation, a complex interaction between globalising and localising tendencies.

In fact there are potentially more ways of conceptualizing the interaction of the global and the local; Gibson-Graham, for example, discusses six (see Box 2.3). There has been much talk of the prospect of globalization from above and globalization from below; if one accepts that globalization is a dialectical process (that it is local and global at the same time) the binary of globalization from above or below becomes redundant – it happens both ways at the same time.

Conclusion – new geographies

According to a number of geographers, there exists considerable confusion with respect to the conceptualization of the causal power of

---

Box 2.3

Global and local: six ways of conceptualizing interaction

Gibson-Graham (2002) goes further than Crang (2013) or Thrift (2000), arguing that there are in fact six ways that the global and local, and the interaction between them have been conceptualized in geographic writings:

1 The global and the local do not exist – they are just ways of 'framing' things.
2 The global and the local each get their meanings from what they are not; that is, in opposition to the other.

3 The global and the local offer different points of view concerning social networks.
4 The global is the local – all global things have local expression. Multinational firms are actually multi-local.
5 The local is the global – the local is where global processes interact with the surface of the Earth.
6 All spaces are glocal – the global is constituted by the local and vice versa.

'globalization'. In particular, in many studies, the lines between globalization as discourse, process and outcome are blurred. In this context Dicken notes (2004, p. 7): 'The problem is that . . . material processes are themselves enmeshed within a web of *discourses* from which material processes need to be disentangled – although, of course, the discourses themselves influence material processes and outcomes.' In this context, according to Yeung (2002), globalization does not have determining power in and of itself, and this idea is similarly refuted by a number of commentators, including Urry (2003) and Hay and Marsh (2000). Rather, its importance is arranged around two crucial themes. First, it reconstitutes scale. As we have seen in the above discussion, globalization leads to a collapsing of scales and increased interpenetration between them. This does not mean, as Thrift (2002) mischievously argues, that scale does not exist; but it has become more fluid. The dominant process in this respect then is that in the lived experience of humans the local grows (becomes more extensive) and the global shrinks (is compressed), leading to new 'glocal' outcomes. This requires that governments, businesses, people – agents in general – have to transform their activities and responses to cope with this. Second, there is a geography of globalization as discourse (i.e. how it is used – and abused – by different social groups to justify certain interventions and readings). In short, globalization is historically, spatially and politically contingent. To respond to it effectively in any given situation, it has to be understood within the context of that situation (Jones, 2009).

These arguments remind us that geography does indeed matter. The constraints of geography are changing as space and scale are transformed; this creates new geographies of old and new places, which are undoubtedly harder to understand than ever before.

Further reading

Castells (1996): This is the classic book on the network society which has been very influential in subsequent geographic thought. See also Stalder (2006).

Crang (2013): This chapter provides a useful overview of geographers' concepts of global–local interaction.

Gibson-Graham (2002): This chapter explores the links between the global and the local in a more detailed fashion than Crang and provides compelling arguments for glocalization.

Harvey (1989): It is in this book that the idea of time–space compression is most completely explored and used as a means of explaining capitalist expansion across the globe. See also Harvey (2007) and Harvey (2011).

Held *et al.* (1999) Chapter 1: This is the classic chapter on the three theses of globalization and should be a starting point for anybody attempting to understand the competing perspectives that exist. See also Held and McGrew (2007a, 2007b).

Herod (2003): This chapter provides an excellent and readable review of critical notions of scale as used in human geography.

Little (2013): An excellent chapter on the concept of space as employed in geography.

Rosenberg (2005): In this article, a 'post-mortem' is provided for globalization theory. This should be read critically but provides a stimulating piece.

Scholte and Robertson (2007): This is one of the best references dealing with globalization that is available, and it deals with many of the issues covered both in this chapter and this book in general.

Sparke (2013): This book is excellent on the relationship between globalization and geography – see especially Chapter 8 on space.

3 ▶ Globalization and time – historical transformations

- Competing global histories
- A hyperglobalist view – globalization as unprecedented
- A sceptical view – globalization in question
- A transformationalist view – global transformations
- Waves of globalization – a spatial framework
- Conclusion – old myths and new waves

Competing global histories

Interpreting the history of globalization is deeply contested, and different perspectives are greatly influenced by discourse, ideology and definition. Few works tackle the subject through the use of empirical evidence, although Held *et al.* (1999) Hirst and Thompson (1999) and Hirst *et al.* (2009) are notable exceptions. Below, four historical 'frameworks' are outlined. We begin by discussing a hyperglobalist and sceptical viewpoint respectively. This is followed by an analysis of two transformationalist histories – one sociological and one based in political economy. This section illustrates that the battleground over defining and conceptualizing globalization, let alone measuring it, is fraught. The chapter concludes with a simplified historical framework for interpreting the different waves of globalization that will be followed in the remainder of the book and asks to what extent might we be moving into a distinct contemporary wave following economic upheavals globally in the late 2000s. Underlying the discussion there are two questions: Is globalization new? What has driven it over time?

A hyperglobalist view – globalization as unprecedented

In 1990 Kenichi Ohmae, a business guru from Japan, wrote a book entitled *The Borderless World* which, if nothing else, served to elevate the debates concerning the nature of globalization to the popular stage. Since that time, the management analyst, who is head of a collective which seeks to reform the interventionalist Japanese government system, has written a number of books that have been influential beyond the management world. These works seek to support the neoliberal argument against government intervention. In *The End of the Nation State* (1995), the author develops an argument against the state and offers a chronological interpretation of history. He begins the book by claiming that the end of the Cold War was a watershed in the history of the global economy and society. Contrary to Fukuyama's (1992) proclamation that the collapse of socialism led to the 'end of history', he argues:

> nothing could be further from the truth. In fact, now that the bitter ideological confrontation sparked by this century's collision of 'isms' has ended, larger numbers of people from more points on the globe than ever before have aggressively come forward to participate *in* history. A generation ago, even a decade ago, most of them were as voiceless and invisible as they had always been. This is true no longer: they have entered history with a vengeance, and they have demands – economic demands – to make.
>
> (Ohmae, 1995, p. 1)

He goes on to argue that such demands cannot be met by nation-states that 'no longer possess the seemingly bottomless well of resources from which they used to draw with impunity to fund their ambitions' (p. 2). Should these new participants in history turn to international bodies, such as the UN, they will find that they are nothing more than collections of nation-states. Economic groupings such as OPEC, APEC or the EU are much the same and, according to Ohmae, incapable of managing and meeting the demands of the 'new global citizen'.

Underpinning Ohmae's argument against nation-states is the increased mobility of what he calls the four 'I's: Investment, Industry, Information technology and Individual consumers. He argues that each of these has become increasingly footloose and pervasive and that, only by tapping into the new flows that have emerged, can welfare be maximized. He argues, then, for a world where the newfound mobility of the four Is is allowed to work in global space:

> Taken together, the mobility of these four I's makes it possible for viable economic units in any part of the world to pull in whatever is needed for development . . .. This makes the traditional 'middleman' function of nation-states – and of their governments – largely unnecessary. . . . If allowed global solutions will flow to where they are needed without the

intervention of nation-states. On current evidence they flow better precisely because such intervention is absent.

(Ohmae, 1995, p. 14)

Running through Ohmae's analysis is an argument for a new geography which transcends what he terms the 'cartographic illusion'. In the place of nation-states, Ohmae argues for 'region states' – 'where real work gets done and real markets flourish' (p. 5). Such units are not defined in terms of their formal political borders, as this 'is the irrelevant result of historical accident' (p. 5). Rather, it is essential that they be 'the right size and scale to be the true, natural business units in today's global economy. Theirs are the borders and the connections that matter in the borderless world' (p. 5).

As noted above, neoliberal thought greatly influences Ohmae's argument, and the work has been seen as the epitome of hyperglobalist claims in that it imagines a world where flows of capital, people, information and goods are unrestricted. It is a normative argument, in that it suggests that such governance (or lack of) is preferable and will maximize welfare, for 'where information reaches, demand grows, where demand grows, the global economy has a natural home' (p. 25). This is the recipe to get countries on what Ohmae calls the 'ladder of development', harking back to Rostow (1960). In terms of the policy implications that flow from his argument, he offers the following:

> The evidence, then, is as exhaustive as it is uncomfortable: in a borderless economy, the nation-focused maps we typically use to make sense of economic activity are woefully misleading. We must, managers and policy makers alike, face up at least to the awkward and uncomfortable truth: the old cartography no longer works. It has become no more than an illusion.

(Ohmae, 1995, p. 20)

Ohmae's chronology of the rise of globalization is straightforward and somewhat simplistic. He divides history into two periods – the industrial age and the information age. The first, which corresponds to the nineteenth and twentieth centuries, was driven by nation-states, national sovereignty and centralization. This led to flows that were disrupted by borders and protected domestic capital above international capital. The centrality of the nation-state grew out of colonial models, and newly independent states adopted this model of governance – what Ohmae terms 'post-feudal relations' (p. 142). Ohmae views globalization as entirely new, beginning in the 1990s. To paraphrase the author, the 'economic' pendulum swung at the end of the twentieth century when the end of the Cold War and the triumph of capitalism paved the way for a new information age reducing the 'cholesterol' in the arteries of the global economy. This new age is driven by private capital and autonomous networks and is characterized by citizen sovereignty. It welcomes foreign capital and is based on governments nurturing entrepreneurship. In terms of the winners and losers between the two periods, Ohmae argues that new region states, such as Hong Kong/ Shenzhen, Singapore/Johor/Batam, Silicon Valley and New Zealand, benefit. The old winners – Japan, the UK, the USA and Germany – all lose out relatively speaking as their strait-jacketed governance systems become ineffective.

Criticisms of the hyperglobalist thesis

Ohmae's arguments have been criticized roundly from many quarters and much of what follows in this chapter, and indeed this book, takes issue with a number of the central points he makes. A few of the more important criticisms are considered below:

- *Idealization or reality* – It is unclear, as Ohmae writes, whether he is talking about an ideal world (in his eyes) of borderless flows or if he is claiming that this is what is actually occurring.
- *Empirical data* – Linked to the above, the use of empirical data to back up his major points is scant and anecdotal. In *The End of the Nation State*, for example, he uses a journalistic article from the *Japan Times* (Peter Osbourne, 23 December 1994) as the main evidence for his argument that New Zealand is a winner in the new borderless world following its neoliberal reforms in the mid-1980s.
- *Unilinear model of development* – Ohmae explicitly argues for a ladder of development where economic growth leads to an ideal end-goal. In this sense, his argument is a-historical and a-geographical.
- *Characterization of the nation-state* – In order to attempt to demolish the nation-state, Ohmae paints a characterization of it that is unfounded. He suggests that national governments are inherently bad, yet no evidence is offered to back this up.
- *Dualistic interpretation of history* – History is reduced to two periods with no particular rationale for the transition between them and with enormous generalization with respect to the conditions in each.
- *Favouring the networked elite* – Ohmae is way too optimistic with respect to the implications of the new information economy for those

that are not in networks. There is no credible evidence to suggest that free markets close the wealth gap. The information gap, for example, is growing; and a policy to reverse this is not suggested.

- *The region state* – The region state is poorly defined. What does it look like? How does it function with respect to other scales? This idea, which is central to Ohmae's argument, is barely developed and the concept of scale is not dealt with coherently.
- *Internal contradictions* – Ohmae talks of the role of governments in nurturing entrepreneurship and of the 'borders that really matter'. The use of national governments and borders to illustrate his argument contradicts his attempted demolition of both.

A sceptical view – globalization in question

Hirst et al. (2009) question the very existence of globalization and some of the claims that have been made in its name, and argue that the world economy has become less integrated over time. Theirs is the sceptical position which offers a salient antidote to the 'hyperglobalist' approach. It is also one which, in the words of one reviewer, is 'systematically grounded in evidence; the literature on globalization contains all too few examples of this' (Perraton, 2001, p. 670). In *Globalization in Question*, Hirst and Thompson (1999, pp. 2–3) make five central points that highlight the exaggeration of claims made by the hyperglobalists:

1 Today's levels of integration fall short of the period between 1870 and 1914 (what is referred to as the classical Gold Standard era). In this respect they argue: 'The present highly internationalised economy is not unprecedented: it is one of a number of distinct conjunctures or states of the international economy that have existed since an economy based on modern industrial technology began' (p. 2).
2 Contrary to the literature from some areas of business studies, truly transnational corporations are scarce. Such companies are embedded nationally, and they do their trade internationally.
3 The vast majority of foreign direct investment (FDI) is concentrated in advanced capitalist economies, and this trend is becoming exacerbated. Integration is not leading to the diffusion of investment from such nations to peripheral ones – apart from a few leading 'emerging nations'.
4 The majority of the world's economic activity remains concentrated in a 'triad' of regional blocs, namely the countries associated with the

North American Free Trade Agreement (NAFTA), the EU and East
Asia including China and Japan. Much global trade flows within and
between these areas and 'this dominance is set to continue' (p. 2).

5 The core economies maintain the ability and wherewithal to regulate
the global economy, its firms, financial markets and other less
powerful nation-states. To do so, they often have to act in cohorts; but
this has been fruitful. The authors refer to the triad powers as the G3
and argue: 'Global markets are by no means beyond regulation and
control even though the current scope and objectives of economic
governance are limited by the divergent interests of the great powers
and the economic doctrines prevalent among their elites' (pp. 2–3).

Hirst and Thompson use two major measures to illustrate that the global
economy is no more open today than it was in the early nineteenth
century. The first is the ratio of trade levels to GDP: in 1914 trade/GDP
ratios were higher in the major economic powerhouses than they were in
1973. By 1995, the major economies of Japan, Netherlands and the UK
were still less open using this measure (while France and Germany were
slightly more open) (see Table 3.1). They do concede, however, that the
USA was substantially more open in 1995 than it was in 1914. They
make the important point that if analysis concentrates on the period after
the Second World War only, then growth in trade openness is steady and
consistent. This, they claim, is underpinned particularly by the rise of the
East Asian countries as trading economies.

Hirst and Thompson argue that 'the evidence also suggests greater
openness to capital flows in the pre-First World War period compared to
more recent years' (1999, p. 27), a finding which is supported by
Grassman (1980) and Lewis (1981). There is, however, evidence to

Table 3.1 *Ratio of trade to GDP, advanced economies, 1914 and 1995 (exports and imports combined)*

	1914	1995
France	35.4	36.6
Germany	35.1	38.7
Japan	31.4	14.1
Netherlands	103.6	83.4
UK	44.7	42.6
US	11.2	19.0

Source: Adapted from Hirst and Thompson (1999, p. 27)

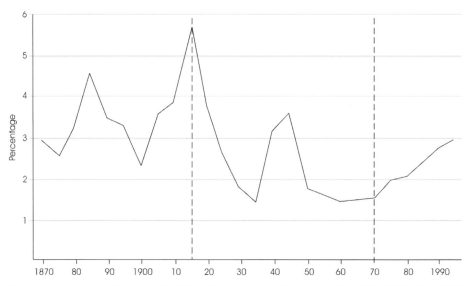

Figure 3.1 *International capital flows between the G7 economies 1870–1995 (% of GDP)*

Source: Adapted from Hirst and Thompson (1999, p. 28)

suggest that, again, since the Second World War, levels of capital integration have indeed risen (see Figure 3.1). In conclusion, building in evidence for decreased per capita migratory moves since the start of the twentieth century, the authors argue:

> [W]e can say that the international economy was in many ways more open in the pre-1914 period than at any time since, including from the late 1970s onwards. International trade and capital flows, both between the rapidly industrialising economies themselves and between these and their various colonial territories, were more important relative to GDP levels before the First World War than they probably are today. . . . Thus the present position is by no means unprecedented.
>
> (Hirst and Thompson, 1999, p. 32)

As Perraton argues (2001), there is a clear normative agenda in what Hirst and Thompson seek to establish through their work. In arguing that current levels of integration are a result of political choice and change rather than technological 'progress', they seek to establish that claims of 'inexorable globalization' are merely a smokescreen to further the neoliberal agenda. This implies that 'globalization' can be reversed and governed. As Perraton argues: 'They repeatedly emphasise links between globalization and advocacy of neoliberalism as the only viable policy option' (2001, p. 671).

Criticisms of the sceptical thesis

There are numerous critiques of Hirst and Thompson's thesis but the most complete is that undertaken by Perraton (2001). While broadly supportive of their work, and in particular their recourse to empirical fact rather than speculation, his central critique is that the methodology is faulty. Hirst and Thompson draw their major conclusions by comparing existing trends with an imagined 'hyperglobalized' economy. Perraton, as well as Held *et al.* (1999), argues that this is teleological and that the evidence should be redirected to provide 'a more nuanced assessment of the impact of contemporary integration' (2001, p. 678) rather than to dismiss the concept of globalization altogether. He argues that, by comparing current levels of integration with the hyperglobalist thesis, the work is:

> if not quite a 'straw man' conception of globalization, GiQ [*Globalization in Question*] still focuse[d] on a particular version of globalization analysis that commands very limited academic support. . . .
> Fundamentally it is unhelpful to study economic processes in terms of single implied end-point.
>
> (Perraton, 2001, p. 672)

The most important criticism is that of the proposition that global levels of integration are not unprecedented. Perraton argues that the measures employed by Hirst and Thompson are too crude to capture the changing nature of the global economy and, in particular, its *qualitative* aspects. One of the most notable trends has been the rise of cross-border transactions in bonds and equities, which have risen exponentially since the 1970s (see Table 3.2). Overall, Perraton (p. 675) argues:

> there is substantial evidence of unprecedented levels of trade, FDI and international financial activity relative to national economic activity. . . .
> Simply comparing crude indicators across time periods misses the changing character of contemporary integration, notably the growth of MNCs and the rise of short-term international financial flows.

Perraton offers five counterpoints to Hirst and Thompson's central claims:

1 Financial and trade integration moves have advanced further than ever before; although integration is not unprecedented, global markets have emerged.

2 True TNCs are rare, and, although they retain a national base (they have to), their mobility has increased making the sharing of the profits – through taxation, for example – within nation-states harder.

Table 3.2   *Cross-border transactions in bonds and equities, 1980–1998 (% of GDP)*

	1980	1990	1998
USA	9	89	230
Japan	8	119	91
Germany	7	57	334
France	5	54	415
Italy	1	27	640
Canada	9	65	331

Source: Adapted from Perraton (2001, p. 674)

3 There are uneven patterns of economic activity, but no bloc is insulated from the rest of the world.
4 Many poorer countries are indeed peripheral and networks bypass them. However, this does not imply that global markets are non-existent. Global markets are not likely to be perfect and therefore incomes are not likely to converge globally.
5 Nation-states still regulate markets but they are increasingly constrained by market shifts and international agreements.

Perraton argues in conclusion that 'global markets have evolved for goods, services and financial assets. As such, this goes beyond "internationalisation" and can reasonably be characterised as globalization' (2001, p. 682). This conclusion is echoed most explicitly in human geography in the work of Dicken (2004) who argues that Hirst and Thompson fail to distinguish between the shallow integration of the pre-1914 period and the deep integration of the contemporary era, stating that 'it is in a qualitative sense that the material processes of globalization must be analysed' (p. 8). To Ash Amin, the argument fails because 'it does not provide any sense of the trends and changes in the world economic system which might genuinely be challenging the balance between national and global influences' (1997, p. 124).

A transformationalist view – global transformations

There are numerous 'transformationalist' views on the history of globalization including an excellent exposition by Waters (2001). Particularly useful alternatives are offered in Rosenau (1990), Harvey (1989), Robertson (1992, 2003) and McGrew (2008). Generally, transformationalist perspectives are more holistic than hyperglobalist or sceptical accounts in that they emphasize the importance of culture and politics as well as economics. Here Held *et al.*'s (1999) chronology is discussed in detail.

The authors take a broadly transformationalist view of the evolution of globalization over time and make two central points. First, they suggest that 'globalization is neither wholly novel, nor a primarily modern social phenomenon. Its form has changed over time and across the key domains of human interaction, from the political to the ecological' (p. 415). Second, they state that the processes of globalization do not unfold according to some unilinear or inherent logic. They say: 'globalization as an historical process cannot be characterised by an evolutionary logic or an emergent teleos. Historical patterns of globalization have been punctuated by great shifts and reversals, while the temporal rhythms of globalization differ between domains' (p. 415).

Premodern globalization – pre 1500

This period began around 9,000 to 11,000 years ago with the rise of distinct centres of settled agrarian civilization in Eurasia, Africa and the Americas. These civilizations developed relatively long-distance trade and, crucially, the projection of power across space – although these two things were limited by transport and communications technology. During this period, it was the Eurasian realm that was characterized most by interregional and/or intercivilizational flows, while Oceania and the Americas remained largely autarkic.

Globalization during this period was evident in four areas: military/ political empires, world religions, migratory movements and – to a lesser extent – trade. In the case of empires, the earliest known example was that of the Sumerian empire (3000 BC), and subsequent empires differed greatly in their extent and endurance. In contrast to the later periods of 'global' empires, most were relatively small, although some went on to form important regional-level organizations which set the stage for encounters in the subsequent centuries – including the Indic civilizations, the Han Dynasty, China and the Roman Empire. Early world religions formed during this period too, crossing imperial boundaries with universalized messages and texts – thus unifying many disparate cultures. The most important religious development at this time was the rise of

Islam in the sixth century which some see as the first globalized world religion.

Migratory movements of those in nomadic empires including the Germanic people (who eventually settled in Western Europe having conquered the Roman Empire in the fifth century) and later the Mongols caused major upheavals through conquest during this period. Held *et al.* make the point that such movements, creating devastation and plague in their wake, did not leave as great an imprint as the slow diffusion of settled agriculture into hunter-gatherer societies which took place throughout the period.

Some forms of economic interaction such as trans-Eurasian and Indian Ocean trade assumed importance in this period. For example, Muslim traders from the Indian subcontinent established an enduring presence on the east coast of Africa. It is of course highly difficult to quantify such flows. What is important though is that such flows represented conduits for the diffusion of cultural and social ideas, as well as technology, as interaction among elites intensified. As such, one-off interactions were often spectacular, but sustained political and military interactions and extension of political control were difficult to sustain. In this sense, the degree to which global interactions were institutionalized and regulated remained minimal. According to the authors this period could be termed one of 'thin globalization' which was limited in its global extensity.

Early modern globalization – 1500–1850

The sixteenth century is the period that many see as the starting point of the 'rise of the West' (Held *et al.*, 1999, p. 418). It was during this period that Europe developed the technologies and institutions that facilitated modernity and led, eventually, to the establishment of the European global empires. The period is sometimes referred to as the mercantilist phase of colonialism and witnessed the first stage of the diffusion of merchant capitalism across the globe. In the early part of this period, the most notable expansion was that of the Spanish and Portuguese into the 'New World' in search of precious metals and minerals and other natural resources to trade (see the section on Latin America in Chapter 7). By the mid-1500s, virtually all of South and Central America was colonized (see Map 3.1). Held *et al.* are sceptical of 'conventional accounts', arguing that European expansion was opportunist and fragile and that much of the innovation which facilitated this was in fact imported and

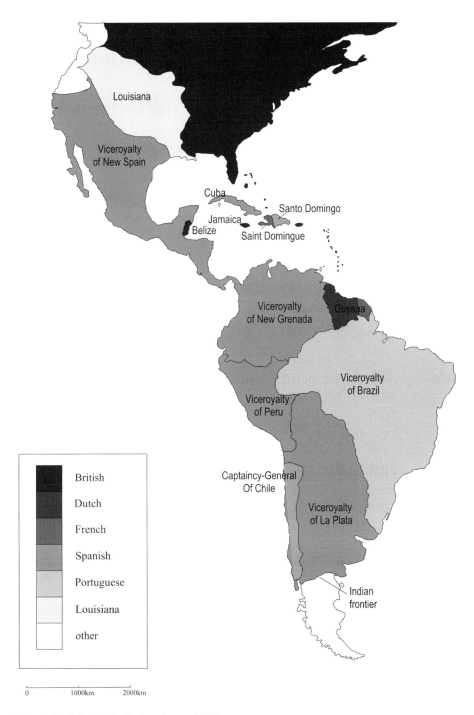

Map 3.1   Colonial Latin America, *c.*1780

adapted from other regions. From a geographical point of view, they also emphasize that the European 'global' empires were in fact spatially partial; Africa and Asia remained largely untouched. However, the new global reach of Europe, together with changes in technology and politics in the heartlands, led to major transformations across the world. As Europe's political and military empires flourished, new forms of economic globalization were forged. Examples included the trading organizations of the Dutch and English East India companies, which were early precursors to TNCs. In general, European expansion into other lands took the form of nodes between which 'ran the thickening sinews of global economic interaction' (p. 419). Particularly important in terms of global economic flows during this period was the Atlantic slave trade (see Map 3.2), which only began to dwindle starting at the end of the seventeenth century as the European heartlands were industrialized and colonial corporations were replaced by formal government. This trade was made illegal by many European nations and the USA by the 1820s.

As in the premodern period, the largest interactions were sporadic and concentrated at the levels of elites. This is evidenced by the fact that by the end of the era the Spanish and Portuguese empires in Latin America and the British and French empires in North America had gained independence. In other words, 'spectacular European thrusts' could be made but it was difficult for the Europeans to retain full control over their colonies. Notwithstanding this the cultural impacts of colonialism, especially in the form of the diffusion of the Catholic Church, were profound (Gwynne and Kay, 2004) (see Plate 3.1).

The infrastructure which facilitated this period of globalization remained much the same as it had been in the premodern period. By the end of the eighteenth century, railways, iron ships, canal and road building were more extensively developed – although their existence did not often stretch beyond the borders of the heartlands of the Western empires themselves. The velocity of these flows, however, was ultimately limited by the speed of shipping and other less advanced technological transport forms such as the horse. The institutionalization and regulation of global relationships, as in the premodern period, remained limited. There was a nascent system of diplomacy in Europe between states, and the Roman Catholic Church increased its influence – but the ability of such institutions to regulate the periphery was relatively minimal.
As in the previous period, one-off encounters often had devastating effects but long-lasting impacts were often yielded by less spectacular interactions.

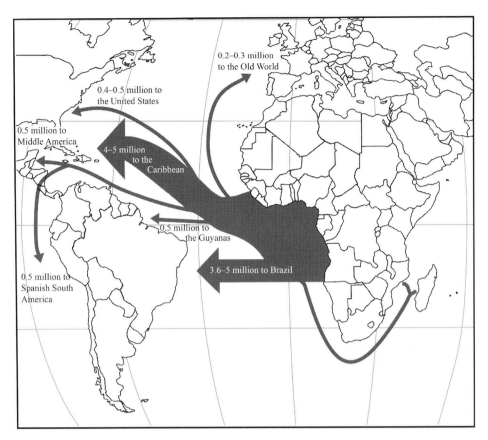

Map 3.2 *The Atlantic slave trade, c.1500–c.1820*

Arguably, however, the greatest impact of the new global relationships was felt in Europe itself where the wealth from the New World, transatlantic migration and the stimulation of inter-imperial rivalry led to new institutional power rivalries and a military technology race. Indeed, it is during this period that the seeds of the great European conflicts of the twentieth century were sown. During this period, too, the development of formal states, with demarcated territories and fixed borders, began in Europe (see Chapter 5 for details). The creation of the nation-state was to become one of the major defining processes which would underpin the subsequent period of globalization according to Held *et al.*

Plate 3.1 *Waves of globalization in Santiago, Chile*
A colonial church in Santiago, Chile, reflected in a modern building
Source: Photo by authors

Modern globalization 1850–1945

This period 'witnessed an enormous acceleration in [the] speed and entrenchment' (Held *et al.*, 1999, p. 421) of global networks and flows principally under the control of the European powers. The era has sometimes been referred to as industrial colonialism – based on the expansion of the industrial core party in search of raw materials for their new production techniques (Daniels and Lever, 1996). Four processes in

Europe, which to a large extent were underpinned by events in the previous period, were instrumental in this respect: the development of capitalist economies; the rise of advanced weapons technology; naval technology; and the rise of powerful state institutions. Of these, the rise of capitalism based around the industrial revolution was of greatest importance. These changes were put into motion to the effect that, by the mid-twentieth century, European powers had 'exploded' across the globe taking in Africa, much of Southeast Asia directly and making important controlling inroads into East Asia – an area from which Europe had been formerly excluded. This was the age of imperialism. As global trade and investment increased massively, the extensity, intensity, velocity and impact of globalized flows increased. There were large migrations across the Atlantic, and the end of the slave trade led to the export of indentured labour from India to many parts of the world. This was underpinned and facilitated by the rise in transport and communications technology – especially more advanced railway systems and the telegraph.

It was the global reach of the European and American empires and their political and military operations that characterized this period (see Map 3.3). Given communications advances, political influence could be exerted at great distance. This led to economic globalization at

Plate 3.2 *A colonial-era building in Nairobi, Kenya*

Source: Photo by the authors

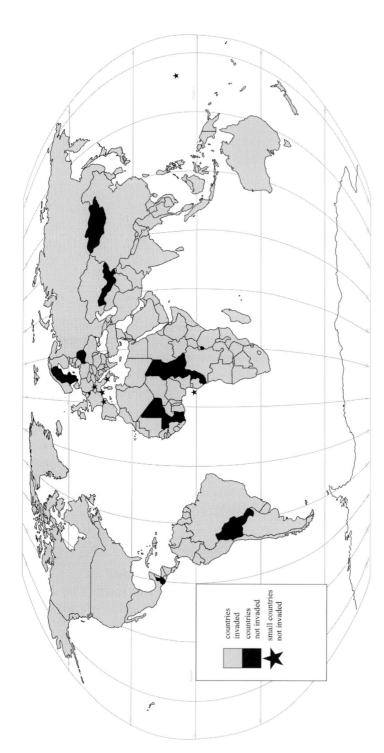

Map 3.3   Countries invaded by Britain

Note: This map is based on one which appeared in *The Telegraph* newspaper in 2012 (www.telegraph.co.uk/history/9653497/British-have-invaded-nine-out-of-ten-countries-so-look-out-Luxembourg.html, accessed 2 September 2013). This in turn was based upon the book *All the Countries We've Ever Invaded: And the Few We Never Got Round To* by Stuart Laycock (The History Press, 2012). Although this is a fairly tongue-in-cheek look at British intervention in other countries and the map is not an accurate representation of the places affected by British imperialism (a whole country is included even if only a small part of it was invaded), it does show how global was the reach of British military influence and interference over the past two centuries and more.

unprecedented levels as trade and financial investment, especially between the European core and its colonial periphery, increased. There was, relative to the current period, limited trade between richer nations. Furthermore, the massive migrations from Europe were paralleled by large migrations across the world from other societies, including the Japanese to Latin America and Indians to the Pacific Islands under the indenture system. Migrations within and to the core stimulated the expanding industrial complex.

In the cultural realm, during this period, the emphasis shifted away from the diffusion of Western religious discourses to the diffusion of secular discourses including science, Marxism and nationalism which eventually filtered down to the masses and transformed lives everywhere (see Chapter 5). Great advances in transport and communications technology made possible partly through the industrial revolution underpinned all of this. The unit costs of transport fell due to the extensive development of railroads and the steamship, and transoceanic telegraphy reduced communications times enormously. Thus elites at a distance could be kept well informed of events far away. Increased literacy and the advancement of the mass media meant that some sections of the population developed a global consciousness (Robertson, 2003). What set this period apart from the earlier period in particular was the roles of institutions in underpinning globalization processes. For example, migratory flows were heavily regulated by states for the first time. In the economic arena multinational banks' rise in importance and the Gold Standard (see Chapter 4) regulated trade and financial sectors. There was a range of other common global standards that evolved in this period, emanating from the West, in order to facilitate increased global interactions. This bears much in common with the model of modernization suggested by Rostow (1960).

The period of modern globalization was shattered by the First World War, which came about partly due to the rising rivalry between the European powers which imperialism both reflected and perpetuated. The end of the war led to a collapse in global trade and investment, and the Great Depression which followed saw an end to the Gold Standard and other systems of global regulation. The period of openness was then replaced by imperial preference in Western empires and autarky in the USSR and Germany. Ironically then it was this imperial war, which inspired unprecedented imperial war efforts, which laid the seeds of the destruction of the great empires.

Decolonization led to a new order – which arguably pervades until today:

> In their place [the European empires] arose a hegemonic USA which would establish the formal and informal structures of global governance alongside which a renewed wave of globalization, supported by new technologies and infrastructures of interaction, would cross a world dominated, not by amorphous empires, but by territorially demarcated nation-states.
>
> (Held *et al.*, 1999, p. 424)

Contemporary globalization 1945–

Held *et al.* (1999, pp. 424–425) argue that contemporary globalization is a distinctive phase in a range of ways. Counter to sceptical theses, they argue (p. 425):

> in terms of the extensity of global linkages, the appearance is one of catching up. . . . But we argue that in nearly all domains contemporary patterns of globalization have not only quantitatively surpassed those of earlier epochs, but have also displayed unparalleled qualitative differences.

Innovations in transport and communications technology have been extraordinary, the emergence of institutions of global governance is unprecedented, and there has been a 'clustering' of patterns of globalization across many domains. New migratory patterns, such as refugee and asylum seeker movements, have been established. There has been a major globalization of environmental impacts of industrial production and consumption as well as concern about it (see Chapter 8). The velocity of social-cultural flows has increased enormously as communications technology has advanced. The rise of global tourism to become a major economic sector, as well as unprecedented levels of migration, are testimony to the changes in facilitating technologies (see Plate 3.3). In the previous periods of globalization, cultural interactions were often elite-to-elite, but there has now been a proliferation of popular culture. The dominance of English as the major *lingua franca* across the globe is an element of this cultural shift which also sets it apart from the other periods, together with the increased role of private capital in cultural diffusion to the masses.

From a geographical point of view, one of the most important aspects of Held *et al.*'s transformationalist argument is that simple

Plate 3.3 *Waikiki Beach hotels, Hawai'i*
Honolulu and Waikiki Beach became one of the most famous international tourist destinations
following the rise of cheaper air travel in the 1970s. Most tourists come from mainland
USA or Asia.

Source: Photo by the authors

global dichotomies such as 'North–South' and 'core–periphery'
have been eclipsed by more complex patterns of economic
distribution and political power relations. These flows are uneven,
however:

> This pattern of interconnectedness is unique to this era. That said, the
> enmeshment of different states and of different social groups within those
> nation-sates, and their relative levels of control over those flows, remains
> highly uneven.

> (1999, p. 427)

Held *et al.* (see 2007a), describe the current era as 'thick globalization',
which is distinct from previous phases in a range of ways (see Box 3.1).
Notable points are that the current period is characterized by decreased
Westernization compared to modern globalization, territories are more
fixed relative to the past, and governments play a bigger role in national
affairs than ever before. These points contradict hyperglobalist arguments
and exemplify the benefits of a long-run historical perspective.

Ultimately, the authors do not see the phenomenon unfolding along a predetermined path – underlying these transformations is no unifying logic, but rather a collection of forces that may or may not converge at any given time.

Box 3.1

Distinguishing features of the current era of globalization

- *Spatio-temporal* – Unprecedented extensity, intensity, velocity and impact of global interactions and flows in all domains.
- *Organizational* – Unprecedented worldwide social, political and economic regulation of power relations through new institutions of multilayered and multilateral governance.
- *Reflexivity* – A worldwide elite and popular consciousness of global interconnectedness as never before. National elites versus transnational social forces are the main groups of contestation as opposed to the coercive empires of the past and their subjugated colonial territories.
- *Contestation* – A new global politics of agenda setting, coalition building and multilateral regulation often designed to contest globalization. This was an internal imperial affair up until this era.
- *Regionalization* – Regional blocs have replaced imperial blocs. These blocs are not autonomous but link into and mutually reinforce each other.
- *Westernization* – Until the current period, globalization was largely synonymous with the spread of Westernization.

Arguably, today's forms of globalization are less Eurocentric.
- *Territoriality* – Contemporary globalization is associated with a different kind of politics of territoriality. Territorial control and governance has always been at the core of globalization. The threats to territory are not just military or external; territories are now *more* fixed.
- *State form* – Compared to earlier phases of globalization, states are characterized by 'big government' (despite recent reforms which have sought to reduce this) where states spend a large proportion of GDP and have responsibility for management and welfare of populations. States are thus more visibly impacted by globalization today.
- *Democratic governance* – Individual states are democratic (or most claim to be) while global governance structures have limited elements of democracy. Combining democratic ideals of national governance with the global organization of social and economic life raises unique dilemmas.

Source: Adapted from Held *et al.*
(1999, pp. 430–431, and 2007a)

Reflections and criticisms of the global transformations argument

Held *et al.* provide a very detailed and historically rich interpretation of the evolution of globalization. In doing so, they seek consciously to avoid repeating the well-trodden histories of the development of Europe and how this has underpinned what we have come to know as globalization. In this sense, to a certain extent, they decentre the debate, shifting the focus away from Europe and to the other parts of the world which have simultaneously, and in very different ways, been exposed to the forces of globalization. Further to this, their approach is multidisciplinary, combining commentary, analysis, measurement and interpretation of economic, social, technological and other forces of change and how these have combined to create the grand sweep of global history. This is a very ambitious undertaking – one that is bound to be fraught with difficulties regarding overgeneralization. However, they do escape much of the determinism of other interpretations from single disciplines.

A number of criticisms can be made of the thesis, however. There is no force underlying globalization and no theoretical rationale for the increased velocity, intensity and extensity of global interactions. Whilst there is no inherent or unilinear logic posited by Held *et al.*, on the other hand there is a sense of inevitability in the transformations they discuss which is not theorized. A further criticism refers to the focus on Europe and Europeans as the central agents in globalization. Chronologies from the remainder of the world would have a very different perspective, and there remains a gap in the geographical literature in terms of alternative chronologies of globalization in this respect. Their definition of globalization is also very broad; one might argue that it effectively captures everything. Every migratory movement, every trade flow becomes evidence of 'globalization' under this schema. Geographers should be critical too of their dichotomous use of the terms 'global' and 'local' which, as was argued in Chapter 2, need greater critical reflection in order to gain analytical power. Notwithstanding these criticisms, there can be little doubt that the periodization set out in *Global Transformations* is the most complete and coherent interpretation of globalization that exists, and one which coherently captures its nature at the turn of the millennium (see Box. 3.1).

Waves of globalization – a spatial framework

There can be no doubt that we are in an accelerated period of globalization. However, that globalization is not fully complete is also obvious, and whether it ever will be is doubtful. In this sense, it is perhaps more accurate to talk of a globalizing world as opposed to a globalized world. The processes and agendas that are leading to this are contradictory, dialectical, complex and heterogeneous. This is why mapping the impacts is both a necessary and highly challenging task for geographers, and one which has only just begun. In undertaking this endeavour it is important that spatial elements be placed to the fore and that the relationship between space and time is emphasized. This approach offers a novel perspective on the chronologies and theories of globalization that have been discussed in the literature, providing a geographical take which we argue is sorely required.

A framework of the waves and phases of globalization is offered below which draws together a number of the ideas discussed so far in this book. Particular emphasis is placed on transformationalist accounts influenced, partly, by world systems theory and, to an extent, regulation theory, and by the periodization work of Robertson (2003), McMichael (2013), Held *et al.* (1999, 2007a) and Firth (1999). For this framework, the definition of globalization developed in Chapter 1 is used.

The wave framework begins in the early 1500s with the rise of the Hispanic empires. This is not to suggest that processes resembling globalization did not exist before that point or that some civilizations, given limited horizons, believed they were acting in a 'globalist' fashion (Robertson, 2003). These were, arguably, forms of nascent globalization that did not attain the stretching of systems and networks that took place with colonization of the New World. In general, the velocity of interaction was greatly impeded by the low level of technology and the poor infrastructures that existed. The Romans and the Incas, for example, had roads that only extended to the edge of their respective empires, and canals in China were in their infancy. In general, then, the vast majority of the world's population remained isolated.

The framework sees the process divided into two *waves* (dates approximate): colonial globalization (the first wave), *c.*1500–*c.*1945, and postcolonial globalization (the second wave), *c.*1945–. Each of these waves is broken down into two further sub-waves referred to here as *phases.* The first wave proceeds through the mercantilist phase (*c.*1500–*c.*1800), and the industrialist phase (*c.*1800–*c.*1945). The second wave moves through the modernization phase (*c.*1945–*c.*1980), and the current neoliberal phase (*c.*1980–) (see Table 3.3). Between each of these waves are major restructuring events that disrupt the dominant – though not ubiquitous – world order, an approach which builds on ideas from the regulation school (see Chapter 5). As the waves proceed, time–space compression intensifies – though not evenly across time and space. As each world order replaces another, a dominant spatial configuration is forged that involves a new relationship between time and space.

In the first wave, the transition between mercantilism and industrialism was caused by major shifts in Europe, based partly on processes unleashed by mercantilist colonialism that led ultimately to the industrial revolution and the consolidation of capitalism. This cultural shift had profound economic and political impacts. It led to intensified competition between newly industrialized powers and the globalizing search for resources and locations for the investment of surplus capital. It is during this period that the 'core–periphery' structure of the world economy theorized by dependency and World Systems thinkers crystallized. There existed a relatively affluent industrialized core and a relatively impoverished periphery from where raw materials were sent (see Chapter 4). There was a range of impacts across the periphery of the spread of colonialism, according to the particular contingency of place. There were major politico-military impacts (the decline and destruction of many non-European states and the modernization of those not directly colonized), trade and investment impacts (especially the locking in of

Table 3.3 *Waves of globalization – a framework*

Wave	Period (approximate dates)	Restructuring crisis
Wave 1	Colonial globalization (c.1500–c.1945) Mercantilist phase (c.1500–c.1800) Industrialist phase (c.1800–c.1945)	Industrial Revolution Great Depression and the Second World War
Wave 2	Postcolonial globalization (c.1945–) Modernization phase (c.1945–c.1980) Neoliberal phase (c.1980–)	Oil Crises

many regions as primary product suppliers), migratory impacts (especially to North America but elsewhere too) as well as cultural impacts (principally the spread of Christianity in some areas and the circulation of Western secular discourses in others).

It was during the interwar period that the first attempt at a truly global system of governance was attempted in the form of the League of Nations established in 1919. However, imperial rivalry meant that it was not as effective as had been hoped. It was unable, for example, to prevent the Second World War, which arose due to continued rivalry in Europe and the mistakes made during the post-First World War settlement. It was this second 'global' war that finally led to the collapse of the European empires and the first wave of globalization. Weakened and exposed by the war, the colonial powers were unable to prevent a tide of decolonization from taking place. Ironically, a generation of independence leaders and their movements employed concepts of nationalism diffused by the colonizers themselves.

The major historical moment defining the start of postcolonial globalization and the rise of the second wave was the Second World War, leading to the end of empire and the emergence of the Cold War. Military power relations between the two superpowers, the USA and the USSR, took on global proportions (see Chapter 5), pulling in many of the world's new nation-states to one side or the other. At the same time, the League of Nations' successor, the United Nations, was established; and this set a new precedent in terms of global governance. Working in parallel with these trends, the Bretton Woods institutions (see Chapter 4) – the World Bank, the IMF and the GATT (General Agreement on Tariffs and Trade) – have fostered economic globalization. Following the 1973 oil crisis in particular, these institutions played a crucial role in the spread of free-market neoliberalism across the world. The deregulated governance model has diffused globally – even to the last remnant of communist societies in China and Vietnam, for example – propagated by GATT's successor, the WTO. TNCs benefitted and flourished in this new economic environment and have been key agents in the diffusion of neoliberalism; in this sense, the core–periphery structure of the world was disrupted by more complex systems of inclusion and exclusion where borders became less important and networks of power and privilege and flows of capital, goods and ideas came to dominate.

The two World Wars, then, punctuated by the Great Depression, led to the crisis in the colonial world order and eventual decolonization. This marked the shift to a postcolonial period where control was affected through new cultural and economic means rather than through overt political ones, fostered in part by the development of globalized political institutions which were designed to support this hierarchy. In the first period of the second wave – the modernization phase – state developmentalism promoting TNCs was the primary motor for the diffusion of the culture of capitalism (McMichael, 2013; Daniels and Jones 2013). A restructuring crisis, brought about by the oil hikes of the 1970s, culminating in the debt crisis, led ultimately to the rise of global neoliberalism. The end of the Cold War seemed to affirm the dominance of the democratic, capitalist and generally economically liberal state, which has been propagated as the desirable form of governance by a host of global institutions. The age of empires had come to an end as we saw a worldwide system of nation-states overlain by regional, multilateral and global systems of governance and regulation. Even the world's only superpower and hegemon, the USA, did not practise overt colonization – through formal political means at least. The fact that patterns of globalization are not driven by coercive institutions of empire, sets the current wave apart from earlier periods.

In the neoliberal phase, the power of the state is challenged by global institutions that seek to further the penetration of capitalism and also by civil society that grows increasingly frustrated by the state's ineffectiveness in a multi-scalar world. At the core, then, is the diffusion of the culture of capitalism that seeks to reduce turnover time of capital through the development of technologies of time–space compression.

It should be stressed that this is not economic determinism – capitalism is a system that has a specific *culture* which produces economic outcomes and political change. The ideologies of capitalism and the traits which comprised it evolved from the Enlightenment period in Europe. It is also highly variable in its spatial expression, as the recent rise of alternative capitalisms in East Asia, Latin America and Eastern Europe attest (see Gwynne *et al.*, 2003). The outcome of each period is to create a system of imperialistic power relations (Harvey, 2003, 2005, 2011) – the first wave involving explicit territorial imperialism and the second, more insipid less spatially explicit neo-imperialism (these ideas are discussed in subsequent chapters). In the first wave, asymmetries in power were articulated spatially largely between nation-states. The second wave is much more complicated spatially, involving partially disembedded networks of power which are less bounded by territorial space. Globalization then comprises both agendas and processes and is fuelled by asymmetries in power and well-being. Cores, semi-peripheries and peripheries still exist; they are just not as neat as they once were. We now see more core *within* the periphery in the form of networked elites and peripheries within the core in the form of those who are marginalized from networks. Returning to the concepts of Chapter 2, within the system of capitalism, disembedded networks have become more prominent in the second wave and as a consequence the mosaicking of space has become increasingly fine-grained and complex.

There is no reason why the process of globalization cannot be reversed – although there are powerful vested interests that are likely to resist any attempts to do so as the recent clashes between powerful state interests and the anti- and alter-globalization movement seems to indicate. But globalization does not have its own inexorable trajectory – it is the result of collective human actions, agendas, desires and perceptions. Although globalization to date has been driven by capitalism, to move to a post-capitalist globalization would require new 'globalized imaginations' that are only now beginning to form.

Reducing the almost infinite complexity of the world to such periodization is a risky business, but such frameworks allow us to untangle the inherently messy business of history. The terminology outlined above will be referred to at various points in the book – although the periodization is not *tested* in a formal sense and provides a framework only. Much of what follows deals with the second wave of globalization – and especially the neoliberal phase. This is because the intensity, extensivity and velocity of globalization processes are unprecedented in this period and also because this is the period that we live in and need to know more about if we are to rebuild globalization in a way that promotes equality and sustainability rather than threatening it.

Has the so-called GFC (see Chapter 7 for a critical geographical analysis) led to the initiation of a third wave of globalization? Following the collapse of financial institutions across the OECD in particular, there was a concerted intervention by rich-world governments that led to the bailing out, through subsidization, of numerous banks and other financial capital institutions. This was combined in some countries with a discernible shift in the late 2000s away from pure neoliberal tenets to a more mixed-economy, arguably Keynesian, approach to economic regulation. The ideas of neostructuralist theorists (see Chapter 7) are relevant here in this regard, and some thinkers have suggested the advent of a new regulatory regime (see also Murray and Overton, 2011a). In this book we take the position that whilst there has been a short-term rise in the intervention of the state and that in this sense there has been a move beyond pure neoliberal policy, this has been driven largely to service the needs and interests of capital and to perpetuate the free-market accumulation process. In this sense, whilst the relationship between capital and state has changed, there has been no seismic shift in regulation. We believe, then, that the model that best describes globalization in the early 2010s is still neoliberal in essence, irrespective of some shifts in practice. Notwithstanding, it is important to note that the GFC did provide an opportunity to reconfigure the relationship between markets, states and individuals in the context of globalization. Disappointingly this opportunity was not seized, in the short term at least; although ongoing conflict manifested most recently in outcomes as diverse as the street protests of Brazil, the riots in Greece and the continued effects of the 'Arab Spring' suggest that it may be too early to completely reject the possibility of moving to a new wave of globalization as a result of recent political-economic upheavals.

Conclusion – old myths and new waves

We began this chapter by asking the question: is globalization new? This idea has been effectively dispelled above and in the literature in general. It has been clearly shown as a collection of processes and outcomes which are 'manifested very unevenly through both time and space. It is neither an inevitable, all pervasive, homogenizing end-state nor is it unidirectional and irreversible' (Dicken, 2004, p. 8). Rather, it is 'multiple non-linear interconnected *trajectories*' (Hart, 2001, p. 655). This chapter adopts the position that the depth and breadth of the interconnections are indeed unprecedented, although the roots of that interconnection can be traced back to the first global European empires. The logic underpinning the expansion of the World System has changed little, being that of capitalist accumulation through time–space compression, although the conditions and facilitating technologies have evolved beyond recognition. The globalization of today is not as new as the popular imagination might believe, but it is certainly qualitatively very different.

With respect to the idea that we may be entering a new wave of globalization, we suggest that the interventions of governments and international institutions designed to arrest the GFC have had little more impact than delaying the crisis – a crisis which, as will become apparent from the remainder of this book, we see as comprised of economic, development and environmental concerns each of which is insolvable without resolution of the other. As such, mini-crises bubble continuously and rise to the surface from time to time and will continue to do so. In this conception, then, globalization to date has been driven by the logic of capitalism and the need for that system to constantly expand its spatial frontiers. It is debatable whether globalization can proceed without capitalism and whether we can envisage a post-capitalist world that is globalized. This debate is revisited in the conclusion.

In the last two chapters, we have touched upon a range of ways of looking at the concept and history of globalization. It is hoped that, at the very least, two common myths have been dispelled. The first is that globalization has been cast over the world like a blanket. This is clearly not the case as was argued extensively in Chapter 2; it is a net which does not affect everybody equally or make everywhere the same – there are holes in the net and some parts are woven more finely than others. However, everything *is* increasingly defined in relation to it. The second myth is that globalization is new. Clearly, it is not unprecedented – but there are important distinctions between the waves that we have identified (see Box 3.1). These historical and geographic points are returned to throughout the remaining chapters of the book, which now moves on to deal with shifts in the three spheres of human experience – economy, polity and culture – and to analyse some of the challenges presented by such transformations.

Further reading

Friedman (2007): This classic book has a very positive view on the impacts of globalization and should be read in conjunction with something more critical.

Held *et al.* (1999): The chronology explored in this chapter builds from Held *et al.*'s classic work which is comprehensive and coherent.

Hirst and Thompson (1999): This study is one of the few that sets out to empirically test the concept of globalization. The sceptical viewpoint it takes has been influential.

Hirst, Thompson and Bromley (2009): This is the third edition of Hirst and Thompson (1999) which adds a third author and updates the analysis substantially. It is worth reading both the second and third editions as the former remains a classic (its core arguments are discussed above).

Jones (2010): This book discusses the key thinkers in globalization according to Andrew Jones and provides an excellent resource for further study on thinkers including Giddens, Castells, Wallerstein and Friedman among others.

McMichael (2013): This book is one of the only texts that explicitly combines globalization and development. The chronology used is especially helpful and the case studies are illuminating and deal with the manifold crises that the world faces.

Ohmae (1995): This is the classic hyperglobalist account and reading it will reveal the roots of many popular arguments concerning the positive impacts of globalization.

Robertson (2003): Robertson's three-waves argument deals with the evolution of the global consciousness from a multidisciplinary perspective.

Stearns (2009): This provides an excellent historical account of globalization that reaches back many centuries.

Part II

Globalization in three spheres

4 ▸ Globalizing economic geographies

- New economic geography and spatial divisions of labour
- The evolution of the global economy
- Post-World War II – new divisions of labour?
- Flexible accumulation – the newer international division of labour?
- Chains and networks in the new global economy
- Contemporary patterns in the global economy
- Transnational corporations
- Regional economic integration
- Conclusion – a transformed economy

The global economy has grown in complexity and breadth almost without interruption since the industrial revolution. As it has done so, it has been conditioned by and has conditioned economic geographies of change that have the power to fundamentally alter people's lives for better or for worse in specific places in all four corners of the planet. This change has become bewilderingly rapid and intricate, simultaneously integrating and marginalising individuals and societies at increased velocity. In this chapter, we trace the evolution and nature of the globalising economy. We begin with a consideration of the crucial role of circuits of capital and spatial divisions of labour both of which underpin the historical evolution of the global economy and have conditioned the differentiated patterns we see today. Having considered the role of changing modes of regulation, we consider the pivotal influence of the contemporary transnational corporation. Finally, we discuss the increased tendency towards regionalization that we witness in the contemporary global economy. A number of key questions underlie our discussion here: What imperatives underlie the rapid expansion of the global economy? How do commodity chains and divisions of labour integrate economic activity? What are the implications of the growth of TNCs economically and politically? Does regionalism imply a move towards broader globalization or a move away from it? Is the global economy becoming more uneven or more homogeneous?

New economic geography and spatial divisions of labour

Geographers have defined economic geography in many different ways, but it essentially refers to the *spatiality of the economy*. The most straightforward, and yet surprisingly rich, definition is offered by Roger Lee in *The Dictionary of Human Geography* (2000a, p. 195) as 'geographies of peoples' struggle to make a living'. Stutz and De Souza (1998, p. 41) offer another useful definition, seeing economic geography as 'concerned with the spatial organisation and distribution of economic activity, the use of the world's resources, and the distribution and expansion of the world economy'. In the 1990s there was something of a revolution in the sub-discipline. Critics charged that the old economic geography was economically deterministic; ahistorical; overly descriptive and quantitative; and incapable of dealing with moral and ethical questions (see Barnes, 2009; Bryson, 2012; Coe *et al.*, 2009; Daniels and Jones, 2012; Peck and Yeung, 2003). Reflecting broader changes in human geography in general, and in particular the *cultural turn*, the sub-discipline now attempts to offer holistic explanations for change; is more sensitive to history and the uniqueness of place; is more willing to engage with the cultural roots driving economics; and incorporates qualitative analysis to make sense and interpret activity. Most importantly, economic geographers are now incorporating ethical analysis into their work (i.e. the normative 'rights and wrongs' of change). This is an important step in the evolution of human geography, which allows it to contribute to globalization debates more effectively.

The globalizing economy and spatial divisions of labour

We need a solid historical backdrop for the study of the contemporary world of economic globalization. In this chapter, we use the concept of *spatial divisions of labour*. A division of labour in general refers to the separation of tasks in the production process. There are two generally recognized divisions of this nature: (1) *social divisions of labour* – where different social groups concentrate on particular activities, and (2) *technical divisions of labour* – where distinct stages in the production of one good/service are separated. Social divisions of labour have been evident in society from prehistoric times; for example, in precolonial Fijian society different tribes were charged with different socio-economic functions such as fishing, hunting and oratory. Technical divisions of labour were greatly advanced from the industrial revolution onwards when technology allowed the establishment of the first 'production lines'. Geographers and others have identified four other division-of-labour types, and these often overlap with each other: *gender divisions of labour, cultural/racial divisions of labour, international divisions of labour* and *spatial divisions of labour* (Painter, 2000). Suggested first by Massey (1984), spatial division of labour refers to the concentration of specific stages, nodes or sectors of any given production chain or complex in different geographic locations. As capitalism has expanded outwards so divisions of labour have become increasingly stretched across space, and, in many sectors, new globalized divisions of labour link together localities at different ends of the world. These globalized divisions of labour are both a product of globalization and play a role in reproducing it. A subsequent section on the evolution of the global economy traces the evolution of spatial divisions of labour through the various waves of globalization as defined in Chapter 3.

The evolution of the global economy

By tracing spatial divisions of labour and how they have influenced the circuit of capital across time and space through history, we gain a better understanding of the patterns and issues that define the contemporary world economy. The periods discussed below draw on the chronology of globalization developed in Chapter 3.

Mercantilist roots 1450–1880

The 'classic' period of growth in the world economy was in the nineteenth century, but a visible world economy evolved earlier than that. Before the colonial wave of globalization (i.e. before the late 1400s; see Chapter 3) trade was largely localized, although some regional networks existed. With the emergence of Hispanic colonialism, the first world economic system was forged – based initially on mercantilism. This led some localities to specialize in the production and distribution of particular items based on comparative advantage – i.e. the items they produce relatively cheaply (see argument for free trade later in this chapter). The main actors in this were merchants who became the richest class and who acted as intermediaries between nodes in the core and periphery. The expansion across the Atlantic to fuel trade underpins the evolution of colonialism there and lies at the core of the evolution of the global market.

Industrialist expansion 1780–1900

Based on the increased rural to urban migration that took place in Europe in the 1600s, a cheap labour force was made available. This, combined with efficiency gains and the resulting surplus in agriculture, stimulated the industrial revolution. This development further fuelled rural to urban migration and thus set a virtuous circle of capitalist growth in motion. The industrial revolution involved increased levels of specialization, and divisions of labour were set up in factories. These employed new forms of technology and pursued *economies of scale*. Capitalism needs these economies of scale, which are realized when increasing the volume of production lowers unit costs, because *growth* is absolutely central to the system's survival. Industrial complexes and chains were extended across space forging a worldwide division of labour which largely favoured accumulation in the ruling core. Space-shrinking technology was used to search out cheap labour and resources and to foster flows of trade and investment that reduced capital's turnover time. By the early nineteenth century, Great Britain was far and away the world's greatest economic power – due partly to the legacy of colonial history and the fact that the industrial revolution began there. An important point to reiterate is that capitalism is a cultural system – other societies have ways to manage and regulate people's need and desire for accumulation. It is not inherent that people wish to make ever-increasing profits; the *growth ethic* emerged out of capitalism during this period and continues to this day.

The twentieth century – colonial divisions of labour?

The principles of capitalist expansion altered little between the eighteenth and nineteenth centuries, although the pace did. In the late nineteenth century, the economic power of the USA expanded enormously based on cheap resources, immigrant labour, innovation and – in particular – the emergence of *global corporations*. These organizations were unprecedented as they were single firms producing both within and outside their home country. As such, the USA became the leader in terms of *foreign investment* which was funding an ever-increasing proportion of economic output. A virtuous circle of outward foreign investment and profit repatriation by core economies led to the further skewing of an already highly concentrated global economic system, and this led to a concentration in political power also. In 1914, for example, 71 per cent of total manufacturing output was concentrated in just 4 countries (and 90 per cent in 11) – the most powerful being Germany, France, the UK and the USA. Dependency theorists, and other neo-Marxists, have argued that the early evolution of the global economy produced a core–periphery system underpinned by a *colonial division of labour*.

This saw industrial production located in the core and raw material production based in the periphery (Kay, 1989). Although this simplifies the world system greatly, it does capture the essence of global economic structure until at least the mid-twentieth century. The Great Depression and the Second World War were the major events which led to economic restructuring. The former led to a collapse of the liberalized economy and a retreat into protectionism, and the latter was followed by an attempt to return to free trade and reconstruct the world economy along modernist lines.

Post-World War II – new divisions of labour?

The geography of the world economy shifted profoundly after the Second World War (1939–1945). A number of processes stand out, including: the rebuilding of industrial capacity following the war; the evolution of new technologies and products through the war effort which underpinned subsequent growth; the wave of decolonization across the world from 1945 onwards; and the new international division of labour. However, the major economic imperative to emerge in the West following the war was the pursuit of *free trade* regulated through the General Agreement on Tariffs and Trade (GATT) (now the WTO). This was attempted largely as a means of avoiding the worst implications of the Great Depression, which had led to a massive collapse in world trade and investment flows. The rise of Japan as a reformed economic power was also an important process in the post-war economy, growing rapidly to become one of the largest in the world with overwhelming influence in its own region. In the socialist bloc, there was an attempt to remain aloof from capitalist development and moves to internationalize socialist systems between like-minded countries. During the modernization phase, which lasted until the late 1970s (Chapter 3), a strong role for the state was envisaged. In the context of the poorer world, aid and loans for industrialization were seen as the major component in stimulating growth. The oil crises of 1973 and 1979 altered the nature of the global economy profoundly. Neoliberal policies first adopted in Chile, and then by the UK and the USA, were diffused across the poor world in the wake of the debt crisis provoked by the oil price hikes. This recession led to the eventual triumph of neoliberalism, which has become the hegemonic global economic paradigm ushering in a new phase of globalization, as argued in Chapter 3.

The new international division of labour (NIDL)

One of the most important trends in the post-Second World War economy was the rise of the new international division of labour (NIDL). Lee (2000b, p. 552) defines this as '[a]n emergent form of worldwide division of labour associated with the internationalisation of production and the spread of industrialisation'. As argued above, until the end of the Second World War, the global economy was characterized by a colonial division of labour, revolving around a basic core/periphery dichotomy in trade and investment. The NIDL evolved particularly in the 1960s/1970s as profits in the core declined and companies increasingly invested and located abroad pursuing cheap factors of production (land, labour, capital) and new markets. Frèobel et al. (1980) argue that three things facilitated the NIDL:

- developments in transport and communications technology;
- developments in processing technology that can standardize production and use unskilled labour;
- the emergence of a world reservoir of labour.

To the above list should be added the increased mobility of capital from the neoliberal reforms of the 1980s. Some have argued that the powerful control which TNCs have gained over poor-world nation-states is a form of neocolonialism – that is to say, economic dependence combined with political sovereignty (Murray, 2004a). In this context, Frèobel et al. bemoaned the impacts of the NIDL in 1980 (quoted in Lee, 2000b, p. 553):

> So far export-oriented industrialisation has failed to achieve any
> improvement in the social conditions of the population of many of the

developing nations, not even as far as their most fundamental needs such as food, clothing, health, habitation, and education are concerned.

Arguably, a new semi-periphery has been added to the world economic structure through the operation of the NIDL (Taylor and Flint, 1989). The major recipients of investment during the forging of the NIDL were those that became known as the NICs (newly industrializing countries), particularly Taiwan, Hong Kong, Singapore and South Korea. These countries experienced rapid economic growth in the 1980s and early 1990s based in part on the spread of capital from the USA and Japan. Figure 4.1 shows the economic growth record of the Asian 'tigers' as compared to three non-core regions. The 'tigers' had enjoyed significantly higher rates of growth until the 1997 crash; this interruption proved temporary however. During the first decade of the 2000s, the economies recovered and then survived the financial crisis that hit Europe and North America in 2007–2008 almost unscathed (see Chapter 7). Significantly

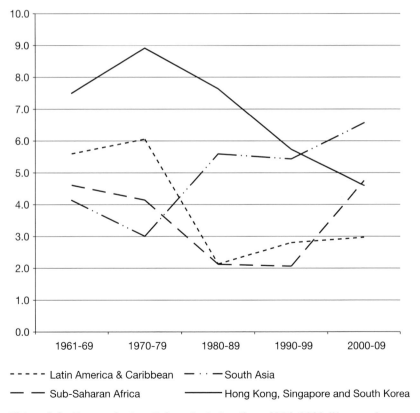

Figure 4.1 *Economic growth in selected regions 1961–2009 (% annual growth of GDP, based on local constant currency)*

Source: Calculated from World Bank data (http://data.worldbank.org/ accessed 3 September 2013)

though, the other regions, especially South Asia, are now experiencing similar or higher rates of growth. Theorists are divided over what lies behind the so-called 'Asian Miracle'. Some have argued that open economic policies based on neoliberal and neoclassical ideas underpin development success there. Others have argued that strong state interventionism, regulating capital and trade flows and designing 'competitive advantage', are the explanatory factors. As Rigg (2002) notes, it is probably the combination of the two in different ways in different places that really lies behind the heady economic ascent of the region.

Based on the high rates of economic growth and reductions in absolute poverty in East Asia and some other semi-peripheral countries (such as Chile), it could be argued that Frèobel *et al.*'s (1980) critique of the NIDL is now inapplicable. However, a range of persistent negative impacts has been posited which include:

- rural to urban migration and explosive urbanization and industrialization;
- environmental degradation and pollution resulting from industrialization;
- cultural shifts – particularly towards potentially unsustainable Western-style consumerism;
- increased economic vulnerability as economies open themselves up to investment flows.

The latter criticism became especially relevant following the Asian Crash beginning in 1997 that some blamed on open-economy and neoliberal policies (Rigg, 2002). On the other hand, continuing state regulation in Asia arguably helped such countries largely escape the effects of the 'global' financial crisis ten years later.

There is some evidence pointing to a second wave of NICs emerging in Southeast Asia and, to a certain extent, Latin America based on successor waves of the NIDL. In East Asia, for example, outward foreign investment by TNCs from first-wave NICs has played an important role in the economic development of countries such as Malaysia, Thailand and Indonesia. Despite the diffusion of industrialization to the NICs made possible through the NIDL, it should be noted that the majority of investment, trade and growth takes place within and between the core economies of Japan, the USA and Europe. Furthermore, the NICs have now been joined by China. China is now the world's second largest economy and has, to a large extent, taken on the mantle of the world's fastest-growing industrial region in an extension of the NIDL. However, although China has become integrated into the global economy, it has done so with a high degree of local – and state – control over its development. Chinese corporations have joined Western companies as the world's largest, and Chinese investment in Western economies and developing country extractive industries has been highly significant. Perhaps we should now suggest that the global triad has been transformed with Japan being joined by China and the tiger economies to form a broader East Asian corner of the triad. Yet, elsewhere, the semi-periphery is left largely outside of this triad. As we consider later in this chapter, the *newer international division* has the potential to further marginalize economies in the semi-periphery and periphery.

Flexible accumulation – the newer international division of labour?

The rise of 'flexible accumulation' is complicating the global economic map further, giving rise to a *newer international division of labour*. This restructuring has potentially profound spatial and socio-economic implications for the unfolding of globalization.

From Fordism to flexible accumulation

'Fordism' is the name used to describe the 'regime of accumulation' (see discussion in Chapter 5) that dominated from the Second World War until the 1970s and underpinned the NIDL. Fordism involves the mass production of standardized goods, pursuit of economies of scale, and a rigid division of labour that sees different parts of the task clearly demarcated. This conceptualization is based on the production techniques first introduced by Henry Ford in car manufacture. A linked concept is 'Taylorism' (see Daniels and Jones, 2012). The 'going abroad' of many TNCs was based on the search for low-cost factors of production, especially labour, for the assembly part of their chains. In the late 1960s there was a crisis of Fordism, and firms experienced a fall in profits due in part to rising wages and competition from low-cost economies. Further to this, tastes were shifting away from mass-produced items towards niche products. Lay-offs of labour to cut wage bills together with outflows of FDI led to a process of deindustrialization in the core. In the place of Fordism – some argue – a more flexible regime of accumulation has been forged, sometimes termed 'post-Fordism' (see Amin, 1994;

Malecki, 1991). Among other things, this regime has the following characteristics, developed largely in Japanese corporations and later diffused to Europe and North America:

- more versatile computer programmable machines;
- labour that is more flexibly deployed;
- vertical disintegration of large firms;
- greater use of contracting and strategic alliances;
- just-in-time production;
- closer integration of product development, marketing and production;
- more flexible labour training.

Combining the above characteristics, flexible accumulation caters for niche markets, is often consumer driven and places product innovation at the core (see Box 4.1). Post-Fordist production is strongly consumer driven in that consumers can order a custom-designed product

Box 4.1

Nike – flexible shoes?

There are numerous firms we could use to explore the concept of flexible accumulation – Toyota, Hyundai, Mitsibushi, for example. The Nike case illustrates a number of important points made by sceptics of the flexible accumulation thesis however. Nike is both Fordist and flexible simultaneously. For example, in the case of the former, Nike's *Air Max Penny* basketball shoes are made up of 52 component parts from five different countries, touched by 120 hands during production, on a clearly demarcated global production chain where timing is extremely important. Nike is notorious for searching out cheap labour pools. Nike has 150 Asian factories and over 350,000 Asian workers. Wages in China are reported in some cases to be as low as 11c per hour (Nike's chairman earned a salary of US$2.7 million in 1997). These are truly Fordist

characteristics indicating the operation of the NIDL. But elements of flexibility are also exhibited. For example, product research is extremely important and Nike aims to produce a new shoe every week to cater to niche markets. To do this, it employs a just-in-time innovation structure where it can buy in expertise at short notice. This entails the use of short-term subcontracts that are often given to firms located close to the R&D headquarters in Oregon, USA. These are flexible characteristics, which suggest a post-Fordist approach. In sum, Nike, like many other global corporations, displays elements of both 'regimes of accumulation' as is evidenced in Figure 4.2, which shows Nike's 'commodity circuit'. Examples of this nature suggest that 'flexibility' arguments may in fact be exaggerated.

Box 4.1
continued

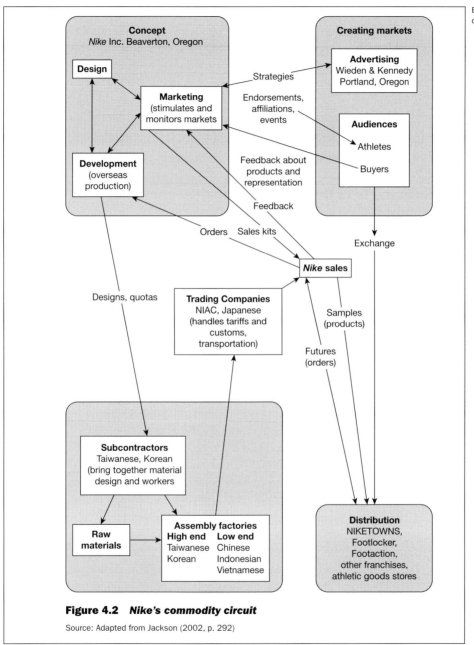

Figure 4.2 *Nike's commodity circuit*

Source: Adapted from Jackson (2002, p. 292)

(for example, specifying the colour, engine size, upholstery, trim and optional extras of a particular car) and this order is relayed directly through to the production line so that this 'unique' product is assembled and delivered to the specific customer. It has replaced the old Fordist approaches to uniformity, economies of scale and mass production with large and centralized industrial locations (such as Detroit for automobiles). Correspondingly, R&D and proximity to market are now more important elements in locational decisions of many TNCs than traditional factors such as resources and labour. This has led to the evolution of a number of industrial districts in advanced capitalist countries, such as Third Italy in Northern Italy and Motor Sport Valley in the UK (see Box 4.2).

Spatial impacts of flexible accumulation – new industrial spaces

New industrial spaces is a term coined by Scott (1988) to refer to the growth of economic nodes of flexible production which have arisen due to the shift to post-Fordist/flexible accumulation. Such areas are subject to the processes of re-agglomeration and, as such, are thought to reverse the NIDL. In new industrial spaces, economies of scope and scale act to cluster activities as the relative priority of the factors of production is altered. Bryson and Henry (2001) discuss four overlapping examples of such spaces:

1 *New industrial districts* – 'Dense agglomerations of, normally, small and medium-sized firms specialising in the high quality production of a particular good or service' (p. 368). Examples include Jutland in Denmark (furniture, shipbuilding), Småland in Sweden (metalworking) and 'Third Italy' in Northern Italy (textiles, ceramics, shoes) (Malecki, 1991). These are also sometimes referred to as Marshallian districts and have been the subject of work by geographical economists such as Krugman (1998).

2 *High technology areas* – These are also sometimes referred to as 'sunbelt' areas. They are described as '[a] diverse range of technopoles, corridors and innovative complexes, often in previously un-industrialized areas, which have grown through their capture of the "sunrise" high technology industries of information technology, electronics, pharmaceuticals, and R&D' (Bryson and Henry, 2001, p. 369). Examples include Japan's technopolis programme, Boston's Route 128, the UK's M4 Corridor (London to Swindon), Motor Sport Valley, Oxfordshire, UK (see Box 4.2) and 'Wellywood' film production in Wellington, New Zealand. The most well-known example however is Silicon Valley, Palo Alto, California (see Plate 4.1). A former agricultural region, in 1950 it employed no

Plate 4.1 *Google Headquarters, Palo Alto, California*

Source: Photo by the authors

more than 800 manufacturing workers. Today nearly 400,000 high-technology workers in such companies as Hewlett Packard, Intel, Apple, Google and Facebook are employed.

3 *Flexible production enclaves within old industrial regions* – A range of economic developments have taken place in more traditional industrial regions. Bryson and Henry argue: 'Often rocked by the decline of previously core "sunset" industries (shipbuilding, coal, steel, heavy manufacturing) some alternative economic growth has been achieved through firms and sectors applying flexible production techniques' (Bryson and Henry, 2001, p. 369). Examples include South Wales, which restructured away from coal and steel towards car production based on Japanese and Korean investment in the 1980s. The US

Midwest has been largely restructured through what is known as 'transplant' investment from Japan.

4 *World cities* – As the advanced economies have moved largely to service-based activity, global cities have become the 'pinnacle of the global urban hierarchy once again' (Bryson and Henry, 2001, p. 369). Major world cities such as Tokyo, New York and London (perhaps now joined by Shanghai) are the control centres of the global economy in a world city network where the links between such sites far outweigh the links to other cities. They house the headquarters of major global corporations as well as providing financial and business services that manage the global corporate world. The district of Canary Wharf in London, Wall Street in New York and downtown Los Angeles are the physical manifestations of this process. Such centres also include the 'servicing' industries – cleaners, couriers, guards, nannies, maids and so on – that service the new global elite as well as clustered cultural and retail outlets that service the same population. The world city hypothesis and its structure is outlined in Box 4.3.

The rise of flexible accumulation has implications for theses of globalization. Supporters of the idea of a new regime argue that it is leading to a reconcentration of investment away from the periphery to the core, resulting in a reversal of the NIDL (Gwynne *et al.*, 2003). For some, this is a bad thing, as the NIDL diffused industrialization and created growth in the Third World (Chandra, 1992). Critics of the notion of flexible accumulation argue that much production is still carried out using Fordist techniques or something that closely resembles them. In many sectors, and in relevant parts of commodity chains, 'going abroad' remains a central strategy. This is particularly the case in labour- or natural-resource-intensive industries, such as garments in the case of the former and fruit production in the case of the latter. Bangladesh, for example, has become a major centre for garments manufacture, largely due to the very low wage economy there. It employs about four million people, mostly women, and the country is second only to China as a producer of garments. Although most garments factories in Bangladesh are locally owned, they are export oriented and produce Western brands of clothing through subcontracting networks. Communications possibilities over the World Wide Web add another dimension to the global division of labour, as skilled labour in low-cost economies such as India can be employed in sectors such as computer programming to produce for markets anywhere. In general, work remains to be done on the empirical outcomes of the newer international division of labour.

Box 4.2

Sunbelt: Motor Sport Valley, UK

Henry has undertaken research on the rise of an important high-technology sunbelt area in the UK known as Motor Sport Valley. This industrial cluster is located in a 70-kilometre radius around Oxford in south-east England. It employs around 50,000 people in hundreds of small- and medium-sized firms. As Bryson and Henry (2000) argue: 'Motor Sport Valley has become the place in the world to produce your motor sport product – no matter how large the company and the global choice of locations they offer'

(p. 371). Three-quarters of the single-seat racing cars used in over 80 countries are UK built. This area is home to the vast majority of world rally teams including Ford, Mitsubishi, Hyundai and Subaru. Over 50 per cent of the supply chains that won the major rally championships of 1995 were sourced in the Valley. These functions cluster, taking advantage of localized R&D, production technology, expertise, flexible labour and parts to produce highly advanced and differentiated products.

This research would do well to focus on social impacts. The casualization of the workforce and the rise of flexible labour markets in the core, and the impact in terms of socio-economic change in the Third World, require greater consideration, for example.

Chains and networks in the new global economy

Two useful ways of thinking geographically about the organization of the globalizing economy are commodity chains and networks. The first to elaborate the chain concept was Porter (1990) in his exposition of the 'value chain'. Gereffi (1994, 1996), a sociologist, has been the leader in seeking to further our understanding of commodity chains. He argues that the concept of the 'global commodity chain' (GCC) is the best organizational unit through which to map the complexity of the global economy. The approach provides a new paradigm for the investigation of the global economy – one which transcends nation-states and other formerly assumed 'containers' of economic activity. This new framework:

> permits us to more adequately forge the macro–micro links between processes that are generally assumed to be discretely contained within global, national and local units of analysis. The paradigm that GCCs embody is a network-centred and historical approach that probes above

and below the level of the nation-state to better analyse structure and
change in the world economy.

(Gereffi, 1994, p. 2)

Any commodity chain is characterized by four features: (1) the flows of
commodities from each point in the node (an input–output structure);
(2) the utilization of labour at each point on the node – what is sometimes
referred to as the relations of production (for example, feminization, use of
unskilled labour and so on); (3) the mode of production at each node, in
technological and scale terms (governance); (4) the geographic location of
the nodes (a territoriality). From a regulation theory point of view, some
geographers (Goodman and Watts, 1997; Guthman, 2009; LeHeron, 1993;
Whatmore, 2002, 2009) have argued that we must also consider the forms
of regulation that exist between the nodes.

Gereffi distinguishes between two types of commodity chain that link
together various parts of the production system – the *buyer-driven chain*
and the *producer-driven chain*. These have been globalized to the extent
that they have stretched so that different nodes are located in various
national territories. There are a number of problems with the concept of
Gereffi's chains, but they do provide a useful starting point for analysing
the global economy. Below, we distinguish between producer- and
consumer-driven commodity chains (see Figure 4.3).

Producer-driven chains

These are found in industries where TNCs generally control the mode of
production and the overall direction of the chain. This tends to be the
case in technology-intensive industries such as computers and motor
vehicles where a high proportion of costs are associated with advanced
capital. Gwynne *et al.* (2003) use the example of the global motor
industry, which 'provides a good example of a producer driven chain,
with multilayered production systems that involve thousands of firms'
(p. 167). In such systems, assemblers are linked with component
suppliers at different tiers of organization. The Japanese production
system in the late 1980s, for example, involved the linking of one
assembler to an average of 170 first-tier suppliers, 47,000 second-tier
suppliers and 31,600 third-tier suppliers – all linked through contracts
(Hill, 1989). The rise of flexible accumulation methods, as noted above,
means that these suppliers are often located within the core. However,
there are still examples of component suppliers in lower wage cost
economies – as was more traditionally the case under Fordism.

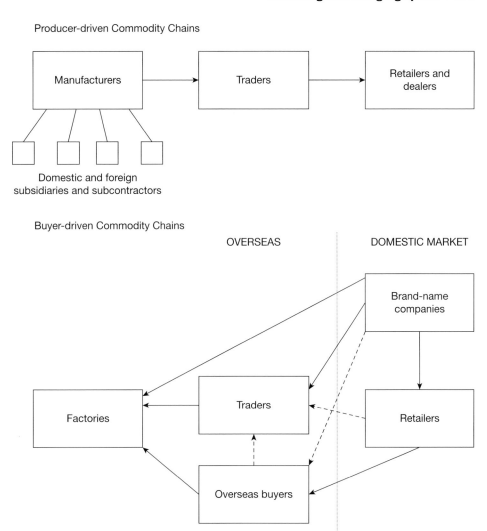

Producer-driven Commodity Chains

Buyer-driven Commodity Chains

Figure 4.3 *Producer- and buyer-driven commodity chains*

Source: Adapted from Gwynne *et al*. (2003)

Buyer-driven chains

These are different from producer-driven chains in that the agents downstream control the development of the chains, leading to the development of 'decentralized' networks where subcontracting at different stages in the chain is more common. Commodities such as toys, garments and footwear fit into this model. As Gwynne *et al*. (2003, p. 168) say, 'essentially they are manufacturers without factories, at the centre of a highly flexible and global network of production distribution and marketing'. According to some authors, these kinds of systems offer greater scope for emerging market economies, with Taiwan and Singapore in the 1980s being good examples. Such systems are generally geared towards labour-intensive industries, and thus production is often located in lower wage economies (such as the example of garments in Bangladesh). As products are altered and go through cycles, the required factors of production (land, labour, capital, knowledge) change (see Gwynne, 1990 on the Product Life Cycle model), and this can shift the optimal location of production. The TNCs in control of such chains are able to strictly direct, through the subcontracting systems, the production and supply of components, and they are able to capture much of the surplus value that is added in the chains and to pass on the risk to the subcontractors. Critical for these companies is the necessity to build and maintain global brands as their key resource: marketing is arguably more important than production in securing long-term profit. According to Gereffi, this type of commodity chain is becoming dominant in the world economy – although this has not been shown empirically in any satisfying way.

Beyond chains?

Dicken *et al.* (2001) and Dicken (2012) are critical of Gereffi's concept of global commodity chains – despite 'the very substantial contribution of his work to our understanding of the structures of the global economy' (p. 98). They argue that the concept offers only a partial framework of analysis dealing with a limited number of commodities. The input–output nature of the chains remains relatively unexplored, and their linear nature is overemphasized. Further, the regulatory aspects in between the nodes of the chain – the institutional context – is under-analysed (Whatmore, 2009). Gereffi accords too much power to TNCs in governing the chains, *vis-à-vis* the nation-state. Some geographers then suggest that a more useful concept is that of the 'network'. This is based on the following view of the global economy (adapted from Dicken *et al.*, 2001, p. 97):

1 *The economy is constituted of spaces of network relations.* Networks allow us to build relational views of the global economy – that have to do with the relations between agents within them.
2 *Agents act intentionally and this has effects on power structures.* The way social actors – which include individuals, households, firms,

industries, states, unions – behave has what may be referred to as 'intentionality' – there are motives behind actions.

3 *The relationships between actors are 'embedded' or take place within particular spaces.* This does not imply that they take place within territorially delimited space but network relationships do have a spatiality.

Summing up their argument, Dicken *et al.* (2001, p. 97) claim:

> [T]he global economy is thus made up of social actors engaged in relational networks within a variety of 'spaces'. The analytical lens we adopt can thus vary widely. It may be geographical, it may be sectoral, and it may be organisational. It may be some combination of these.

They go on to argue the case for actor-network theory which better addresses the three characteristics above. This theory is useful for assessing the globalizing economy in three main ways:

1 *Networks ground globalization* – They provide a real time-and-space face to the process which is sometimes seen as universalizing and inevitable. It is played out in networks by the intentional action of individual actors with certain motives. It has both discursive and material outcomes but it is an agenda as well as a process. This also means that globalization can be resisted, as Whatmore and Thorne (1997) point out in their work on 'fair trade' coffee production networks.

2 *Networks allow space for ethical analysis* – The rights and responsibilities of all network participants can be considered – be they human or otherwise. In addition, because network theory explicitly increases the range, or reach, of our individual actions it means that we can take responsibility for outcomes that are distant. It also provides justification for representing distant others by virtue of our links to them (Corbridge, 1993).

3 *Actor-network theory transcends the problems of grand theory* – Unlike Gereffi's commodity chains, actor-network theory does not make grand claims about the structure of the global economy and does not suggest any overriding logic. This takes economic geography beyond the impasse of thinking about the economy in a structuralist way.

In reality the functioning of the global economy is somewhere in between the structural relationships that Gereffi suggests and the 'intentionality' or subjectivism of actor-network theory. As is often the case with the major

Box 4.3

Networks in action – world cities

Geographers argue that cities are the principal points through which the processes of globalization have been articulated and resisted. Armstrong and McGee (1985, p. 41) proposed that '[c]ities are the crucial elements in accumulation at all levels and the locus operandi for transnationals, local oligopoly capital and the modernising state'. The concept of *world cities* refers to a tier of cities that dominates global economic and political affairs, acting as controlling nodes on a network of flows of capital, goods, people and knowledge (Friedmann, 1986; Sassen, 2000, 2011). The idea originates in the work of Geddes (1915), was later pursued by Hall (1984), and is now associated with Knox and Taylor (1995), Knox (2011) and Sassen (2011) in particular. As 'basing points' for global capitalism, world cities function as major manufacturing centres, provide advanced service systems, operate transport and communication hubs and house TNC headquarters. They also play a major role in the diffusion of cultural symbols, especially those associated with capitalism and Westernization. In sum, they are 'both the cause and effect of economic, political and cultural globalization' (Knox, 2002, p. 329).

In summing up the early debate on world cities, Friedmann (1996) notes that by the 1990s there were a number of points of convergence with respect to the research on world cities:

1. World cities articulate economies into the global economy and serve as nodes in the global complex.
2. There is a space of global capital accumulation represented by world cities; most areas are off this network however.
3. World cities are large, urbanized areas of intense social and economic interaction.
4. World cities can be arranged hierarchically based largely on their economic power; this is determined by their ability to attract investment.
5. The controlling strata of people in world cities have a shared vision and form a transnational capitalist class. They have a shared vision of the effective functioning of the global economy, consumerism and cosmopolitanism. They often clash with more localized groups around them.

World cities have been classified into a hierarchy based on the political/economic power they command and the extent of their articulation into global networks. The most complete classification is that undertaken by the Loughborough Globalization and World Cities Research Group (see Knox, 2011). This separates world cities into three tiers: Alpha, Beta and Gamma. Alpha world cities include Los Angeles, Chicago, New York, London, Paris, Frankfurt, Milan, Sydney, Mexico, Beijing, Sao Paulo, Hong Kong, Santiago, Tokyo and Singapore. Beta cities

Box 4.3
continued

(a) San Francisco

Box 4.3
continued

(b) Dubai
Plate 4.2 *World cities: (a) San Francisco and (b) Dubai*
The ranking of world cities is reviewed often and in the latest, at the time of writing, both Dubai and San Francisco had entered the ranks of alpha cities.

Source: Photographs by the authors

Box 4.3
continued

include Montreal, Rome, Cairo, Vancouver, Manchester, Birmingham, Auckland, Lima, Rio de Janeiro and Osaka. Gamma world cities include Glasgow, Hanoi, Doha, Belgrade, Quito, Porto, Wellington and Durban. Under this classification there are relatively few world cities in South Asia, the Middle East, or Africa, illustrating that population size is not the principal determinant of world city status. Over the last 20 years of research on world cities, perhaps the most notable trend has been the rise of East Asian (especially Chinese) and Latin American cities through the index. Although the world city is an important concept in globalization studies, at least two things need to be built into future studies. First, differentiation within world cities themselves; second, change in large non-world cities – or what Rennie-Short (2004) refers to as 'Black Holes'.

problems in human geography, geographers are grappling with the relationship between structure and agency. Dicken *et al.* (2001) talk of 'attempts to draw together structural understandings of power relations with relational approaches to human agency' (p. 107), while Larner and LeHeron (2002) and Le Heron (2007) discuss the possibility of 'post-structural political economies'. This debate will continue to be an important one in economic geography.

Contemporary patterns in the global economy

The global economy transformed rapidly over the past two centuries as capitalism spread across the planet in its search for profit. During the neoliberal phase its structure in terms of the proportional role of different sectors has shifted fundamentally. Across the world, primary (exploitation of natural resources) and secondary sectors (the manufacture of value-added products from primary goods) are shrinking in terms of their relative contributions to both total employment and GDP. The tertiary (retail and wholesale trade, transport and personal services) and quaternary sectors (banking, finance, business services, media, insurance and administration) now form the most important component in many of the world's economies and all the 'advanced' ones. When the sources of world output are divided into agriculture, industry and services, the overwhelming importance of the latter becomes apparent. In 2011, services accounted for 66 per cent of the value of total world output. In developing countries, services are playing an increasing role, rising from 30 per cent of total income in 1965 to 51 per cent in

2011. In developed countries, the contribution of services to output has risen from 54 per cent to 74 per cent in the same time period. Agriculture is now of relatively little economic importance, representing 4 per cent of income for the world as a whole. In developing countries, agriculture remains important, though declining, contributing 9 per cent to total income in 2011 (see Table 4.1). Daniels (2001) outlines six further central trends in the contemporary global economy:

1 *A marked increase in the power of finance over production* – Capital moves almost uninterrupted across the globe and has the power to transform companies, economies and even entire regions almost instantly.
2 *Knowledge is a vital factor of production* – Skills and the production of an educated workforce have become major policy priorities.
3 *Technology and its diffusion have become transnationalized* – Managing technology has become essential for economic success and the technology gap which exists further compounds the global welfare gap.
4 *Global oligopolies have grown* – Successful TNCs have come to dominate production and marketing in any given industry (e.g. Microsoft or McDonald's).
5 *Transnational institutions have arisen to regulate the global economy* – These include the UN, IMF and WTO.
6 *The 'wiring' of the world* – This has led to more rapid cultural flows, what Amin and Thrift (1994) call 'deterritorialised signs, meanings and identities'.

Globalization, and the increased role of networks in particular, has made the world economy more uneven than ever. The discussion of global investment and trade patterns that follows illustrates this asymmetry clearly.

Geography of foreign investment

As previously discussed, the NIDL changed the geography of foreign investment, leading to significant outflows from the core. In general terms the range of recipient countries has broadened, and while North America, Europe and East Asia are the main destination regions for FDI, investment in Latin America and Southeast Asia is growing. Notwithstanding the broadening in the range of recipient countries, at the

Table 4.1 Share of world output by sector, 2000 and 2011 (% of world output)

	2000				2011			
	Agriculture	Industry	Services	Total	Agriculture	Industry	Services	Total
Developing economies	61.9	27.2	17.3	21.7	75.3	45.8	27.1	34.9
Transition economies	3.3	1.5	0.9	1.1	4.4	4.1	2.9	3.4
Developed economies	34.8	71.3	81.9	77.1	20.3	50.0	69.9	61.8
Africa	7.8	2.3	1.3	1.8	10.1	3.3	1.9	2.7
South America	10.0	7.0	5.9	6.4	10.1	8.3	7.3	7.7
Asia	50.5	34.7	25.5	29.0	57.4	42.2	27.8	33.5
Oceania	1.9	1.3	1.5	1.4	1.7	2.2	2.5	2.4
North America	10.2	28.2	38.4	34.4	6.7	17.9	29.9	25.3
Europe	16.3	25.1	26.5	25.8	9.7	21.9	27.6	25.1

Source: Calculated from UNCTAD data

aggregate level, inflows into 'developed' countries (based on UNCTAD – United Nations Conference on Trade and Development – classification) accounted for a consistently high and growing proportion of the total stock (57.5 per cent in 1980 and 62.3 per cent in 2012) (see Table 4.2). In this sense, there has been a reconcentration in global investment patterns. Approximately 70 per cent of outflows from the 'developed' world stay within the 'triad' grouping of the EU, USA and Japan. Overall, the 'developed' countries accounted for 79.1 per cent of all FDI stock outflows in 2012, which represented a fall from the figure of 1980 (86.9 per cent). This fall was attributable to the fact that outward investment from a handful of 'emerging' Asian economies has increased, including Hong Kong, Singapore, Taiwan, South Korea, Malaysia and especially China in the past decade. Asia's share of outward FDI has increased significantly from just 6.6 per cent in 1980 to 18.2 per cent in 2012. In general, however, outflows from the developing world outside Asia remain very low, Latin America and Africa standing at just 5.5 per cent in 2012. Africa's position in this respect has worsened significantly over the past three decades; the continent now accounts for only 0.6 per cent of global FDI outflows and 2.8 per cent of inflows.

Geography of trade

Trade has become an increasingly important component in national economies as borders have opened across the world. From 1950 to the present, merchandise trade increased approximately fifteenfold, while output increased only fivefold. Trade flows are concentrated however. In 2002, for example, a small number of large countries accounted for an overwhelming proportion of world exports, the most important being the USA (10.7 per cent), Germany (9.5 per cent) and Japan (6.5 per cent). Interestingly, a decade later China had emerged as the world's largest exporter (with 11.2 per cent of world exports), followed by USA (8.4 per cent), Germany (7.7 per cent) and Japan (4.4 per cent). The world's 20 largest importers and exporters and their shares of world trade are shown in Table 4.3, and the geographical distribution of exporting countries is illustrated in Map 4.1. Taken together, the top 20 countries account for approximately 70 per cent of world merchandise imports and exports. Despite globalization trends, countries still tend to trade with territories that are located closest, so much of global trade is

Table 4.2 Foreign direct investment by region, 1980 and 2012

| | Inward FDI stock | | | | Outward FDI stock | | | |
| | 1980 | | 2012 | | 1980 | | 2012 | |
	$US bill	%	$US bill	%	$US bill	%	$US bill	%
Africa	41.097	5.9	629.632	2.8	7.584	1.4	144.735	0.6
Latin America	41.790	6.0	2,310.630	10.1	47.512	8.7	1,150.092	4.9
Asia	218.690	31.3	5,060.621	22.2	36.231	6.6	4,289.477	18.2
Oceania	27.139	3.9	716.891	3.1	4.850	0.9	448.229	1.9
North America	138.348	19.8	4,570.442	20.0	239.178	43.6	5,906.953	25.0
Europe	230.849	33.1	8,676.610	38.0	213.567	38.9	11,192.494	47.4
Transition economies	–	–	847.854	3.7	–	–	460.760	2.0
Developing/transition countries	**296.280**	**42.5**	**8,592.377**	**37.7**	**71.720**	**13.1**	**4,920.116**	**20.9**
Developed countries	**401.633**	**57.5**	**14,220.300**	**62.3**	**477.203**	**86.9**	**18,672.623**	**79.1**
World	**697.913**		**22,812.68**		**548.922**		**23,592.739**	

Source: Calculated from UNCTAD data

Table 4.3 The world's 20 largest merchandise exporters and importers, 2002 and 2012 (US$ millions)

Rank	2002						2012					
	Exporters			Importers			Exporters			Importers		
	Country	$US bill	%	Country	$US bill	%	Country	$US bill	%	Country	$US bill	%
1	USA	693.9	10.7	USA	1,202.4	18.0	China	2,048.8	11.2	USA	2,335.4	12.6
2	Germany	613.1	9.5	Germany	493.7	7.4	USA	1,547.3	8.4	China	1,818.1	9.8
3	Japan	416.7	6.5	UK	345.3	5.2	Germany	1,407.1	7.7	Germany	1,167.4	6.3
4	France	331.8	5.1	Japan	337.2	5.0	Japan	798.6	4.4	Japan	885.8	4.8
5	China	325.6	5.0	France	329.3	4.9	Netherlands	655.8	3.6	UK	680.4	3.7
6	UK	279.6	4.3	China	295.2	4.4	France	569.1	3.1	France	673.7	3.6
7	Canada	252.4	3.9	Italy	243.0	3.6	South Korea	547.9	3.0	Netherlands	590.7	3.2
8	Italy	251.0	3.9	Canada	227.5	3.4	Russia	529.3	2.9	Hong Kong	554.2	3.0
9	Netherlands	244.3	3.8	Netherlands	219.8	3.3	Italy	500.2	2.7	South Korea	519.6	2.8
10	Belgium	214.0	3.3	Hong Kong	207.2	3.1	Hong Kong	493.4	2.7	India	489.4	2.6
11	Hong Kong	201.2	3.1	Belgium	197.4	2.9	UK	468.4	2.6	Italy	485.9	2.6
12	South Korea	162.5	2.5	Mexico	173.1	2.6	Canada	454.8	2.5	Canada	474.9	2.6
13	Mexico	160.7	2.5	Spain	154.7	2.3	Belgium	446.3	2.4	Belgium	434.8	2.4

Table 4.3 *continued*

Rank	2002 Exporters Country	$US bill	%	2002 Importers Country	$US bill	%	Rank	2012 Exporters Country	$US bill	%	2012 Importers Country	$US bill	%
14	Taiwan	135.1	2.1	South Korea	152.1	2.3	14	Singapore	408.4	2.2	Mexico	380.5	2.1
15	Singapore	125.2	1.9	Singapore	116.4	1.7	15	Saudi Arabia	386.0	2.1	Singapore	379.7	2.1
16	Spain	119.1	1.8	Taiwan	112.6	1.7	16	Mexico	370.9	2.0	Russia	335.4	1.8
17	Russia	106.9	1.7	Switzerland	83.7	1.3	17	Taiwan	301.1	1.6	Spain	332.2	1.8
18	Malaysia	93.3	1.4	Malaysia	79.9	1.2	18	United Arab Emirates	300.0	1.6	Taiwan	270.5	1.5
19	Ireland	88.2	1.4	Austria	78.0	1.2	19	India	293.2	1.6	Australia	260.9	1.4
20	Switzerland	87.9	1.4	Australia	72.7	1.3	20	Spain	292.2	1.6	Turkey	247.6	1.3
Top 20		4,902.5	75.8		5,121.2	76.8	Top 20		12,818.8	70.0		13,317.1	72.1
World		6,455.0			6,693		World		18,324.6			18,469.3	

Source: Calculated from UNCTAD data

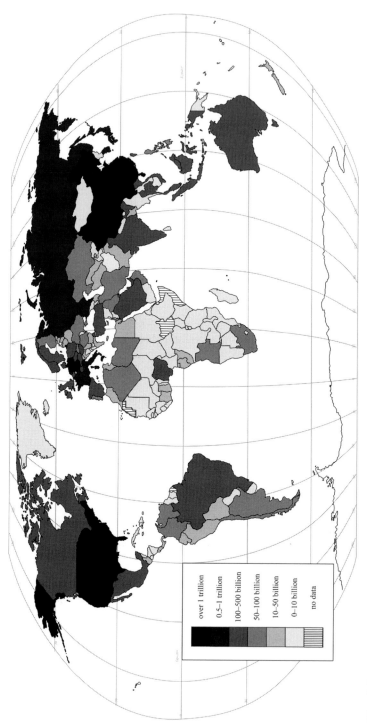

Map 4.1 Merchandise exports by country, 2012 (US$)

over 1 trillion
0.5–1 trillion
100–500 billion
50–100 billion
10–50 billion
0–10 billion
no data

Source: Calculated from UNCTAD data

intra-regional. Western Europe is the largest trader, although two-thirds of that trade takes place within the region. Asia is the second largest with around 50 per cent of its trade being intra-regional. North America is the third largest region conducting about 40 per cent of its trade internally. In terms of the relative importance of exports and imports, the largest surpluses are found in East Asia, the Middle East and some European nation-states. The USA has by far the largest trade deficit on the planet and this is growing. In 2012 USA and China had almost identical aggregate trade volumes ($US3,882.7 billion and $US3,866.9 billion respectively) yet this involved a trade deficit for the USA of $US788.1 billion and a trade surplus for China of $US230.7 billion.

As noted above, services have grown in proportional economic importance over recent decades. Services exports grew at around 11 per cent per annum between 1989 and 2000 while merchandise exports grew at 10 per cent. Trade in services is also concentrated, although not to the extent of merchandise exports. Approximately 65 per cent of service exports come from 15 countries. The balance of service imports and exports is different geographically to that of merchandise exports, with the USA being a major surplus exporter and Japan being a major services importer. When trade in goods and services is taken together, a clear regional concentration emerges. However, the geographical picture is changing. In 2002, Europe accounted for 42.2 per cent of total goods and services exports and 40.3 per cent of total imports but, a decade later, these figures had fallen to 35.4 and 33.8 per cent respectively. The Asia Pacific, by contrast, emerged during this time to become the world's most dynamic trading region, accounting for 39.3 per cent of global exports and 38.0 per cent of imports in 2012 – although this is reduced significantly if Japan and China are taken out of the calculation. The surplus in service exports from the USA is not great enough to offset the deficit in goods trade, with North America accounting for 12.1 per cent and 15.1 per cent of total global exports and imports respectively. Table 4.4 illustrates also the diminutive proportional role played by Latin America, Africa and the transition economies of Eastern Europe in global trade.

Table 4.4 *International trade in goods and services by region, 2002 and 2012*

Region	2002				2012			
	Exports		Imports		Exports		Imports	
	US$ bill	%	US$ bill	%	US$ bill	%	US$ bill	%
North America	7,900	15.9	1,640	21	2,741	12.1	3,306	15.1
Latin America	414	5.2	408	5.2	1,286	5.7	1,315	6.0
Western Europe	3,336	42.2	3,147	40.3	8,008	35.4	7,383	33.8
Transition economies	379	4.8	358	4.6	992	4.4	826	3.8
Africa	173	2.2	165	2.1	693	3.1	728	3.3
Asia-Pacific	2,097	26.5	1,913	24.5	8,869	39.3	8,311	38.0

Source: Calculated from UNCTAD data (http://stats.unctad.org accessed 30 September 2013)

Summary – the global economic triad and shifting economic patterns

In terms of economic flows, there is considerable evidence for the existence of a 'global triad'. However, it is important to remember that such macro trends hide important variations across regions and within countries. When the geographical scale of analysis is reduced, we see that most investment and trade actually takes place between specialized industrial spaces. These *clusters* of economic activity increasingly define the global economy and imply even greater spatial concentration of activity than national macroeconomic figures suggest. Examples include Europe's 'vital axis' from the English Midlands to Northern Italy, the urban growth corridors of East Asia (e.g. the South China Economic Zone) and the US–Mexican border. In analysing and describing the evolving global economy, it is important that we are able to see beyond national borders and look also at networks within countries and firms.

Summarizing the most important geographical trends in the world economy in 2000 over the three decades prior to the turn of the millennium, Dicken (2003, p. 45) outlines seven points of note:

1 Japan's rise to the second largest economy and third largest single exporter;

2 the dominance of the USA as the world's largest economy despite intense competition from Japan over the recent past;

3 the uneven performance of the Western European economies;

4 the rise of a number of East Asian NICs to become global players in merchandise exports;

5 the heady rise of China over the past few years to become a major economic power;

6 the weak performance of most 'developing countries' except for a few exceptions, such as Chile;

7 the appearance of a number of transitional economies in Eastern Europe following the collapse of the USSR.

Given the dramatic changes in the global economy in the past decade, we can see how some of these trends have continued but others have been reshaped. We would emphasize in particular the rise of China to usurp Japan and threaten USA as the world's largest economy. China, along with Brazil, Russia and India (together constituting the BRIC countries) have become the new growth centres of the global economy as both Europe and North America have faltered, especially in the wake of the financial crisis of 2007. We might also point to the way the commodities boom of the early 2000s has led to some gains being made in Latin America, Australia and Africa, though these remain fragile and dependent on the continued growth of China in particular.

Transnational corporations

One of the most controversial debates in globalization studies surrounds the rise and role of transnational corporations. TNCs are loathed by the anti-globalists and seen as saviours by pro-globalists. They are defined as:

> A firm which has the power to co-ordinate and control operations in more than one country, even if it does not own them . . . typically involved in a spider's web of collaborative relationships with other legally independent firms across the globe.
>
> (Dicken, 2003, p. 7)

In the above sense, Schoenburger (2009) argues that TNCs cannot be merely reduced to foreign direct investment and what is crucial is the ability to command and control production complexes and chains. This may be achieved through direct ownership of the different parts of the production complex, but may also involve subcontracting as well as other less formalized means of influence. Such enterprises operate at a worldwide scale and account for an increasing proportion of the world's flows of trade, investment, knowledge, skills and technology transfer. Between one-fifth and one-quarter of all production in the advanced capitalist economies is undertaken by TNCs, and a large proportion of world trade is intra-firm exchange (over 50 per cent in the case of USA–Japan trade, for example). In sum, TNCs have the following characteristics (though there are many different types):

- *Functional integration* – in terms of R&D, production, investment, marketing and sales at a global scale. Essentially, within a TNC you find the location of various parts of the production network or complex in different global locations. In the past, this spatial division of labour may have happened at the national scale only.
- *Global sourcing* – an ability to take advantage of geographical differences in the distribution of factors of production (labour, capital, resources and knowledge) and state policies (regulations, tariffs and taxes).
- *Geographical flexibility* – an ability to switch its resources and key operational facilities to different locations.

Overall, TNCs evolve through either vertical or horizontal integration – and some may evolve through both:

- *Horizontal integration* occurs when a given company attempts to gain greater control of the market for its product by taking over or expanding production at the same stage of production that comprises its core business. This may come about through purchasing companies at that same stage or linking to them through less formal means. Firms will usually prefer to trade abroad than invest directly, and they will do the latter only when the benefits of having a base location overseas outstrip the benefits of exporting there. Such advantages may include cheaper labour costs or reduced tariffs. Often, companies expand in this way in order to gain larger markets (Gwynne *et al.*, 2003). Horizontal integration and expansion of this nature is responsible for the growth of 'global consumerism' through the expansion of service TNCs – such as fast food/beverage corporations like KFC, Starbucks, McDonald's and Pepsi (see Plate 4.3). However, horizontal integration has also driven the global expansion of the following sectors: motor vehicle industry

(Nissan, General Motors, Ford); international accounting services (PricewaterhouseCoopers); newsagents and booksellers (Waterstones and WHSmith); retail outlets (Walmart); and, banking (Lloyds, HSBC).

● *Vertical integration* occurs when companies attempt to gain control over different stages in the production process either upstream (supply) or downstream (distribution or outlet). Vertical integration is driven by TNCs' attempts to reduce risk. Some vertical integration takes the form of direct ownership of the different stages of production, as this can reduce the 'transaction cost' of contracts. In the fruit sector, for example, Del Monte owns orchards in Chile, packing and shipping facilities in Chile and Holland, refrigeration plants in the UK, and its R&D and headquarters are in the USA. Other examples of vertical integration include the oil industry as well as the aluminium sector. Many TNCs, in the face of the increased flexibilization of the global economy, attempt to reap the benefits of vertical integration while dissipating risk by developing contract-based networks. This greatly reduces fixed costs and makes reaction to markets more flexible and rapid. The Nike case in Box 4.1 illustrates this strategy.

TNCs have a relatively long history and are rooted in the expansion of the colonial trading companies. The USA was the leader in TNC development and invested in Latin America and Asia early in the twentieth century. As noted elsewhere, TNC growth blossomed in the 1960s driven by the NIDL. The rise of TNCs has been facilitated by two major factors during the neoliberal phase of globalization. First, the increased mobility of capital in deregulated markets clearly favours these institutions. Second, progress in communications and transport technologies has made corporate decision-making possible at increasingly long distance. Data and control have become centralized, while operations have been increasingly decentralized. In this way, corporations have been able to reduce their risks while increasing flexibility.

The 'home' locations of the world's largest TNCs are concentrated in the global triad – the USA, EU and Japan (see Map 4.2). If we look at the world's largest companies by sales volume in 2013, we can see that both the USA and Western Europe are each home to approximately 30 per cent of the largest TNCs, while Japan is home to 15 per cent. Over the past three decades there has been some diversification in terms of the home location of TNCs as non-Japanese East Asian TNCs have grown in

Plate 4.3 *McDonald's in Kuwait*

Source: Photo by the authors

particular (Chinese and other Asian companies account for 15 per cent of the world total). The range of destinations for TNC investment has broadened considerably over the past 50 years however. There has also been a diversification in the nature of TNCs in terms of their social and cultural organization. In South Korea, the *chaeabol* based on the *zaibatsu* of post-war Japan represents a particularly state-linked form which differs from the Western models (see Gwynne, 1985; Gwynne *et al.*, 2003). Japanese *keiritsu* are characterized by 'transactions conducted through highly symbolic alliances of affiliated companies based around long-term relationships within "families" of related firms and founded on

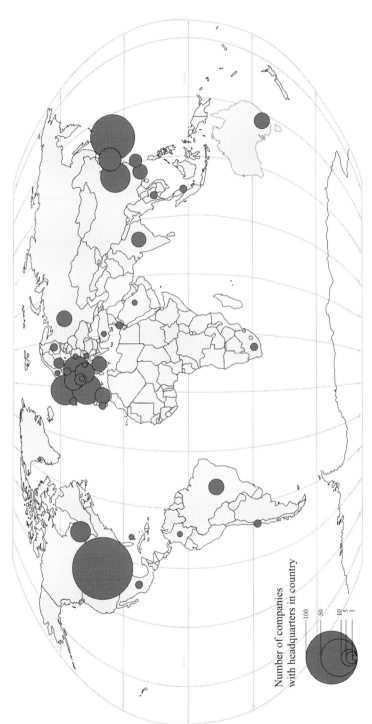

Map 4.2 Home location of the world's 500 largest TNCs, 2012

Number of companies
with headquarters in country

100
50
10
5
1

Source: Forbes 500 (www.forbes.com accessed 3 September 2013)

Note: TNCs are ranked by annual sales volume.

highly complex links' (Lee, 2000c, p. 853). Such organizations may be centred around a bank or a central trading company (e.g. Mitsubishi) or around a parent company (e.g. Sony). It could be argued that the vertical and horizontal linkages of such corporations make restructuring difficult and that this partly explains the inability of Japanese firms to react quickly to the 1997 Asian crisis. By contrast, many of the Chinese companies that now appear in the global list are virtually state-run enterprises.

Controversies in the geography of TNCs

As argued previously, TNCs have been central players in the evolution of globalization since the Second World War. In terms of sheer size, and thus economic command, a number of facts warrant mention. Many of the 50 largest TNCs have a higher turnover than the majority of the world's countries, while the larger ones are greater in absolute size than some medium-sized countries. In 2012, Walmart recorded turnover that was greater than the national income of Austria (see Table 4.5). In the same year, there were 13 TNCs larger in economic size than New Zealand. The ten largest TNCs have a total income greater than the world's 100 poorest countries. In general, the largest TNCs have the ability, by virtue of their vast size, to stabilize and/or destabilize national economies.

There are many controversies surrounding the role and impacts of TNCs both in the West and the Third World, though particularly in the case of the former. Two issues are dealt with below: the race to the bottom, and the 'embeddedness' or otherwise of TNCs.

The race to the bottom? Effectively, states bid for the investment of TNCs, often offering economic incentives to attract them. To remain competitive, states must maintain, among other things, low tax regimes, flexible labour markets (and low levels of unionization) and flexible environmental laws. TNCs can take advantage of these conditions in two ways: first, in their production location decisions; second, through 'transfer-pricing'. The latter sees TNCs allocating sales and profit returns in ways that exploit local conditions, such as low taxation, irrespective of the actual sales and profits at the real point of production. States have invested billions of dollars in attracting TNCs through, for example, the provision of infrastructure as well as the establishment of tax-free export

Table 4.5 *The economic size of TNCs and nation-states compared, 2012 (country GDP or company annual sales in $US billions)*

1	USA	15,685	31	Austria	400
2	China	8,227	32	South Africa	384
3	Japan	5,960	33	Venezuela	382
4	Germany	3,400	**34**	**BP**	**371**
5	France	2,613	35	Colombia	370
6	United Kingdom	2,435	36	Thailand	366
7	Brazil	2,253	37	UAE	360
8	Russian Federation	2,015	38	Denmark	314
9	Italy	2,013	**39**	**PetroChina**	**309**
10	India	1,842	40	Malaysia	304
11	Canada	1,821	41	Singapore	275
12	Australia	1,521	42	Chile	268
13	Spain	1,349	43	Hong Kong	263
14	Mexico	1,177	44	Nigeria	263
15	South Korea	1,130	45	Egypt	257
16	Indonesia	878	**46**	**Volkswagen Group**	**254**
17	Turkey	789	47	Philippines	250
18	Netherlands	772	48	Finland	250
19	Switzerland	632	49	Greece	249
20	Saudi Arabia	577	50	Israel	243
21	Sweden	526	**51**	**Total**	**241**
22	Iran	514	52	Pakistan	231
23	Norway	500	**53**	**Toyota Motor Corporation**	**225**
24	Poland	490	**54**	**Chevron**	**223**
25	Belgium	484	**55**	**Glencore International**	**214**
26	Argentina	475	56	Portugal	212
27	**Walmart Stores**	**469**	57	Ireland	210
28	**Royal Dutch Shell**	**467**	58	Iraq	210
29	**Exxon Mobil**	**421**	59	Algeria	208
30	**Sinopec-China Petroleum**	**412**	60	Kazakstan	202

Source: Sales figures are from the Forbes list (www.forbes.com/ accessed 3 September 2013); GDP figures are from the World Bank (http://data.worldbank.org/ accessed 3 September 2013).

processing zones (EPZ). It could be argued that this erodes the sovereignty of nation-states *vis-à-vis* TNCs.

The embeddedness of TNCs? Linked to the above is the issue of the supposed 'placelessness' of TNCs. To some, TNCs are so powerful and pervasive that they are increasingly bypassing nation-states. The argument of hyperglobalists is that such firms operate on a 'global surface' (Dicken, 2004). There has been some groundbreaking work in geography, informed by a number of the discussions concerning scale and space (see Chapter 2), which seeks to go beyond this simplistic notion. The first counter to the 'placelessness' argument is built on quantitative analysis. If firms were truly 'global' it would be expected that the majority of their assets and employment be located outside of their country of origin. A useful measure in this case is the transnationality index (TNI), which calculates a combined ratio of foreign to total assets, sales and employment; although these measures cannot always capture the true extent of the control that large firms have over global production complexes. In 2012, 9 of the largest 15 TNCs had TNIs greater than 60 per cent, and 6 of those exceeded 75 per cent (see Table 4.6). This seemed to represent a large and increasing degree of global reach. Yet TNCs remain remarkably tied to their home base and to their national identity (for example Nokia from Finland or Volkswagen from Germany). Quantitative measures provide only a partial picture of 'embeddedness' or otherwise. The particular configuration of economic, political and cultural traits of both home and destination states can have an important influence on the transnationality of TNCs. Although TNCs are global in scope, they often take on some of the characteristics of the host economy or society in order to further their business – Japanese firms in the UK, for example, have adapted Japanese production practice to suit local labour forces and markets. Marketing is also increasingly differentiated to reflect local conditions – as in the case of the Kiwi Burger in New Zealand, which comes with beetroot! There is certainly a move among large globalized corporations to insert the 'local' into their operations. Slogans such as 'your local global bank' used by HSBC exemplify this. Overall, many TNCs are classic examples of glocal organizations.

Costs and benefits of TNC diffusion

There are highly divergent opinions with respect to the costs and benefits of TNC activity for both home and, especially, host countries. Gwynne *et al.* (2003) separate theoretical opinion on the net impacts on the latter into two camps, as explored below.

Neoclassical approaches – Lewis was the first to write of the benefits of TNC investment in peripheral countries in the context of his influential modernization paradigm in the 1950s (Lewis, 1955). He argued that TNC investment would promote technology transfer, better labour

Table 4.6 *Transnationality of the world's 15 largest non-financial TNCs, 2012[a]*
($US millions)

	Corporation	Home economy	Industry[c]	Assets		TNI[b] (%)
				Total	Foreign	
1	General Electric Co	USA	Electrical and electronic equipment	685,328	338,157	59.7
2	CITIC Group[d]	China	Diversified	514,847	71,512	43.8
3	Volkswagen Group	Germany	Motor vehicles	409,257	158,046	78.3
4	Toyota Motor Corporation	Japan	Motor vehicles	376,841	233,193	66.0
5	Royal Dutch Shell plc	UK	Petroleum expl./ ref./distr.	360,325	307,938	76.4
6	Exxon Mobil Corporation	USA	Petroleum expl./ ref./distr.	333,795	214,349	77.7
7	EDF SA	France	Utilities (Electricity, gas and water)	330,582	103,015	54.4
8	BP plc	UK	Petroleum expl./ ref./distr.	300,193	270,247	83.8
9	GDF Suez	France	Utilities (Electricity, gas and water)	271,607	175,057	90.2
10	Chevron Corporation	USA	Petroleum expl./ ref./distr.	232,982	158,865	58.1
11	Total SA	France	Petroleum expl./ ref./distr.	227,107	214,507	52.1
12	Enel SpA	Italy	Utilities (Electricity, gas and water)	226,878	132,231	62.8
13	Vodafone Group Plc	UK	Telecommunications	217,031	199,003	60.6
14	Daimler AG	Germany	Motor vehicles	215,408	99,490	58.5
15	Walmart Stores Inc	USA	Retail and Trade	193,406	84,045	76.4

Source: UNCTAD

Notes:

Companies are ranked by total assets.

[a] Preliminary results based on data from the companies' financial reporting; corresponds to the financial year from 1 April 2012 to 31 March 2013.

[b] TNI, the Transnationality Index, is calculated as the average of the following three ratios: foreign assets to total assets, foreign sales to total sales and foreign employment to total employment.

[c] Industry classification for companies follows the United States Standard Industrial Classification as used by the United States Securities and Exchange Commission (SEC).

[d] Data refers to 2011.

practices and managerial know-how, and increase domestic income and foreign exchange. Although Lewis favoured domestic capital, he argued that, in its absence, it is better to have foreign capital which would lead to improvements in consumption and education through raised incomes.

Lewis' neoclassical assumptions have been criticized by authors such as Jenkins (1987). The idea that in the absence of foreign investment there may be no investment at all is seen as true for some sectors only. This may be the case, for example, in capital-intensive industries with high entry and running costs – such as mining. In contrast, in sectors where smaller-scale operations are more prevalent such as agriculture and services (e.g. hotels, supermarkets) foreign investment could have the effect of crowding out domestic capital. Furthermore, in protected markets, such as those operating import substitution industrialization policies, the entry of large foreign firms can lead to what economists term an 'oligopsonistic reaction' leading to higher prices and lower wages.

The neoclassical approach has evolved from a macroeconomic perspective to one which has become more microeconomic in its focus, particularly at the scale of the corporation itself. Led by authors such as Dunning (1988) and Vernon (1977), a diffusionist model has evolved. Diffusion theory argues that the diffusion of technology has been of particular benefit for host countries (see Malecki, 1991). In general, then, these approaches 'see TNCs as efficient allocators of resources internationally and as providing benefits for the host and home nations' (Gwynne *et al.*, 2004, p. 153). Benefits in terms of trade, foreign capital inflows, employment and linkage effects are emphasized. It is this perspective that underpins neoliberal ideas which are supportive of TNC FDI in institutions such as the WTO.

Radical approaches – These approaches have their roots in Marxist and neo-Marxist theory. Broadly speaking, such thinking sees TNCs as agents of imperial and neo-imperial power attempting to control resources, capital, labour and markets to the advantage of elites in core nation-states. These processes are accentuated where host countries are economically small. The *neo-imperialistic* or *dependency* perspective has been influential in academic, activist and socialist policymaking circles. With respect to dependency thinking on the role of TNCs (see Kay, 1989), there are at least three ways in which TNC investment is linked to the underdevelopment of the periphery:

1 *TNCs are the main vessels through which surplus capital is appropriated back to the core.* There is an unequal exchange that

characterizes this process. Some have extended this analogy to culture where TNCs act as the main lines of the diffusion of Western cultural models, designed to promote consumerism and the growth of capitalism.

2 *Monopolies and oligopolies tend to spread from the core countries to the periphery.* Large companies that have gained significant market share face little competition in relatively small economies and are thus able to capture significant proportions of economic activity, reducing the space for local capital.

3 *The emergence of what Frank (1969) called a 'comprador' class.* This 'class' is dependent on TNC patronage and acts as a conduit for its demands. Such elites will influence national decisions on the basis of what is required for the TNCs rather than the host nations.

Overall, Lee (2000) and Schoenburger (2009) argue that assessing the balance sheet of TNC activity is a difficult analytical task. This is partly because it is a counterfactual problem (i.e. we do not know what would have happened in any particular case had the TNC not invested). For the circuit of capital as a whole it is uncertain whether TNCs make it more or less efficient. From the perspective of peripheral countries, however, Lee is quite unequivocal, claiming:

> Their [TNCs'] influence is malign insofar as at the same time that they connect economies into the global circuit of capital and so increase supply lines and markets and introduce new technologies (which have both positive and negative developmental effects), they displace control over the making of history and geography from people struggling to make their living and define their identity in such economic geographies by imposing a particular understanding and measure of value and progress.
>
> (Lee, 2000c, p. 853)

Notwithstanding all of the above criticisms, and the current popularity of a populist dependency-type approach evident in the works of journalists such as Michael Moore and John Pilger, it is the neoclassical approach that is hegemonic in policymaking. Left-leaning critics might argue that this is the case because TNCs have purchased enormous influence over the policy agendas of Western governments. Those on the right argue that the role of TNCs is largely beneficial and that poorer countries are not liberalized enough to make the most of what TNCs can bring. The arguments of the right are based upon theories that assume perfectly competitive markets. In reality, the imperfect competition of monopoly capitalism is becoming increasingly pervasive through the activities of TNCs, and even neoclassical analysts would argue that this does not bode well for society.

Regional economic integration

The thesis of economic globalization is made controversial, to sceptics at least, by the recent increase in regional economic integration. The WTO estimates that by 2012 there were over 170 Regional Trade Agreements (RTAs) in force, covering close to 40 per cent of world trade. These have continued to grow in number and coverage. The nature and scope of such agreements vary but essentially they represent attempts among often (though not always) neighbouring countries to reform the rules that govern trade and investment flows such that preferential access be granted to signatories. This usually involves lowering tariff barriers and non-tariff barriers such as subsidies and environmental controls. In this sense they are liberating and discriminatory at the same time. Examples abound including the European Union, the North American Free Trade Agreement, the MERCOSUR, the ASEAN (Association of Southeast Asian Nations) and APEC (Asia Pacific Economic Cooperation) (see Table 4.7). One of the largest and most interesting blocs that is emerging is the Trans-Pacific Partnership (TPP). This involves 12 economies in the Americas and Asia (including USA and Japan), some with hitherto highly protective trading regimes. Although still under negotiation, the TPP seeks a more liberalized trading agreement than under APEC and its membership is likely to increase (Kelsey, 2010). It sits alongside several important bilateral free-trade agreements in the region, such as that between Australia and USA (2004), New Zealand and China (2008) and Chile and many other countries in the Pacific Rim (Murray and Challies, 2010).

The agreements vary in size and coverage, and in some cases – especially in Latin America – membership overlaps. Dicken (2003, p. 146) identifies four 'waves' of regionalization: during the second half of the nineteenth century; after the First World War; from the 1950s until the 1970s; and the most recent wave following the end of the Cold War. We must be careful not to see regionalization as a homogeneous process. The nature of integration in Europe is very different to that which has taken place in ASEAN, for example (see Map 4.3); the latter is often seen as a purely economic body which does not intervene in the politics of member states. The free-trade agreement of 1992 was not prevented, for example, by the poor human rights records of numerous governments within the grouping. There are thus 'alternative regionalisms'.

Table 4.7 *Important regional trading blocs*

	Formation	Type	Population 2010 (mill)	GDP 2010 ($USbill)	GDP per capita 2010 ($US)
Andean Community	1969 (restarted 1990)	Customs union	100.9	528	5,232
APEC	1989	Pre-free-trade area	2,752.4	35,536	12,911
ASEAN	1967 (ASEAN) 1992 (ASEAN free-trade agreement)	Free-trade area	593.9	1,883	3,171
CARICOM	1973	Common market	16.8	66	3,924
COMESA	1994	Common market	436.2	502	1,150
EU	1957 (European Common Market) 1992 (EU)	Economic union with limited political union	506.7	16,233	32,037
MERCOSUR	1991	Customs union	245.4	2,570	10,475
NAFTA	1994	Free-trade area	461.3	17,031	36,917
SADC	1980	Pre-free-trade area	250.7	561	2,239

Source: World Bank (http://data.worldbank.org/ accessed 3 September 2013)

Classical trade analysis allows for two broad outcomes in terms of the impacts of regional integration:

- *Trade diversion* – where, due to the formation of the new agreement, trade with a former partner outside the bloc is reduced and replaced by trade with a partner inside it.
- *Trade creation* – where, as a result of the agreement, growth creates new trade or replaces domestic production.

An RTA may be considered a 'success', in purely economic terms, where creation outstrips diversion. In general, RTAs shift flows of trade and investment in ways that have significant geographical impacts on the ground as networks connecting places are created or destroyed. There are at least five different types of RTA according to different levels of integration:

- *Free-trade area* – where the barriers to trade between members are reduced but each country retains sovereignty with respect to external parties in terms of trade policy.

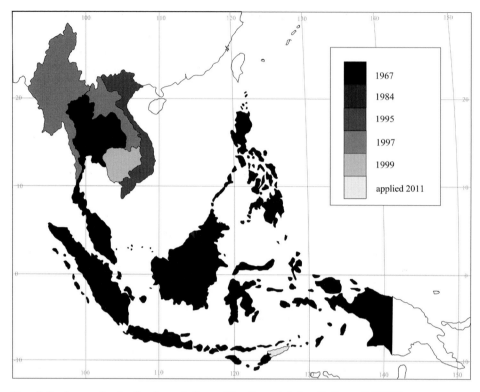

Map 4.3 *Ascension to the Association of Southeast Asian Nations (ASEAN)*

- *Customs union* – a free-trade agreement plus a common external policy towards non-members.
- *Common market* – a customs union plus the free movement of factors of production – labour, capital, resources, knowledge – between members.
- *Economic union* – a common market plus the harmonization of economic policies, including monetary and fiscal policy. Such things become governed by extra-national bodies.
- *Political union* – an economic union plus harmonization of social policy and the surrendering of national sovereignty to the regional body.

Most agreements are of the *free-trade area* type, and thus much of the regionalism that is occurring has been 'open' regionalism. Examples of free-trade areas include NAFTA, EFTA and AFTA-ASEAN. Customs unions include the examples of ANCOM and CARICOM. In contrast, MERCOSUR is a common market. The EU represents a different example, and is a customs union with elements of economic and political union also (see case study in Chapter 5). Of the various regional 'blocs' three stand out as 'megaregions' (McMichael, 2013). NAFTA, the EU and the APEC account for two-thirds of all global manufacturing and three-quarters of world exports. Of the three megaregions, the EU is the least reliant on trade with the outside world. Some critics have labelled it *Fortress Europe*, around which protectionist walls, especially in agriculture, are built. In the case of Europe, and to a lesser extent in the case of other examples, the issue of regional integration throws up tensions with regard to the sovereignty of nation-states in conducting their economic affairs.

Regional integration has inspired an enormous literature that deals with the political, economic and cultural impacts of such change. Thinkers are divided as to whether integration is a defensive reaction to globalization and other regional integration moves, or a precursor to further global interconnectedness. The discriminatory aspects of the formation of RTAs contravene GATT and WTO rule in principle, but article XXIV allows for the formation of such groups under certain conditions. As already pointed out, most of the world's trade takes place between three major groupings: EU, NAFTA and East Asia. APEC, however, arose out of a particular concern among potential members that regional blocs were eroding the potential trade benefits of globalization. Does regionalism confirm or deny the thesis of economic globalization? To sceptics such as Hirst and Thompson (1999), the regionalization of the world economy is seen as further evidence that globalization does not actually exist. Furthermore, they argue, it is a process that compounds the core/periphery nature of the world economy. For others, it represents a *creeping globalization*, which will eventually encompass all regions. Individual countries are under great pressure to take part in such ventures in order to face the enormous economies of scale and scope that existing regional groupings possess. In South America there is an active process of integration taking place (see Box 4.4). Chile is currently the most aggressive country in the world in terms of forging new bilateral and multilateral trade and integration agreements.

Impacts of regionalism case study – NAFTA and the maquiladoras

The North American Free Trade Agreement was signed in 1994 by the USA, Canada and Mexico. Under the agreement, flows of goods, capital

Box 4.4

Regional integration in action: from MERCOSUR to UNASUR in South America

Regional integration is a theme that has a long heritage in Latin America. Following the conclusion of independence in the 1820s there were some efforts to create a unified Spanish America. Following the Great Depression, and the theories proposed by the structuralists to explain its devastating consequences, the idea of economic integration was once again raised. Structuralists argued that it was the nature of the insertion of the Latin American economies into the wider economy – as resource peripheries established during colonialism – that explained relative impoverishment. Crucially, the structuralists wanted to stimulate industrialization through import substitution (ISI) in order to break out of the primary product export trap. This would require economies of scale that regional economic integration would create – fostering autonomous capitalist, as opposed to dependent, development. Furthermore, such agreements would be geared towards restructuring the insertion of the economies into the global economy as manufacturers. Efforts towards this end included the Latin American Free Trade Area, The Andean Pact and the Central American Common Market (see Gwynne and Kay, 2004; Munck, 2011; Murray and Rabel, 2008). Many such models were not especially successful, which, in part, led to the eclipsing of structuralist ideas by more radical 'semi-autarkic' dependency theories (see Chapter 7) (Kay, 1989). In the early 1990s, however, a more 'successful' example of regional economic integration was constructed on the Continent. The *Mercado Comun del Sur* (Common Market of the Southern Cone) was established in 1991 to facilitate trade between member countries Argentina, Brazil, Paraguay and Uruguay. During the 1990s intra-MERCOSUR trade grew by a multiple of five and trade barriers were removed consistently. In 1995 a common market of the four countries came into effect and a cooperation agreement with the EU was signed.

In the same year, Chile and Bolivia expressed their interest in joining and have since been admitted as associate members. MERCOSUR has perpetuated the rise of neoliberalism in the region, where it is now hegemonic. When the agreement was set up in 1991, the era of ISI had ended in the continent and governments were committed to outward orientation. Governments appealed to interest groups who might otherwise oppose the negative implications of structural adjustment policies by offering the supposed benefits of a greater regional market. By 2012 MERCOSUR was comprised of six members: Argentina, Bolivia, Brazil, Paraguay, Uruguay and Venezuela (although at the time of writing Paraguay had been suspended due to contravention of the democratic clause of the group due to the impeachment of its president). By that time it has become a full

Box 4.4
continued

customs union (see earlier discussion of regional integration), with plans laid to create a full economic union in the future. This association of full members represented a grouping of over 270 million in population, with a combined GNI of over US$3.2 thousand billion (trillion) – making it one of the largest regional economies in the world. By 2012 the range of associate members of MERCOSUR had expanded to include Colombia, Ecuador, Guyana, Peru and Suriname joining Chile in this status. Mexico and New Zealand were observer members, the latter becoming increasingly interested in the continent over the 2000s (Murray and Challies, 2008) building in particular on a bilateral agreement with Chile signed in 2006. Despite the heralding of the agreement as a success story, there are many imbalances between and within countries in the context of free-trade reform. Marginalized groups (such as peasant farmers, and agriculturalists in general) have voiced considerable concern and protested the impacts of the MERCOSUR widely (Murray, 2004a, b).

An agreement signed in 2004 stated that it was an objective to eventually integrate with Andean Community of Nations (CAN) to create a bloc covering all of South America. In 2008 this became the Union of South American Nations – or UNASUR – which held its first summit in Chile in 2008 and eventually led to the creation of Banco Sur in 2009 – a development bank to rival the World Bank. Both institutions were designed to offset the power of the United States politically and economically and led to the eclipsing of the proposed Free Trade Area of the Americas. The headquarters and parliament were to be located in Cochabamba, Bolivia and the bank headquarters in Caracas, Venezuela. It was intended that UNASUR would eventually supersede MERCOSUR and CAN to become a supranational body emulating the European Union. It is hoped that the UNASUR will be more development focused and address some of the issues of imbalance that MERCOSUR has experienced.

and services are governed by WTO rules. Trade volumes between the three countries have tripled in the decade since the agreement was signed. Mexico has grown rapidly to become the ninth largest economy in the world. Canada and the USA have benefited, it is argued, from cheaper manufactured imports in particular. Spectacular regional economic growth along the US/Mexican border has been promoted, especially on the Mexican side, where hundreds of assembly branch plants (*maquiladoras*) of US TNCs have sprung up in cities such as Ciudad Juarez and Tijuana (see Plate 4.4). The economic transformation of the border zone actually builds on Mexican policy beginning in 1965 when the Border Industrialization Programme was established. Under this plan, capital and parts were imported duty free from the USA to Mexico, assembled into final products, and exported back to the former. This represented one of the earliest examples of an export-processing zone in the Third World. In the 1970s, the factories were largely labour-intensive operations populated by unskilled workers, many of whom were women. With the arrival of high-technology TNCs from the mid-1980s, skill requirements have shifted somewhat (Gwynne *et al.*, 2003). By the end

Plate 4.4 *Maquiladoras*
A shanty town in the foreground is a source of labour for the assembly factories on the Mexican/US border pictured in the middle distance.

Source: Photo by Robert N. Gwynne

of the 2000s the total trade between NAFTA partners stood at over US$950 billion, and investment flows between countries stood at approximately US$750 billion with pro-NAFTA groups such as NAFTA Now estimating that the agreement had created over 40 million jobs between 1994 and 2008 (NAFTA, 2013)

There are many critics of the apparent macroeconomic success which NAFTA has brought to its participants. In the USA, unions from manufacturing sectors, such as car production, have protested the shipping out of jobs to cheaper labour locations in Mexico resulting in large-scale structural unemployment in many 'rust belt' cities of the US North West (such as Detroit and Chicago). Liberal US critics have argued that the rules governing NAFTA have stopped the prevention of the import of products where environmental standards have not been met and where poor labour conditions, such as the use of children, are allowed. In Mexico, in order to gain reciprocal reductions in US protectionism, wage rates were forced down in the 1990s. In 1995, hourly rates were 9 per cent of equivalent rates in the USA. Environmental protection laws were also eased in order to push the deal through. Despite high economic growth in much of the 1990s (aside from the peso crash in 1994), levels of poverty remain stubbornly high. In an absolute sense, it is estimated that some 15 million individuals joined the ranks of poverty through the decade, and income inequality rose considerably. Social conditions on the Mexican side of the border are marginal for many, as crime and deprivation in local economies based on cheap labour exploitation has risen dramatically (McMichael, 2013). Protest in Mexico has been widespread over the last 20 years and was especially intense in the first years of the agreement, the most famous episode being the Zapatista uprising against NAFTA in the traditional agricultural Chiapas region of the South. Indeed, some see this event as the genesis of the anti-globalization movement (see Chapter 5). In the early 2000s, in Southern Mexico, a new 'region state' was posited, the *Puebla de Panama*, which would take advantage of relatively low wages in the periphery of the country. It would function as an export-processing zone linking Mexico with Central America. As such, it was intended that NATFA would lead to the widespread maquilization of the Mexican economy and the diffusion of unskilled and non-unionized employment across the country. This was widely resisted and eventually abandoned. Indeed, resistance to the rolling out of the NAFTA and associated regional integration plans has been influential. According to Hayden (2003, cited in McMichael, 2004, p. 192), there was evidence of an increasingly intense race to retain low-wage competitive advantage:

> Relocating the crisis-ridden maquiladora industry to Southern Mexico, where wages are half those of the Mexican maquilas on the US border, is a desperate effort to prevent the haemorrhage of jobs to China where [the wage] is only one-sixth of the Mexican wage.

Despite these criticisms, free-trade proponents still see NAFTA as a model of regional integration that lays the basis for the wider Free Trade Area of the Americas (FTAA), and the Trans Pacific Partnership agreement beyond this. In the early 2000s it was suggested that this continent of America-wide would include close to 40 and over 800 million people. Many neoliberal Latin American governments were keen for this scheme, which represented an extension of NAFTA, to progress. Based in part on the negative trends that were observed in Mexico – that were characterized by the contradictory free flow of investment and trade but very un-free flow of persons and increasingly draconian measures in the USA to restrict this – a number of countries including Venezuela, Bolivia, Ecuador and Cuba joined the Bolivarian Alternative for America, arguing that the FTAA would lock Latin America in the context of the hemispheric economy into supply of cheap labour and resources. Crucially, regional powerhouses Chile, Argentina and Brazil also opposed the FTAA, at least as it was proposed in the early 2000s. The advance of South American integration has meant that the idea of FTAA is, for now, stalled and any future integration that might take place will do so under more equal power relations between the North and South of the continent of America. Sceptics might argue that these trends indicate a tendency towards regional integration and the carving out of new economic blocs; hyperglobalisits might argue in response that such trends represent the preparing of the ground and movement towards a more globalized economy.

Conclusion – a transformed economy

Circuits of capital and spatial divisions of labour have been stretched through the process of economic globalization, leading to increasingly complex and hybrid outcomes that result in the coexistence of Fordist and flexible modes of production. This evolution is tied up with the shift from the colonial division of labour, through the NIDL to one based on flexible accumulation and the ascendancy of neoliberalism. TNCs have been among the most important, and controversial, driving forces of this transition. The evolution of the globalizing economy is summarized in Box 4.5.

The transformations outlined in Box 4.5 have led to increased unevenness across the global economy. Patterns of exclusion/inclusion are more complex today than they were in the past as networks of power have replaced simple core–periphery dichotomies. Understanding these new patterns requires incorporating both chain and network concepts. At the same time as 'global' flows have increased, powerful regional economic blocs have evolved which, to some, undermine the thesis of economic globalization. Notwithstanding this, as the case study of globalized agriculture illustrates, economic flows have intensified, become more extensive, move at increased velocity and have more profound impacts for those on the networks. Ultimately this has led to a global economy that is more asymmetric than ever, and it is to this inequality that we will return in Chapter 7, which further explores the unfolding world economy and its inherent instability and unevenness

Box 4.5

Summary of the transforming global economy

- Time–space compression, and the desire to decrease the turnover rate of capital, has led to a rapid expansion of the global economy over the past two centuries.
- The colonial division of labour faced a crisis sometime following the Second World War.
- It was replaced in 1960/1970 by a NIDL which saw increased FDI to the periphery. This created a global semi-periphery.
- TNCs were the main actors in the forging of the NIDL – they have brought costs and benefits.
- The expansion of TNCs has been facilitated by the shift to outward-oriented, free-market (neoliberal) development policy in the Third World.

- More recently there has been a further complication in the processes that created the NIDL – of note is the rise of NIC investment elsewhere in the periphery and to the core.
- The crisis of Fordism has led to a more flexible regime of accumulation.
- Flexibility has spatial implications which may lead to an exacerbation of the already concentrated pattern of global economic activity.
- However, in reality, the global economy is characterized by a mixture of Fordism and flexibility. The only truly supportable fact is that the global economy is far more extensive and complicated than it has ever been before.

Further reading

Coe (2013): Another excellent chapter from the 'Birmingham School' *Human Geography* text – but in this case by somebody from the Manchester School of Economic Geography – that covers commodity chains and divisions of labour comprehensively.

Coe, Kelly and Yeung (2009): This is rapidly becoming a classic in economic geography and serves as an excellent introduction.

Cumbers (2009): This provides a very good entrance point to the debates concerning regional integration from a geographical perspective.

Daniels and Jones (2013): This is a very good overview chapter that presents a number of the arguments visited in this book.

Dicken (2012): This is by far the most comprehensive book on the geography of the global economy and is used across the social sciences. A measure of its globalization is that it is translated into Japanese and it is in its sixth edition.

Gwynne *et al.* (2003): This text looks at the concept of 'alternative' capitalisms and adopts a regional geography perspective in comparing and contrasting experiences from Russia, East Asia and Latin America.

Yeung (2009): This entry in the *International Encyclopaedia of Human Geography* is an excellent overview of the issues and trends in studies of economic globalization. See also Dicken (2009) in the same collection on TNCs and globalization from a geographical perspective.

⬤5 Globalizing political geographies

The rise of challenges from both above and below the scale of the nation-state have brought into focus the issue of where political power resides and can be exercised in the light of globalization. The process of globalization has given rise to successive world orders that provide the broad contours of political power. One of the outstanding responses to the increasingly rapid transformation of these contours has been the consolidation of a global civil society movement that, while ebbing and flowing over recent decades, has on the whole grown in its extent. Whether it has been effective has been the site of considerable debate however. This chapter investigates the shifting nature of political globalization and has a range of key questions underpinning it: How is the role of the nation-state being transformed? How is globalization both forged by and forging so-called 'world' orders? And, what is the nature and effectiveness of the anti-globalization movement that has arisen in response to the political challenges of these global times?

New political geography and globalization

The nature of political geography has changed markedly over the past 50 years. Early political geography was dominated by traditional geopolitics, which focused on the formation of, and interaction between, nation-states and the expression of political order in space. Geopolitical ideas were used to justify the expansionist ambitions of some states and, as such, often informed a colonial mindset (see Barton, 1997 on Pinochet's geopolitics in Chile). During the quantitative revolution in geography in the late 1960s to early 1970s, positivist tools were used to describe and analyse the playing out of formal politics in space – electoral geographies, political parties, the business of government were modelled in a spatial manner. This work played an important role in our understanding of the spatiality of the polity. With the post-positivist shift, the rise of radical and critical geography, and the cultural turn, the nature of political geography has been transformed. This is reflected in two main developments in the sub-discipline. First, the impact of formal politics (e.g. government, democracy, elections, transnational governmental organization) on communities and individuals has been analysed in a more critical way. Furthermore, a critical geopolitics school has emerged which seeks to challenge accepted ways of understanding governance and governability in world politics (see Ó Tuathail, 1996, 2002, 2006; Painter, 1995; Painter and Jeffrey, 2009; Sidaway and Mamadouh, 2012). Second, informal politics, sometimes referred to as grass-roots politics, have been increasingly researched. By focusing on the politics of the home, the body, the mind, the street and the community as practiced outside of the formal sphere, the idea that politics is personal has entered into contemporary political geography (Cloke *et al.*, 2005). Thus geographers have been actively researching themes such as citizenship, national identity, nationalism and new social movements, focusing on both the concrete and imagined outcomes of political discourse at various scales.

The above changes are related to the evolution of globalization in important ways. The broadening of the sub-discipline comes in response to the increased complexity of political geographies in a globalizing world. It is apparent that the stretching of social relations across space is giving rise to new networks through which political demands and power are transmitted, and these may be both formal (for example, transnational governmental institutions such as the UN) and informal (for example, grass-roots political communities such as the anti- and alter-globalization movement). Further, the critical perspective that has evolved in political geography enables researchers in the field to grapple with moral questions concerning how best to manage, regulate, respond to and resist globalization. The world is at a crossroads and political geography is in a strong position to analyse the political choices ahead.

The rise and fall of the nation-state?

The role of the nation-state is being contested. Critics from across the political spectrum have come to question the relevance and durability of the concept as the principle organizing unit of globalizing society. Authors such as Ohmae (1995) and Castells (1996) suggest, albeit in very different ways, that it is being eroded both from 'above' and 'below' (see Sidaway and Grundy Warr, 2012). The argument underlying each of these analyses is a 'scalar' one: the nation-state is viewed as too small to cope with global economic, political and cultural flows and too large to represent effectively the needs and desires of local and trans-local communities. On this basis, hyperglobalists have predicted the *end* of the nation-state. Geographers have, on the whole, responded to such arguments in a balanced way; while acknowledging the transformed nature of political governance, arguments predicting the death of the state are seen as exaggerated.

Nation-states are such a central part of most of our lives, albeit to different degrees, since we have grown up in a world divided by neat, seemingly impervious, borders. Behind these borders, we are often led to believe, exist *inherent* national sociocultural traits. Questioning the role and survival of the state rocks the very foundations of what many have been led to believe – through discourses of nationalism – is their very identity and place in the world. The nation-state, however, is a relatively new concept in the history of global society, and it is the product of the European political imagination. The concept diffused across the world, through the processes of globalization itself, during the colonial wave when Enlightenment notions of the state as the 'natural' unit of society proliferated. This was consolidated during the postcolonial wave after the Second World War, which saw the creation of many newly independent modernist nation-states. The fact that the nation-state is threatened by the very thing that propagated it is a paradox (Waters, 2001). If we understand the roots of this *discourse*, we can better understand how and why notions of the state are shifting in the light of contemporary globalization, and we are better placed to accept that changes in governance practices might actually be *required*.

The concept of the nation-state

The nation-state may be defined as:

> A complex array of modern institutions involved in governance over a
> spatially bounded territory. It claims sovereignty over that territory and
> generally protects that sovereignty through a monopoly on the means of
> violence (military).
>
> (Adapted from Johnston *et al.*, 2009 and Daniels *et al.*, 2012)

There are two processes involved in nation-state formation. First,
state-building sees the state centralize power over its territory,
penetrating civil society in order to construct the political and social
infrastructure to implement decisions. Second, *nation-building* involves
elites using power gained through state-building to create a national
identity and/or culture, and thus a notion of *citizenship*. Nation-states are
not to be confused with nations, where the latter refers to a community
of people that shares a common sense of 'togetherness', sometimes
expressed as a common culture. As such, nations can transcend nation-
state borders; and there are few nation-states, if any, which contain one
homogeneous nation. For example, the Kurdish nation stretches across
the borders of Iraq, Turkey and Syria. Under contemporary globalization,
the diffusion of nations through migration can lead to *diasporas*. For
example, the Tongan nation exists largely outside of Tonga itself and
comprises populations in Australia, New Zealand, the UK and the USA
(Lee, 2004). As such, most nation-states are home to multiple nations.
The UK, for example, is home to English, Scots, Welsh and Irish, other
smaller groups (such as Cornish and Manx) as well as a multiplicity of
immigrant diasporas which may or may not continue to practise their
national culture (see Chapter 6).

The evolution of the nation-state

The concept of the nation-state is only around 300 years old. Prior to this
there existed a range of organizational forms governing territory.
Compared to today's system, these were much less rigidly organized in
space (see Calhoun, 2007; Sidaway and Mamadouh, 2012). Across
Europe in the sixteenth century, the feudal state dominated, where
aristocratic landholdings were loosely integrated in a relationship to
barons and a superior monarch, sometimes bounded in empires. By the
early seventeenth century in Europe, monarchical power had been

consolidated, power centralized in bureaucracies, and professional armies were established to impose the will of the state. Such states were often absolutist and 'depended on the monopolisation of control over a particular territory in which it could raise both armies and taxes' (Waters, 2001, p. 97). Some have argued that England and Holland represented the first examples of constitutional states where the powers of the monarch were limited through the collective interest of the emergent capitalist bourgeoisie (Mann, 1988).

The emergence of the modern nation-state and a system of international relations is often argued to have been established by the Treaty of Westphalia in 1648. This was signed at the culmination of the Thirty Years War in which the Austria–Hungary Catholic Empire sought to impose its religion on the protestant states of Northern Europe. The Treaty established three major points:

1 *The ruler is emperor in his own realm* – states should not be subject to external control.
2 *The ruler determines religion* – the state has internal sovereignty.
3 *Hegemons should not be allowed to arise* – a balance of power should be struck between different states.

The Treaty enshrined territorial sovereignty, borders and citizenship. This ushered in a period where diplomacy replaced war as the principal form of interstate relations. By the late nineteenth century, the modern industrial state, based on models from north-western Europe, had been firmly established and had been exported to North America through colonization. This transformation culminated in a set of interrelated processes (Waters, 2001):

1 The *centralization* of state power to overcome internal resistance.
2 The *autonomy* of state power and sovereignty to redefine tradition and resist external dominance.
3 The *broadening* of the political community through constitutional reform and the extension of suffrage.
4 The development of the *symbols of nationhood* – flags, anthems and national languages.
5 The *activization* of the state to move beyond military function into areas of daily regulation.

Underlying the formation of the modern state lay the rise of the transnational industrial bourgeoisie which, based on Adam Smith's neoclassical ideas, sought *laissez-faire* economics combined with an enabling state. This class achieved its economic and social goals through the constitutionalization process and the consequent shift away from absolutist control. The bourgeoisie also used the state to expand control over new colonies and to control resource flows to its benefit, beginning the eventual globalization of the nation-state concept largely to the advantage of the growing land- and capital-owning elite.

Trans-state governance from above and below

Until recently, the global political economy was not much more than the sum total of its nation-states, which were the regulating agents of change at virtually all scales. Today's polity is more like a 'web of interdependence', as Dicken (2011) calls it. This involves three basic types of institution:

- national governments;
- international organizations (e.g. UN, WTO);
- transnational organizations (e.g. NGOs including TNCs, environmental groups, international protest groups).

The contemporary political system is far more complex than the old system and it involves states 'as merely one level in a complex system of overlapping and often competing agencies of governance' (Hirst and Thompson, 1999, p. 183). In this 'world system of differentiated power relationships which operate at different levels of geographical analysis' (Dicken, 2011, p. 80), states are *permeable* and global governance is *polycentric*. As such, we are living in a less predictable 'multipolar' world where power is exerted from above and below the scale of the state. It seems paradoxical that today's world has more nation-states than ever (there are currently 193 member states of the United Nations), and this number may rise further. It is possible, however, to see this as part of a broader dialectical reaction to political globalization, as discussed further into this chapter.

The rise of global governance

A factor that has influenced theorization of the demise of state power is the rise of global 'regulation' practised above the level of the state. Of late, this issue has become one of enormous public interest as the nature and practices of global institutions such as the WTO, the IMF, the World

Bank and the UN have been widely criticized. Later in this chapter we look at one of the most important manifestations of this – what became termed the 'anti-globalization' movement – which has since moved to broaden in its scope (Economist, 2013). Here we trace the rise of institutions of global governance, which have proliferated since the end of the Second World War (1939 to 1945). It is argued that, to date, such entities in fact represent the will of the individual nation-states involved in their construction, particularly a small handful of very powerful states. It is this latter point that has given rise to concern over the 'undemocratic' nature of such bodies. Indeed, it was over ten years ago that Roberts (2002, p. 143) claimed the current malcontent is so overwhelming that we are possibly 'witnessing and participating in the establishment of a new order of global regulation'. Whether this has evolved is debatable although there has been a shift to a more multipolar world, as we shall see. Large-scale street protests building in places as diverse as the UK, Egypt, Brazil, Chile and Turkey have built upon the legacy of the anti-globalization movement and give the impression that global governance and state monopolies on power are being challenged. Despite some political victories in the Arab World and policy-based reform in places such as Brazil and Chile in response to such tactics, it is highly debatable whether a new order has been created; although we may still be in the throes of such a transition (see later section on the anti-globalization movement).

'Global regulation' is a term that needs carefully considering before we can understand the rise of its formal institutions. The French Regulation School sees capitalist society defined at any given time by an economic 'regime of accumulation' within a broader sociocultural 'mode of regulation'. According to Lipeitz (1987, p. 33), a mode of regulation includes 'institutional forms, procedures and habits, which either coerce or persuade private agents to conform to (the regime of accumulation's) schemas'. Thus a mode of regulation includes not only a system of rules but also social and cultural norms that sustain and reproduce the regime of accumulation. When we consider the evolution of global governance over the past 50 years, it is apparent that it operates as a 'system of rules' which forms part of a wider capitalist mode of regulation designed to uphold successive regimes of accumulation (we reviewed the Fordist and flexible regimes in Chapter 4).

The current mode of regulation has evolved in contested and often uneven ways over the past six decades, and its path has been historically contingent. Two points are worth noting in this regard (adapted from Roberts, 2002):

1 *Trans-state organizations did not begin after the Second World War* – although they certainly intensified. The initiation of the current global governance regime may be dated back to 1920, and two factors were especially important as a result of the end of the First World War. The first was the creation of new nation-states as the Austro-Hungarian, Ottoman, Prussian and Russian empires dissolved, and the second was the establishment of the League of Nations (1920). The system set up after the Second World War built on the failures of its precursors. As such, global regulation reflects both contemporary and historical institutions and power relations.

2 *Global regulation continues to evolve* – and each of the major institutions has been significantly contested since the Second World War. Roberts identifies three major developments that have shaped this evolution: (1) the internationalization of capital; (2) geopolitical shifts, notably post-war independence of former colonies and transition of formerly socialist countries; (3) the recent protest against major global institutions. Some thinkers argue that the terrorist attacks against the USA of 11 September 2001 (9/11) represent a further watershed in the evolution of global regulation as discussed in the geopolitics section. To this could be added recent responses to the GFC which has seen a shift in power away from a traditional economic core to include a broader array of countries, represented in terms of the G20, and also a resurgence in the role of state intervention in order to sustain economic growth.

The eventual victors in the Second World War were instrumental in establishing the post-war order. This was first discussed in 1944 when representatives from 45 countries met at Bretton Woods, USA, in order to negotiate the regulation of the post-war political economy. This conference confirmed the role of the USA as the dominant power given its overwhelming economic size, political influence, and the devastating impacts of the Second World War on its European allies. By 1945, two Bretton Woods institutions had had been established – the IMF and the IBRD (the World Bank). These institutions would attempt to stabilize the global economy and prevent another catastrophic war by assisting in the reconstruction of war-torn economies and the modernization of newly independent nation-states. They were designed in a way that favoured already powerful nations who played the major role in deciding what shape they would take (Young et al., 2007).

Arguably, the most important post-war global institution has been the UN, which came into existence in 1945 (see Box 5.1). There were further key moments in the evolution of the post-war regulatory system. For example, the General Agreement on Tariffs and Trade (see Chapter 4) was established in 1947 to promote global free trade. GATT was intended as a precursor to the establishment of the International Trade Organization, but this was blocked by the USA's continued protectionist stance. It was not until 1995 that GATT was replaced by the WTO (see Chapter 4). In 1960, the Organization for Economic Cooperation and Development (OECD) was formed by the most powerful Western countries building on the success of the Marshall Plan (see Chapter 7), and there are now 34 members.

Box 5.1

The United Nations

The United Nations is one of the most powerful and important global institutions. The seeds of the formation of the UN were laid after the end of the First World War (1914 to 1918) when, through the Treaty of Versailles (1920), the League of Nations was established. This sought to restore global stability in order to administer the transition of the new nation-states carved out of the Ottoman, Prussian, Russian and Austro-Hungarian Empires. The League failed, partly because the USA turned inwards and partly because the policy of shaming Germany laid the seeds for a nationalist-fascist movement there. After the Second World War it was clear that the system needed reform to achieve the twin goals of stability and peace, and the fostering to independence of nation-states from the former French, British, German and Dutch empires. It was clear that the Westphalian state system, which enshrined competition between states and the use of

force (see above) had failed, and developments in nuclear weapons meant that there was a real possibility that a third world war could end civilization. In October 1943, the allied countries (the UK, USA, USSR and China) met in Moscow and agreed to create an improved version of the League of Nations following the war. In 1945, 52 countries – mainly the war victors – met at Dumbarton Oaks, USA and, based on the Atlantic Treaty signed by Churchill and Roosevelt in 1942, established the UN Charter. This was signed in San Francisco on 26 June 1945. The major principles were:

- The use of force and war between states was prohibited except in special cases.

- The Security Council was to have a monopoly on violence and to use this to maintain peace.

- States could use force only as a means of protection.

Box 5.1
continued

A Secretary General heads the UN, a post currently filled by Ban Ki-moon. Its membership currently stands at about 190 and is growing. The two central bodies are the Security Council and the General Assembly. The *Security Council* consists of five permanent members (France, Russia, UK, USA and China – the victors of the Second World War) and ten non-permanent members elected every two years. It is the more powerful of the two major bodies: it is charged with maintaining peace and security, and can call on members to send troops for 'peacekeeping' operations. Each of the members of the Security Council has a veto and thus, during the Cold War in particular, it became relatively ineffective. The membership of the Security Council remains contested. Some nations protest the composition of permanent members as not reflective of current global political realities, for example. The *General Assembly* consists of all members of the UN, each of which has one vote. The assembly debates a wide range of issues related to the UN Charter and holds sessions for matters of special concern (such as the 2002 meeting to discuss the then-impending war in Iraq). For constitutional matters, a two-thirds majority is required, whereas for non-constitutional matters a simple majority is sufficient. Its resolutions are binding (in theory) although its recommendations are not.

Other principal organs of the UN include the *Economic and Social Council* which deals with issues related to human welfare and development; and the *International Court of Justice* located in The Hague, Holland. This court deals with disputes between nation-states, as opposed to individual prosecutions regarding genocide, crimes against humanity and war crimes which are now handled by the *International Criminal Court*. Beyond the principal organs, there is a host of other specialized agencies including the UNDP (United Nations Development Program); FAO (Food and Agriculture Organization); UNESCO (Educational, Scientific and Cultural Organization); ILO (International Labour Office); WHO (World Health Organization); and the International Monetary Fund (IMF). The WTO is independent of the UN, while the World Bank is linked informally.

The UN's record in international conflicts has been one of mixed success. The original aim of the Charter was to create an independent military force, but this has never come to fruition. Instead, the UN relies on calling up troops to enforce 'peace'. The greatest successes have been in the arbitration of conflicts such as Kashmir (1949), Cyprus (since 1964) and the Iran–Iraq War (1980 to 1988). In larger-scale conflicts, the results have been less clear-cut as the UN has relied on the will of the more powerful nation-states providing troops. During the Cold War, the USA used the UN to further its own cause. For example, it cited self-defence as the rationale for the invasion of Korea (1951). The UN has failed to prevent ongoing conflict in places as diverse as Israel/Palestine, the former Yugoslavia, Rwanda, Sudan, Libya and most recently in Syria between 2010 and 2013.

In 2002, the USA and the 'coalition of the willing' flouted UN resolutions and invaded

Box 5.1
continued

Iraq on the basis of the supposed existence of weapons of mass destruction (WMD), which they argued posed a threat to Europe, the USA and allied Middle Eastern neighbours. The most important allied country in the latter region is Israel which was created by the UN following the Second World War to formalize Zionist-inspired Jewish resettlement on their historical holy lands – a move which has led to nearly five decades of Arab–Israeli conflict. The authority of the UN was seriously compromised by the invasion of Iraq. In reality, some have argued, the UN has always been at the whim of the USA – the major budget contributor and hegemonic power. Thus the impartiality of the institution is called into question constantly. The role of the UN in other areas has not escaped controversy; in the case of the IMF the institution is charged with forcing politically unpopular structural adjustment reforms on poorer countries while rarely subjecting richer countries to the same discipline (the UK in 1976 being a notable exception) (see Chapter 7).

The end of the Cold War had an important impact on the UN in that the transition from superpower geopolitics reduced the incentives for liberal-capitalist states to provide financial and military support (Sidaway and Grundy-Warr, 2012). Despite its recent undermining however, the UN has made great advances in many areas since it began. In the development arena the UNDP, while remaining overtly modernist, often provides a relatively holistic counterbalance to the purer neoliberal stance of the other global institutions. Overall, it could be argued that:

> the UN system represents a very clear advance on the Westphalian system and is clear evidence that peace and security is a shared global problem that can neither be left to private treaties between states nor to the dubious intentions of any hegemon.
>
> (Waters, 2001, p. 119)

The withdrawal of two of the Second World War victors, the USSR and China, from the capitalist system in favour of socialist development signalled the start of the Cold War (see section on geopolitics below) which was deeply influential in the further development of global regulatory systems both in the capitalist and the socialist world. Following the collapse of the socialist states post-1989, the institutions that govern the world system are effectively those established at Bretton Woods – the World Bank, IMF, WTO and UN. These institutions have great influence in the design of economic and social policy in their member countries. To some radical thinkers, the overt support of capitalist accumulation practised by these bodies represents a form of neo-imperialism visited on politically independent, but economically subordinate, postcolonial states. As we will see, the rise of China is causing a reconsideration of this viewpoint.

Global regulation from below

The perception of global institutions as undemocratic, combined with frustrations over the capability of the nation-state, is leading to a global civil society movement that seeks to reconstruct and regulate the system from below (Chomsky, 2012; Glasius *et al.*, 2002; Goldsmith and Mander, 2001; Harvey, 2011; Munck, 2006; Reitan, 2006; Routledge, 2013). This has seen the rise of transnational non-governmental organizations (NGOs) representing common interests, shared needs and new globalized identities (see S. Amin, 1997; Bello, 2002). The number of transnational NGOs has risen sharply since the 1960s and generally much more rapidly than the rise in intergovernmental organizations or nation-states (see Figure 5.1). A number of broad dissenting alliances have driven this process, including the global environmental movement (see Chapter 8) and the anti-globalization movement (see below), as well as more contested entities such as the global women's movement (see below). Individual examples include NGOs as diverse as Greenpeace, The Fair Trade Network, Stop the War Coalition, Globalize Resistance, Médecins Sans Frontières, Amnesty International and Oxfam. Although diverse in their perspectives, such bodies generally argue that powerful Western associations such as the G8, G20 and the World Economic Forum are undemocratic, and that Bretton Woods institutions are conduits for the agenda of the dominant elites in rich industrialized countries. These bodies have propagated, it is argued, damaging neoliberal perspectives that have become increasingly deterministic in the daily lives of billions of people across the planet. These NGO networks are organized transnationally and, as such, are sub-state in their constituents but globalized in their extent (Cumbers *et al.*, 2008; McFarlane, 2009).

New forms of representation and lobbying have been developed, most visibly in the street protests of the anti-globalization and Occupy movements, together with other large-scale manifestations such as the Arab Spring and the Chilean Winter, but also in the day-to-day resistance of individuals (for example, those who boycott GM crops in the USA or those who only buy fair trade products in groups such as the *Conscious Consumers* of New Zealand). Elements of this general movement are found in both the rich and the poorer worlds, which in many ways transcends the traditional delimitations between First and Third Worlds and East and West. In this context,

Brecher *et al.* (1993, p. xv) argued over 20 years ago that the emergent 'globalization from below':

> in contrast to globalization from above, aims to restore to communities the power to nurture their environments; to enhance the access of ordinary people to the resources they need; to democratise local, national and transitional political institutions, and impose pacification on conflicting power centres.

This resistance from below represents a new force that challenges the post-Second World War mode of regulation. Recent pieces in *The Guardian Weekly* and *The Economist* (see Economist, 2013) emphasize the new power of networks as a form of resistance and the shifting nature of those that resist in this way, from single- to multiple-issue concerns that transcend traditional political boundaries. The first edition of this book argued that we should expect this sector to become increasingly strident in the future as global income polarizes further and environmental issues come to the fore, and this has most certainly been the case. Whether such decentred and poly-vocal organizations have sufficient power to change the global order is a debatable point; however, the enhanced role of global civil society does, at the very least, signify a watershed in world politics and one which is likely to be increasingly influential in the coming years in ways that both react to and build from globalization (see Chomsky, 2012; Harvey, 2011; Munck, 2006).

Reflections on the nation-state

Broadly speaking, there are two schools of thought with respect to the impact of globalization on the nation-state. The first school, which Waters (2001) terms the 'modernizers', argues that the state is becoming increasingly incapable of coping with the demands of globalization and that its sovereignty is being eroded from above and below. Moderate modernizers such as Held (1995), McGrew (2008) and Waters (2001) argue that although the state is unlikely to disappear as a political form it will become less powerful as transnational social and political flows and grass-roots demands multiply. More radical modernists have argued for some time that the state is just about finished as a meaningful political and economic force (see Castells, 1996; Kindleberger, 1967; Ohmae, 1995).

In contrast, the 'realist' school is much more optimistic about the role and weight of the state in global affairs. Authors such as Albrow (1996), Calhoun (2007), Dicken (2011), Giddens (1985, 1999), Gilpin (1987), Harvey (2001, 2007), McGrew (2008), McGrew and Lewis (1992) and Rosenau (1980), to mention but a few, assert that while the state is being transformed, and governance is being played out at a range of scales from the local to the global, it still fulfils a central regulatory role. Some thinkers point to the fact that the global regulatory regime itself is a product of nation-states and that globalization processes have been facilitated by the existence of states. For example, Robertson (1992) argues that: 'the diffusion of the idea of the national society as a form of institutionalised societalism was central to the accelerated globalization which began just over a hundred years ago' (p. 58).

Although the 'modernizer' versus 'realist' framework is imperfect – there are many overlaps – it does provide a useful way of thinking about the relationship between globalization and the nation-state. The broad arguments of these two camps are used as springboards for the discussion below (see Sparke, 2009c and 2013 for an entrance point).

The powerless state?

In a classic exposition, Held (1991) argues that nation-states will eventually be replaced by a world government. The steps in the argument are as follows:

- Cultural and economic flows are reducing the power of governments at the nation-state scale and flows are becoming more difficult to regulate at their borders. Transnational societal and cultural links are flowering.
- Transnational processes are growing rapidly – this includes TNCs who are increasingly powerful; sometimes being larger in economic size than nation-states.
- Given the above, nation-states are forced to increasingly cede many areas of their traditional sovereignty to international, intergovernmental and regionalized institutions. These powers include defence, security in general, economic management and communications.
- Linked to the above, larger territorial units have evolved to which states are increasingly ceding power: these may be political

(EU), economic (NAFTA), military (NATO) or even broader (WTO, UN, IMF).

- The above leads to a system of global governance which is emerging and which states must increasingly abide by.
- This system provides the seeds of a world government with coercive and legislative power – and this is already beginning.

In this sense then, Held sees the major threat to the state as coming from 'above'. Waters (2001) identifies a number of what he calls external 'threats' to the sovereignty of the state, the most important of which is the *global human rights* agenda. This agenda, established, ironically, by nation-states after the Second World War during the prosecution of war criminals, has provided an important rallying cry for many across the world. It has created a sense that there is an authority higher than the state, together with the institutions to enforce this view. Waters argues that all the cross-societal social movements of the second half of the twentieth century are ultimately rooted in the human rights agenda (see Sidaway and Mamadouh, 2012).

There are those who see processes from 'below' as more troubling for the future of the state (Routledge, 2013; Spybey, 1996; Waters, 2001). It is possible to be critical of the idealism surrounding the celebration of the rise of globalized NGOs however, since it tends to overstate the importance of cross-societal struggles. Intranational differences in welfare are as great as ever, and a wide gulf between Western and Third World populations in terms of needs and wants remains. Castells argues in the *Power of Identity* that the state is actually sandwiched between 'internal' and 'external' threats (2010, p. 304):

> The growing challenge to states' sovereignty around the world seems to originate from the inability of the modern nation-state to navigate uncharted, stormy waters between the power of global networks and the challenge of singular identities.

To 'modernizers', then, the state is in crisis and its powers have peaked, and this decline was clearly visible in the last quarter of the twentieth century. In Western nation-states there was a marked shift away from corporate welfare states to more liberalized and 'downsized' government, often referred to as the 'rolling back' of the state. This 'state weakening' included, and continues to include, privatization, welfare budget reductions, liberalization and the ceding of sovereignty to transnational institutions. Underlying this crisis, claim some sociologists, is the rise of 'post-materialist' society. Witness the recent affront to the power of the nation-state to control information through the WikiLeaks controversy which saw the leaking of thousands of classified diplomatic documents onto the internet. Some have heralded this as an example of the internet serving to inform people with regard to the actions – military, diplomatic, economic and political – of their governments. Others – often governments themselves – have argued that it represents a dangerous precedent that undermines the security provided by the nation-state to its people. There can be no doubt that this example and the more recent case of the exposure of espionage tactics of Western powers over its citizens through the monitoring of Facebook and other social media sites, together with intergovernmental spying including that of the USA on the EU, reveals a growing tension between the opportunity and threat to freedom of information that the internet and its associated technologies offers. It is uncertain where this will take us, but it is certain that it will become an increasing battleground between information liberals and conservatives.

Notwithstanding all of this, the recent response to the GFC is seen by some as evidence against the modernizers' views of the transition of the state. The intervention of governments to restore economic viability and bail out banks and financial institutions, together with a shift away from pure market economics to more Keynesian regulation, is seen as proof that the state has been rejuvenated. Modernizers might argue that, in fact, the fallout of the GFC is that corporations were propped up by governments that acted in unison with them. This represents an extension of the erosion of power of the state and the handing over of collective governance to private concerns that have usurped traditional governments. Under such a perspective, the reaction to the GFC may have seen a temporary shift of power and control back towards the nation-state but only in as much as it was intended to perpetuate the dominance of private capital and the imperative of increasingly rapid and globalized accumulation. Even more, the failure of nation-states to use the financial collapse of the late 2000s as a reason for shifting the nature of governance and restoring balance may actually be seen as evidence of the increasingly weak, decentred, ineffective and co-opted state, controlled at its core by private concerns over property and profit.

The powerful but transformed state

Many argue that the nation-state remains a powerful entity in a globalizing world, simultaneously transforming and being transformed by globalization processes (McGrew, 2008). Some of the arguments put forward by social scientists since the Second World War on the transformed state are reviewed below.

Traditional international relations arguments conceptualize global politics in terms of a snooker ball analogy. Each ball has autonomy but is impacted by the movement of other balls as the game progresses. Although arguably accurate until the end of the First World War, this 'dynamic mosaic'-type approach is unsuitable for the contemporary world. It fails to capture the political networks of power which bypass formal government and underestimates the possibility of alliances of interest. International relations theory has evolved to try and capture the new complexity. Burton (1972), for example, proposed the rise of 'world society' characterized by a 'layering of inter-state relationships with networks or systems relationships between individuals and collectivites that transcend or subvert state boundaries' (Waters, 2001, p. 99). Burton talks of the need to incorporate a concept of 'effective distance' which describes the mesh of relations between places more effectively than territorial space. At the core of his argument, however, the nation-state retains the power to control this process.

Rosenau has been especially influential in terms of the theorization of the transformed nation-state, introducing the concept of 'transnationalization'. This conceptualized a system where new actors and groups were brought into the arena of global politics facilitated by technological advances. This would render the global system an increasingly complex web of interactions between governments and NGOs. The effect of this in terms of the propulsion of inter-societal groups on to the world arena would have a disintegrative effect, causing a: 'transformation, even a break down, of the nation-state system as it has existed throughout the last four centuries' (Rosenau, 1980, p. 2). However, Rosenau concludes that, despite the existence of such corroding processes, the nation-state would remain central as governments generally have the power to adapt, design and implement policy.

More recently, Rosenau (1990) suggested the 'two worlds' of world politics conceptualization. This argued that we are entering a 'post-international' system where chaos and unpredictability rule, and formal nation-state governance is unable to cope with the emergence of truly global problems (such as environmental change) and the demands of an increasingly informed population that is able to articulate its distrust of traditional state power. This has led to a bifurcation of the world system into the 'state-centred' world, where the USA, Japan, the EU and the USSR are dominant (the thesis was written prior to the collapse of communism), and the 'multi-centric' world comprised of interactions between a diverse range of groups at various scales cutting across national boundaries, striving for autonomy from the state. Although attacked, and possibly eroded, the state remains a powerful agent in this conceptualization, however.

Other authors who have written supporting the continued relevance of the state from various viewpoints include Gilpin (1987), Robertson (1992) and Giddens (1985, 1990). Geographers have also entered into these debates. Dicken (2011) sees states performing three important roles with respect to the globalization of the world economy:

- *states as containers* of specific business practices – all economic activity is embedded in broader cultural patterns and processes;
- *states as regulators* of economic activity within and across jurisdictions;
- *states as competitors* where national differences can determine the extent to which nation-states have a competitive advantage.

Numerous geographers have argued that the state continues to play a central role in reproducing the 'mode of production' in the sociocultural sphere (Peck and Tickell, 2002). Nation-states remain active in the reproduction of cultural values and norms: witness, for example, the policy of France with respect to minimum national music content on the radio and television, or China with respect to its policy of regulating internet access in English. Geographical work on nation-state building, discourses of nationalism (see Sidaway and Grundy-Warr, 2012; Kaplan, 2009) and citizenship (Painter and Philo, 1995; Painter and Jeffrey, 2009) has revealed that nation-states are extremely active in the area of identity formation (Goodwin, 1999). For other geographical perspectives on this debate see Taylor and Flint (2000), Smith (1994) and Sidaway and Mamadouh (2012).

Again, the recent resurgence in state intervention in the light of the impacts of the GFC could be interpreted as a reaffirmation of the role of the state – although, as argued in the previous section, it could be viewed as the opposite of this given the inability of states to change the broad mode of governance and regulate financial sectors in anything more than cursory ways. It may be that countries compete more so in the era of accelerated globalization as they seek a distinctive place on the global stage. A quick glance at the advertisements on global news providers such as BBC World, BBC Mundo, CNN and Al Jazeera illustrate countries such as Kazakhstan, South Korea and India vying for business investment as well as tourist earnings. Certainly, in a political sense, the borders of nation-states still play a very important role; in fact during the so-called War on Terror, borders were policed more than ever, ostensibly to manage terrorist threats but also to protect economic growth and political control. During the 2000s, across the EU, in the USA and places such as Australia, it become increasingly difficult to gain access through migration, at a time when global inequalities were arguably more pronounced than ever. Witness the recent cases of global information fugitives Julian Assange and Edward Snowdon – who have sought refuge behind borders (and in the case of the former, behind the walls of an embassy where the national sovereignty of Ecuador is protected) in order to evade capture by those seeking to prosecute them – as evidence of the continued relevance of 'borders' and national sovereignty. Given that many of the technologies associated with globalization were invented in order to stimulate military activity, or at least to support it up, it seems unlikely to us that nation-states will strenuously and physically resist the undermining of information security given their continued monopoly on the legitimate use of violence.

The future of the state

By the early twenty-first century there were more nation-states in existence than ever before in the history of humanity (Penrose, 2009). The remnants of the European empires had disappeared, leaving only very small dependencies such as the Pitcairn Islands, Tokelau and the Cayman Islands scattered across the world. The state system is no longer Westphalian given that nation-states increasingly surrender sovereignty in military alliances and hegemons have been allowed to arise throughout the century. Does the proliferation of nation-states contradict the political globalization thesis? Decolonization and independence, as well as devolution and decentralization, are part of the *dialectical* process of political globalization which creates new space for the assertion of localized identities, ethnicities and cultures struggling against the rise of centralized geopolitics and traditional national ideologies in an attempt to assert their sovereignty. It is this process which is partly behind the rise of nationalist movements we have witnessed over the past 20 years as nations (as distinct from nation-states) seek their place in the globalizing polity. It is likely that the number of nation-states will increase further, probably along ethnic/cultural lines (as is the case within the context of the European Union: see Box 5.2), leading to a proliferation of less powerful states. In the absence of enlightened global political regulation, this multiplication will make the world more chaotic.

Across existing nation-states in the late 1990s and mid-2000s we saw a drift to the political centre-right where legislation from the economic to the environmental, from welfare to workers' rights, is made in the light of the demands of neoliberalizing global institutions. Whilst there was some reaction to this following the GFC and a return to the centre-left from the mid-2000s in Latin America, this general approach has persisted and the underlying mode of governance remains much the same. In fact, in some cases (such as that of New Zealand and the UK), the GFC was used as a reason to increase the shift to the right wing of politics in the name of the imposition of austerity measures that saw rapid declines in government spending, privatizations and other neoliberal-inspired economic and social policies (see Banks *et al.*, 2012). The evolving nature of nation-states allows globalizing agents, such as TNCs, to penetrate boundaries far more easily than ever. But, the global governance system does not supersede nation-states – although it changes the way they interact. As Roberts (2002, p. 157) argues:

> trans-state organisations are not extra-state organisations. They do not mark the eclipse of the state. The formal structures of the post-1945 world have been creations of states and reflect the unequal power relations between states.

The current global 'mode of regulation' is under deep scrutiny since it was designed for a very different world. Global regulation from above is being challenged from below as civil society groups increasingly globalize their concern in an attempt to forge a more democratic and equitable system of global governance.

Whether the forces of transnational civil society can reform the deeply ingrained system remains to be seen, especially in the light of pro-neoliberal and pro-military discourses which have proliferated in the USA since 9/11 and have been confirmed in terms of the reaction to the GFC. The importance of such movements, however, is attested to by the fact that many nation-states have been co-opting some aspects of the movement's arguments. There has been a widespread mainstream adoption of environmentalist ideas, some feminist ideas, and other more specific concepts such as the participatory budget model emanating from urban areas of Brazil.

The human geographical view has tended to fall somewhere in between the 'realists' and the 'modernizers'. For example, Smith (2000a, p. 535) argues:

> The process of globalization, in the form of both the internationalisation of capital and the growth of global and regionalised forms of governance, challenge the ability of the nation-state effectively to practise its claim to a sovereign monopoly over its bounded space and to protect its citizens from external incursion. . . . There is however, general consensus among political scientists that while the powers of the state have been eroded as a consequence, it is a myth to claim that the state has no influence over the impact of such globalising processes.

What we are left with, then, is a multipolar world, where nearly all states are forced to act in concert in order to pursue their will. Multinational military efforts in Europe, the Middle East and Asia, as well as regionalism (see Box 5.2), are testimony to this. At the same time, global institutions are heavily influenced by the desires of powerful nations and seek to co-opt other nation-states to perpetuate the current mode of production, while transnational civil society in places as far flung as Chile, Syria, South Africa and China seeks to resist, transform or revolutionize this process. We are moving away from a state-centred politics to what Held *et al.* (1999) called 'multilayered governance' and, in the light of the technologies that support this resistance, it is increasingly bewildering and unpredictable (Sidaway and Grundy-Warr, 2012).

Geopolitics and world orders

Geopolitics is 'a long-established area of geographical enquiry which considers space to be important in understanding the constitution of international relations' (Smith, 2000b, p. 309). Some might argue that in the era of globalization focusing on international relations in this way is not compelling. The following discussion of the evolution of world orders contradicts that view however. World orders, or geopolitical worlds as Ó Tuathail calls them, 'come together after a general war and subsequently develop until a new crisis or war changes the order of power'; during any given order, the 'distribution of power and

Box 5.2

The European Union

The seemingly paradoxical process of the surrendering of sovereignty by nation-states to transnational institutions combined with the devolution of power and increasing salience of localized politics is best illustrated through the case of the European Union. Some have seen the rise of the EU as evidence that refutes the political globalization thesis (Hirst *et al.*, 2009; Ritzer, 2011). Others see it as a product of globalization, providing a model of what a 'global government' might eventually look like (Held, 1991; Held *et al.*, 1999; Sparke, 2013). Regional integration, focusing on economic processes, was discussed in Chapter 4. The EU has gone beyond the economic, and represents a political, social and cultural phenomenon however.

Attempts at European integration rose out of the ashes of the Second World War and the desire among Western European nations to prevent such a war from occurring again. In 1951, the European Coal and Steel Community (ECSC) was founded by Belgium, West Germany, France, Italy, Luxembourg and the Netherlands. Focusing on the industries that were crucial to the produc-tion of arms, it was intended that the unilateral rearmament of Germany and the escalation of arms production in other nation-states should be discouraged. Buoyed up by the success of this institution, the six members went on to form the European Economic Community (EEC), enshrined in the Treaty of Rome of 1957 and brought into effect in January 1958.

This created a common industrial and agricultural market and laid the ground for economic harmonization in other sectors. At the same time, the European Atomic Energy Commission (EURATOM) was established, focusing on the research and development of nuclear power. These various bodies functioned as separate institutions but were overseen by a European Parliament. In 1966, the Common Agricultural Policy was introduced, which has proven to be the most controversial, and yet binding, of the agreements in force to date.

In 1967 the three institutions – the ECSC, EURATOM and the EEC – were combined to form the European Community. Other European nation-states considered joining at this point, notably the UK which had two applications (1961, 1967) turned down under the veto of French President de Gaulle. Following the latter's retirement, Denmark, Ireland and the UK joined in 1973. In the early 1980s Greece (1981), Spain (1986) and Portugal (1986) joined the EC. In 1986, the *Single European Act* introduced a rule of majority decisions and thus greatly enhanced the powers of the Council of Ministers and the Parliament. It also introduced the goal of removing all barriers to trade by 1992. The Maastrict Treaty, which came into force in November 1991, was the most far-reaching agreement hitherto produced. It confirmed the agenda

Box 5.2
continued

for the removal of trade barriers and set in motion the establishment of a European Central Bank, a single currency and a range of social regulations. The Maastrict Treaty established the European Union, which was joined in 1995 by Austria, Finland and Sweden, augmenting the total number of member states to 15 at this point (see Map 5.1). On 1 January 2002 the Euro single currency was introduced in 12 of the 15 member states of the EU. Some, such as the UK and Denmark, opted out of this system with the possibility of joining at a later date. The largest single enlargement of the EU took place on 1 May 2004 when ten further nation-states, mostly from Central and Eastern Europe, joined. Most recently, Bulgaria and Romania joined the EU in 2007 and Croatia in 2013.

Both the reality and the concept of the EU have been enormously influential in domestic politics across Europe. The desire to join, or not, and issues surrounding the integration of social and economic policy have been influential in national parliamentary elections and shaped political parties. In the late 1980s, the integrationist aspirations of France (under Mitterand) and Germany (under Kohl) clashed with the more isolationist stance of the UK's Margaret Thatcher who argued aggressively against the perceived loss of sovereignty. In general, the UK's Conservative Party has traditionally been 'Eurosceptic' (despite the fact that a Conservative prime minister took the UK into the community), while the Labour Party is broadly supportive – although both have recently hit the middle ground. Euroscepticism in the UK is not just an economic or political issue; it is also a cultural one. The formation of the UK Independence Party (UKIP), whose main policy platforms are to retain the Sterling currency and withdraw from the EU, has struck something of a chord with parts of middle-class England who see the potential loss of the pound as a form of identity erosion.

Across Europe in the late 1990s there was a general retreat into Euroscepticism, both at the level of (some) national governments and the general electorate. The increased enlargement of the EU, and the issue of immigration from new member states, became political 'hot potatoes'. Some argued that this relative disinterest was due to the fact that the majority of the population had no direct recollection of the Second World War, as well as the end of the Cold War, which removed the external 'threat' posed by communism. The fact that the turnouts for European Parliament elections are very low signifies a continued lack of concern, or possibly exasperation, with the EU among the general population.

Significant strains were felt in the EU in the wake of the sovereign debt crisis which began in 2009. Countries such as Greece, Spain, Portugal and Cyprus faced severe financial crises after governments found themselves in danger of defaulting on debt servicing. In response, the EU and the European Central Bank – backed by stronger economies in the region, particularly Germany – bailed these governments out and staved off national bankruptcy but only after insisting on severe austerity measures by the governments. The national economies were saved but there was very strong reaction on the streets, especially in Greece,

Box 5.2 continued

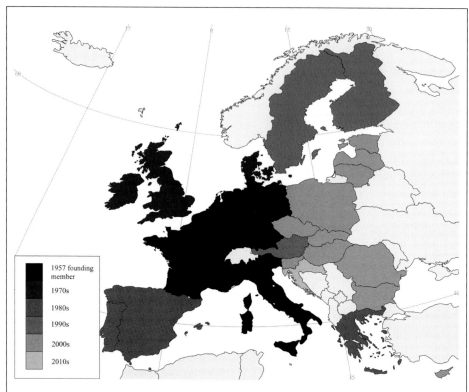

Map 5.1 *Membership of the EU*

to this seeming EU attack on national sovereignty. Withdrawal from the Eurozone and even the EU is a possibility.

As nation-states have increasingly surrendered sovereignty to the EU (albeit with a host of opt-out and veto clauses), we have seen a rise in local autonomy and independence movements across the Continent. These processes are linked, as the new EU creates space for the assertion of such aspirations given the erosion of state power that formerly thwarted such desires. In the UK, political power was devolved in the 1990s to Wales, Scotland and Northern Ireland, where Assemblies and Parliaments have been variously established. In 2014 Scotland will hold a referendum on whether it becomes an independent sovereign state. In Spain, Catalonia and Galicia have achieved significant autonomy (while the Basque Country has not, largely in response to the violent tactics of the armed independence-seeking faction ETA). There are autonomy movements in parts of France, Italy and Germany. In this sense, the evolution of the EU appears to confirm the dialectic process of centralization and devolution that is unleashed by globalization. In some ways, the EU is a laboratory for the potential impacts of globalized politics and what happens when that system enters a political crisis precipitated by an economic one.

configuration of alliances remains relatively constant' (2002, p. 177). The evolution of geopolitics and 'world orders' from 1945 onwards illustrates the increasingly important role of globalization as both a driver and consequence of such 'orders' (see Sidaway and Mamadouh, 2012).

The Cold War

Following the end of the Second World War, many former European colonies gained independence (see Chapter 3). The overstretch of empires, combined with the rise of the USA, led to a watershed in global geopolitics which eventually led to the Cold War. The USA emerged as the richest and most powerful nation on Earth, not least because it had been relatively undamaged during the Second World War. This led to a process of economic expansionism which replaced the political expansionism of the former colonial powers. The USA's post-war foreign policy was squarely aimed at maintaining its dominant position. As Rennie-Short (1993, p. 38) argues:

> The USA has a key role in the world economy and is intimately associated with overseas economies, which it needs for supplies of raw materials, as sites for capital investment and as markets for agricultural and industrial produce. Foreign policy objectives are primarily aimed at maintaining and servicing these links.

In the immediate post-war years, the evolution of a bipolar world, dominated by the USA and the USSR, emerged. The USA attempted, largely through the Bretton Woods institutions, to promote capitalism, wishing to take a leading role in the ideological formation of the post-war order. The USSR, a communist state since 1917, had different ideas, forbidding countries under its influence to accept the Marshall Plan (see discussion below). It occupied much of Eastern Europe and installed satellite polities there, seeking *buffer states* to provide resources for a recovery from its devastating war. The West interpreted this as an aggressive move, summed up by Churchill's argument of 1946 that an 'iron curtain' had descended across the continent. Soon the Western powers argued that communism should be arrested lest it knock over successive countries like 'dominoes'. The resultant Cold War proceeded through a number of phases (Sidaway and Mamadouh, 2012).

In the First Cold War (1947 to 1964) US policy was organized around two pillars – both of which were intended to promote capitalism, contain communism, and thus promote US interests. The first was the Truman Doctrine developed in 1947, which saw the USA effectively appoint itself as the global 'policeman' of freedom, modernization and stability. The speech that launched the doctrine also saw the introduction of the concept of 'development' (see Chapter 7). The doctrine resulted in the development of alliances for the purposes of security – the three most important being NATO (Western Europe), SEATO (Southeast Asia) and CENTO (Central Asia). The policy of the containment of the USSR was explicitly pursued, resulting most explicitly in wars in Korea (1950 to 1953) and Vietnam (1964 to 1975). Smaller-scale interventions – military, political and economic – also took place during this period. These included many interventions in Latin America and the Middle East based on the 'domino theory'. As Rennie-Short argues: 'The recorded interventions are only the tip of the iceberg. Submerged beneath the waves of secrecy lie the endeavours of the CIA to bolster friendly governments, topple unfriendly ones, get rid of troublesome politicians and advance the cause of US interests' (1993, p. 42). Subsequent presidents including Eisenhower, Kennedy, Johnson and Nixon echoed the Truman Doctrine. Notably, President Carter said in 1980 that: 'an attempt by any outside force to gain control of the Persian Gulf region will be regarded as an assault on the vital interests of the United States. It will be repelled by the use of any means necessary, including force' (quoted in Rennie-Short, 1993, p. 43). Indeed, the Truman Doctrine set the tone for US foreign policy until the fall of communism, was then absent during the 1990s and resurfaced – arguably – following 9/11, although the enemy was no longer communism. Its spirit was especially visible in the subsequent invasions of Afghanistan (2001) and Iraq (2003), and George W. Bush's articulation of the 'axis of evil' in 2002, which included Iran, Iraq and North Korea (see below).

The second major pillar of US foreign policy was the Marshall Plan which saw the USA invest nearly $17 billion dollars of aid in Western Europe's post-war reconstruction. A similar approach was taken in East Asia, although not as extensively, under the Colombo Plan. These aid regimes were explicitly intended to bolster capitalism and prevent the spread of the communist 'threat'. Meanwhile, the USSR had set up the Communist Information Bureau in 1947, COMECON in 1949 (Council for Mutual Economic Assistance) supporting communist governments in satellite states, and the Warsaw Pact in 1955 (the equivalent of NATO). Although Stalin's successor, Khrushchev, was more intent on coexistence with the West, and more tolerant of different forms of socialism in the periphery, this period was still marked by irreconcilable conflict.

In the Period of Détente (1964 to the late 1970s) there was a heightened realization by the superpowers that the Cold War was *enormously* risky. This was epitomized by the Cuban Missile crisis of 1962, which almost led to the outbreak of nuclear war due to superpower conflict over the issue of Fidel Castro's Cuban revolution. In 1963, however, a treaty of non-nuclear proliferation was signed. The military-industrial complex in both countries had vested interests and remained a powerful lobby group, stimulating further clashes, most notably in Vietnam. Most US interventions during this period were covert (e.g. Chile in 1973) or economic in nature. The investment role of the USA in East Asia during the late 1960s to late 1970s underpinned subsequently spectacular economic growth there. Both countries increased their global scope and drew in more countries, especially from what was then termed the Third World, into their influence.

During the Second Cold War (late 1970s to mid-1980s), relations between the two superpowers again deteriorated. Tension in the US administration between more 'hawkish' elements and those who wished to reduce the USA's role in global affairs was high. Meanwhile the build-up of the USSR's arms was seen as a threat to the USA. With the election of Ronald Reagan in 1980, there was a return to the more explicitly anti-communist rhetoric of the post-Second World War years. This resulted in direct interventions in Grenada and Libya and covert activity in Nicaragua (the Iran-Contra affair), for example. In the USSR, the military-industrial complex played an increasing role in determining policy there. People began to talk of MAD geopolitics (mutually assured destruction) as an unprecedented arms race took off. The level of hysteria in some Western countries was tangible: in the UK in the 1980s the prospect of nuclear war and the resultant winter were images which hung heavy over the population. In the popular media, Hollywood films, such as *Rambo*, portraying the 'commies' as evil and non-human, were propagated. It was a period of intense and dangerous propaganda, and genuine fear.

The bipolar world order was destroyed by the fall of the USSR in 1989. By the late 1980s, partly because of the costly arms race, the USSR was devastated by economic decline, political dissent and tension. Gorbachev's *perestroika* (restructuring) led to the undoing of the union as centrifugal forces resulted in independence movements across it

(Bradshaw, 2004a). The USA emerged as *the* global superpower, and the world order, it seemed, shifted from a focus on geopolitical to economic forces.

Post-Cold War – the rise of liberal democracy

The 1990s saw the dispersion of policies of neoliberalism and democratic governance across the former Third World to former communist satellite states and to those formerly subsumed by the USSR itself. It seemed the West had won. It is no coincidence that this was the time of greatest rhetoric with respect to the liberating potential of the processes of globalization. Capitalism and liberal democracy, hyperglobalists argued, had succeeded in removing the threat of global conflict. As such, globalization at this point was used extensively as a *discourse* to provide rationale for the extension of neoliberal democracy, but it also intensified as a *process* as an increasing array of economies, especially those in Africa, Latin America and Asia, were opened up to the world's markets.

Democratization is 'political change moving in a direction from less accountable to more accountable government' (Potter, 2000, p. 368). The problem with measuring democratization is that it is subject to competing definitions. The aid and loan powers certainly have fairly fixed ideas with respect to what democratization should look like. For some commentators, the enthusiasm with which Western governments have embraced the democratization of the rest of the world has more to do with its complementarity with free-market economics than with any *moral* superiority. 'Good governance', a mixture of neoliberal economics and liberal democracy, formed the backbone of the so-called 'Washington Consensus', which provided an organizing framework for the post-Cold War order.

There has been a major shift towards democratic governance over the past three decades. In particular, Latin America, Sub-Saharan Africa and the former USSR have moved towards electoral politics and away from dictatorships. The most explicit outlining of this shift is found in the work of writers such as Fukuyama (1992) and Huntington (1991), who herald the arrival of a 'new global political culture' based on liberal democracy founded in the globalization of Enlightenment values. As discussed previously, the shift towards liberal democracy began in the seventeenth and eighteenth centuries in Western Europe and North America, and it intensified through waves of decolonization following the Second World War with the explicit support of the USA and its allies. However, both fascist and socialist dictatorships persisted, the latter supported in many cases by the USSR or China. In the 1970s fascist dictatorships fell in Spain, Portugal, Greece and Turkey, and in the late 1980s there was a widespread transition to democracy in Latin America and some parts of Asia. The end of the Cold War led to a great acceleration in liberal democracy across the world.

The foreign policy of the USA and the rest of the West, together with conditions on loans and aid in the 1990s also strongly incentivized democratic 'good governance'. Most recently we have witnessed a wave of democratization across Northern Africa and the Middle East in the form of the so-called 'Arab Spring' which came about partly in response to the negative impacts of the GFC on Arab populations, together with the rising tide of democratic networking across the globe. Further to this, the winding down of the US-led 'War on Terror' meant that dictatorships that were ostensibly pro-Western in spite of their tendencies to persecution were no longer to be assisted or tolerated by the West. This episode reminds us that democracy unfolds unevenly and imperfectly and is subject to geopolitics and the interests of dominant powers as much as it reflects the will of the people. The continued vigour for democratic governance is illustrated by the proliferation of secessionist movements across the former Third World – for example, in Indonesia, Somalia, Morocco, Sudan and Ethiopia, and also in the West. At a broader scale, we can see that the number of nation-states has increased significantly since 1990 (Map 5.2), largely the result of the break-up of the former Soviet Union and Yugoslavia but also because of the success of some secessionist movements, as in Timor Leste and South Sudan.

Does this mean that the globe has become more democratic for ordinary people? There are many interpretations of 'democracy', and this is greatly complicated by the fact that political impacts on people's lives now come from above and below the scale of the state (see above). There are broad swathes of global society that are not democratic by *any* measure – for example, China, and some states in the Middle East, such as Saudi Arabia and Syria, together with remnants of dictatorships in South East Asia such as Burma, and some parts of Africa where the existence of democracy is debatable such as Mozambique. We should not forget smaller states too such as Fiji where democracy has been

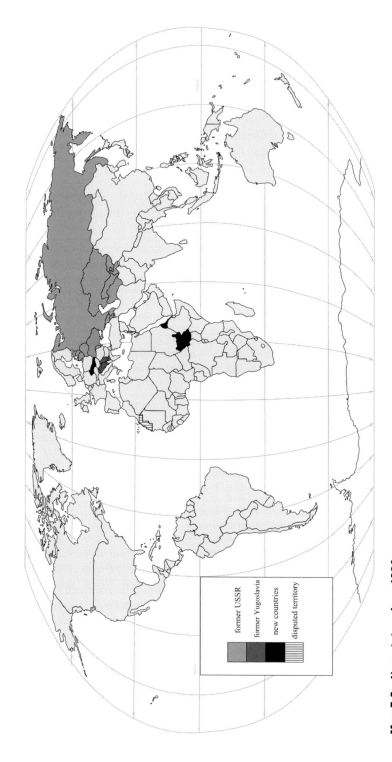

Map 5.2 New states since 1990

legend:
- former USSR
- former Yugoslavia
- new countries
- disputed territory

overturned repeatedly over the past 25 years, mostly by military regimes. This points us to the idea that how we define democracy is contested and various systems exist across the world, each of them with their own costs and benefits. Westminster-style democracy may not be applicable in other political and cultural settings, as has often been argued in the case of the Pacific (see Murray, 2009a). Furthermore, whilst representative democracy may exist in a political sense, other 'unfreedoms' such as in identity politics or media access may be called into question in many places – as has been the case in democracies as diverse as Russia, Singapore and Nicaragua. In truth there is no one definition of democracy, and neither should there be; however, what is for sure is that, as an ostensible ideal, liberal democracy has become the goal for the majority of states across the world and for the most part this has been backed by Western powers and the financial institutions that they largely control. There have been interventions by the West in countries such as Libya, Iraq, Afghanistan and Mali in order ostensibly to safeguard democratic order – although the motives of this intervention are highly debatable and may have more to do with oil, 'terror' and other geopolitical concerns. Fukuyama (1992, p. 45) is more generalist in predicting this shift and ascribed it to the evolution of post-materialist values and neoliberal economics. He is triumphant in terms of his prognosis:

> What is emerging victorious . . . is not so much liberal practice, as the liberal idea. That is to say, for a very large part of the world, there is now no ideology with pretensions to universality that is in a position to challenge liberal democracy.

The rise of democracy is not irreversible however, and rests on the tacit and real support of international institutions. There are regions, however, such as Russia and its hinterland which appear to be going in the other direction.

Were the USA, its allies and linked institutions to drop the favour which they accord democracy and 'good governance' then a major global shift would be likely. This, to an extent, occurred during the so-called War on Terror as we shall see below. The era of 'good governance', it may be argued, has come at a time when people's real choice is more constrained than ever. For example, how do we measure the gradual erosion of power away from nation-states and local populations towards TNCs and supranational institutions that back big business?

The world order after 9/11 in New York – 'War on Terror' to where?

The war on terror instigated by the first George W. Bush Administration (2001 to 2009) following the attacks of 9/11 in New York shaped a new global geopolitical order, which may still be unfolding. The USA and its major allies, including the UK, Spain, Pakistan and Australia, undertook both military and 'diplomatic' offensives in countries including Afghanistan, Iraq, Iran, Yemen, the Philippines, Indonesia, Syria, Sudan and Lebanon in an attempt to neutralize the terrorist networks that were behind 9/11. Intelligence sharing at the nation-state level has also formed a central part of this 'war' and continues to this day. The principal target was al-Qaeda, an alliance of radical Islamic groups that have used terrorist and military tactics for over two decades to defend what they perceive as the oppression of Muslims. The group was formed in 1988 by Osama bin Laden in order to expand the *mujahideen* resistance to the Soviet occupation in Afghanistan. Despite US backing for this movement in the context of the Cold War, the USA and its Western allies (including Saudi Arabia) became the group's principal targets in the 1990s, motivated principally by the West's intervention in the Gulf War of 1990 to 1991 and the ongoing Israeli–Palestinian conflict. The events of 9/11 represented the climax of this conflict and unleashed US efforts designed to remove the threat, although attempts to uproot the network, and to invade Iraq, were ongoing during the Clinton years and before. The rhetoric of al-Qaeda is that the West, led by the USA, is waging a new crusade in the Middle East designed to impose its will and culture, and as such the conflict has far deeper historical roots. The terrorist networks that have evolved, partly in *response* to the war, are thought to be truly worldwide in their extent and represent an amorphous and globalized target for the USA and its allies. In this sense, the war is the first of its kind – fought against a group that is not spatially bounded and that does not identify itself with any one nation-state. The International Institute for Strategic Studies has estimated that al-Qaeda has over 18,000 militants at its disposal spread across 60 countries. Critics have argued that this greatly overestimates the true extent of the network and that such claims should be seen as part of the USA's wartime propaganda.

The rhetoric of the USA during this 'war' was 'you are either with us or against us', an approach which sought to draw new geopolitical lines in ways which echo the Truman Doctrine. Those who clearly demonstrate that they are with the USA could expect grand rewards, as the issuing of reconstruction contracts in Iraq following the 'end' of the war there in 2004 onwards illustrated. Some have argued that the Bush Administration used the war to re-inject life into the industrial-military complex and protect fossil fuel reserves in the wake of declining supplies at home (see Box 5.3). This may explain the administration's great leap from

Box 5.3

The geopolitics of oil

The supply of oil plays a major role in global geopolitics and, since the modernist industrial model of development has been diffused, it has become increasingly important. Many recent conflicts such as the Gulf War following Iraq's invasion of Kuwait (1990 to 1991) and indirect interventions such as the recent US intervention in Venezuela, as well as the wars in Afghanistan and Iraq in the 2000s, have an 'oil link' of some description. The West's tolerance of the deeply undemocratic and anti-humanitarian regime in Saudi Arabia is also illustrative of the power oil has. Demand for oil currently stands at unprecedented levels, and rapid economic growth in large 'emerging' economies in South, Southeast and East Asia lies behind some of this increase. In China alone the International Energy Agency estimates that demand will rise from the current 5.5 million barrels per day (mbpd), to 11 mbpd by 2025. The rise in demand in 2004 was the largest single annual rise in history and pushed OPEC (Organization of Petroleum Exporting Countries) to 96 per cent of its capacity. By far the world's largest consumer both in an absolute and per capita sense, however, is the USA, and this consumption is set to rise.

These developments are having a notable impact on the price of oil which rose to $50 per barrel for the first time in history in 2004 and continued to rise to record levels by the early 2010s, subsequently falling considerably. Some commentators have argued that this was a temporary blip caused by supply pressures in Russia, uncertainty in Venezuela, and insecurity in the Middle East/Iraq given the recent escalation of conflict there. Others argue that the rise in prices represents a more long-term scarcity-induced trend that is unlikely to be reversed. The IEA (International Energy Agency) estimates that the world will require 120 mbpd by 2025 – which is double today's demand.

There has thus been a scramble for new supply locations, as indeed there has been in terms of a wide range of natural resources to continue to fuel economic growth in the USA and China principally but also in countries beyond that. The largest reserves exist in Saudi Arabia (36 billion tonnes) followed by Iraq (15 billion tonnes). As the wars in Iraq and Afghanistan illustrate, there was a campaign to install Western-aligned liberal democracy in the Middle East, arguably motivated in part

Box 5.3
continued

by the desire to secure oil supplies. However, the security situation in the region means that the oil companies turned their attention elsewhere. The USA has become more directly involved in Venezuelan politics and the Mexican sector, for example. There is a major diplomatic offensive underway in the Gulf of Guinea in Africa, which currently supplies 15 per cent of the USA's oil demand. There are ongoing efforts in countries such as Angola, Equatorial Guinea, Nigeria and Chad. The USA itself has also begun the widespread exploitation of tar sands in order to reduce reliance on external countries.

It is in Central Asia, however, where some of the greatest reserves lie – centred around Kazakhstan, Turkmenistan and Azerbaijan. Control of oil in this region is likely to cause a future stand-off between the USA and Russia, the latter seeing it as a natural area in which to expand its military presence in the region as well as its economic links – encouraging the formation of a new association without links to Moscow named GUUAM (Georgia, Ukraine, Uzbekistan, Azerbaijan and Moldova). Moscow has sought to increase its influence through the Collective Security Treaty Organization as a counter to this. These shifts are indicative of the role that energy and, more particularly, oil play in the evolution of geopolitics. In the absence of alternative energy, the continued expansion of the Asian economies and rising demand in the West is likely to result in complex and dangerous geopolitical shifts over the coming decades.

Afghanistan into Iraq, citing unsubstantiated terrorist links and weapons threats as part of the rationale. The invasion of Iraq, for example, was predicated on the existence of weapons of mass destruction (WMD) that could be used to attack the West and its allies in the Middle East. In January 2005, the USA finally admitted that there were no such weapons in Iraq. The invasion of Iraq was also represented as an expansion of the war on terrorism as it was claimed that Saddam Hussein harboured Islamic militants. Again, there was little evidence of this, although it is certainly true that terrorism in Iraq has been inspired by the invasion. The core motivating rationale for the war, radicals argue, is a desire to protect economic prosperity and the spoils of Western-directed globalization (see Mike Moore's film *Fahrenheit 9/11* for a radical exposition of this idea). It should be noted that 9/11 also coincided with the anniversary of the brutal military coup in Chile in 1973 which was backed by the USA in the context of Cold War politics, led to state terror and killed over 3,000 people.

Should the so-called War on Terror be seen as an order in global geopolitics in the post-Cold War world? The Washington Consensus of neoliberalism and good governance was certainly being eclipsed by simplistic notions of 'with us or against us'. If this were not the case, why has the West increasingly turned a blind eye to the most illiberal and undemocratic politics of Pakistan, Saudi Arabia and Indonesia, for example? What is clear is that, as always, the lines around which imagined global geopolitical divisions are drawn are contingent on the desires and needs of those who are most powerful. When, under the new Obama Administration, the support for the War on Terror was gradually withdrawn and the favour with which dictatorial regimes in the Middle East that were anti-Islamic were tolerated by the West reduced, there was a widespread revolution against decades of tyrannical rule that precipitated the Arab Spring.

What is for sure is that the broader War on Terror was used as a means to promote the diffusion of neoliberalism. At a Pacific Island Forum meeting in 2003, Australia cited the war as the rationale for an increased Australian security presence in the region and for the further extension of neoliberal economic reform – which it is argued would reduce the threat of terrorism by increasing prosperity (Murray and Storey, 2003). The challenge for the USA, as the hegemonic global force, is that it must have legitimacy in what it does – lest it fall prey to accusations of isolationism and self-interest. That the will of the United Nations was flouted in assembling the invasion of Iraq, that WMDs were not found, and that some terrorist suspects have been held without trial in the US base Guantanamo Bay for over ten years, has spurred on the global resistance movement which seeks to counter the unregulated and illegitimate actions of the US-led coalition (see also the discussion below). As it stands, however, it is probably most accurate to see the War on Terror as expansion of the second chapter of the post-Cold War order propagated by the world's only superpower in its efforts to spread neoliberalism and the political models that best support this.

The current 'world order' is thus highly debatable. As Ó Tuathail and Jeremy (2009) argue, relative stability in the geopolitical order is punctuated by major crises and wars and, in their wake, new orders form. It is perhaps erroneous to constantly refer to a 'new' world order as by definition the order is never truly fixed – although it may remain relatively stable. The waves of globalization argument adopted in this book would certainly subscribe to that idea; and the orders that are created are more than political – they are also cultural and economic in their nature. At the time of writing, a number of political and economic crises have unfolded, culminating in a growing shift in the geometry of world power. The GFC has had a major impact on the distribution of economic resources across the planet. In relative terms, the neoliberal West has lost ground to the East and, in particular, China and India. Although the West intervened extensively in order to maintain its relative position, it has not restored growth at a level that has prevented a 'catching up' by increasingly important economies. A manifestation of this is the widening of the G8 to the G20, including what have been referred to as the BRICs – Brazil, Russia, India and China – which by 2012 were all in the top ten nations in the world in terms of the total size of GDP. With economic power comes political might and there has thus been a shift away from the hegemonic power of the USA to a more multipolar world, which for the first time includes Latin America in the form of Brazil (and arguably Mexico). In terms of the distribution of wealth, then, the financial crisis has had the effect of closing the gap at the macro level between the West and the rest of the world. However, at the same time, inequality within countries has risen due in part to the austerity measures provoked by the response to the crisis. The world order that is evolving may be characterized by multipolar geopolitics, but it is underpinned by the exacerbation of inequality at the micro level.

At the same time as the GFC has had a major impact, the winding down of the War on Terror and the end of major war in Iraq and Afghanistan has also led to a redrawing of the geopolitical order. As noted previously, this in part precipitated a wave of democratic protest across the Middle East. This wave of democratic change has further stimulated protest and social movements across the world – responding, to some extent, to the impacts of the GFC together with the injustice of the remedies for it. These movements have evolved from the earlier anti-globalization movement as we shall see below. However, what we appear to be witnessing is the further challenging of centralized political power by a global civil society more networked than ever before due to technological change. As noted previously, however, the advent of internet technology and social media also raises the possibility of increased monitoring, and this represents the downside of the so-called Facebook revolutions recently laid bare by the whistle-blowing activity of Edward Snowdon.

Whether this new multipolar structure populated by a networked civil society will make the world more or less stable is indeterminate; but this will turn largely on the relationship between China and the USA which, in the early 2010s under the Obama Administration, was positive – due in part to the fact that much of the debt incurred by the USA in order to bail out its financial institutions is controlled by China which has built up huge surpluses over two decades of spectacular export-led economic growth. The role of Russia will become increasingly important, not least due to its enormous mineral wealth and pivotal role in fuelling the rise of China. However, the evolution of economic change in Brazil and India will also be very important. In all of this the conflict in the Middle East remains crucial; and, as witnessed in the recent stand-off at the UN between China and Russia on the one hand the USA and others with respect to intervention or not in Syria, the Israel and Palestine conflict continues to dominate global geopolitics.

This is a bewildering time to try and draw the lines of geopolitical world orders – as change appears more rapid than ever before, in part due to the technologies that underpin globalization. A number of questions remain salient, such as whether we are seeing a continued evolution away from the Cold War world order that is still being defined. Perhaps more radically, there may be no such thing as a world order any more in these days of postmodern politics and globalized civil society. The latter seems unlikely and we believe that the world is evolving into a post-oil order (see Box 5.3) that is seeing the shift in power and privilege of those associated with fossil fuels to a broader set of geopolitical factors – which, crucially, in the future will include water. This also returns us to our contention in Chapter 1 that we exist at present in the jaws of at least three ongoing 'crises' – the environmental crisis which is bound up with access to and use of oil and other fossil fuels; the inequality crisis which comes about in part due to the broader politics of capitalism associated with the current energy regime; and finally the crisis in capitalism which is increasingly incapable of solving the problems in the other spheres. The world will seem disorganized and chaotic until all three of these crises are addressed in unison – expect greater conflict, increased inequality and economic instability over coming years. We return to this debate in the concluding chapter of this book.

New social movements and anti-globalization

The past three decades have witnessed the evolution of many *new social movements* including, for example, feminism (see Box 5.4), environmentalism (see Chapter 8), gay rights (see Box 6.5), indigenous rights and anti-racism. Geographers have argued that global economic networks have disembedded people from their cultures, histories and

Box 5.4

Feminism, geography and globalization

Feminism is one of the most important examples of a new social movement. An amorphous and contested concept (Pratt, 2009), it has been defined as:

> [A]n awareness of women's oppression on domestic, social, economic and political levels, accompanied by a willingness to struggle against such oppression.
>
> (Wieringa, 1995, p. 3)

There are, however, multiple feminisms which vary, among other things, according to the particularity of place and moment in history (McDowell, 1999; Pratt and Craft, 2013). Is it feasible to argue that there has been a globalization of feminism and that this represents a new social movement? Women's movements actually have an extensive history, with formal activism dating back to the suffragette movement in the UK in the 1800s. It was subsequent to the 1970s, however, that a *global feminist struggle* was articulated. This perspective crystallized in the 1980s with the publication of the book *Sisterhood is Global* (Morgan, 1984). A sense of commonality evolved in terms of analysing and acting against patriarchy across the planet. Arguably, the rise of globalization has provided a space for all women to act from within, while the nation-state had proven adept at excluding and marginalizing women. Even in those democracies where female suffrage had existed for some time, the participation of women in politics was much lower than that of men. It is this reaction to the state which lies behind the formation of the various local, national and transnational NGOs that comprise the broader feminist movement.

The UN, with its framework of universal human rights, has been a particularly important vehicle for the above aspirations. In 1975 it declared the Decade for Women and held a number of conferences subsequently in Mexico (1975), Copenhagen (1980), Nairobi (1985) and Beijing (1995). In 1976 UNIFEM or the UN Development Fund for Women was established. The latter conference was instructive in terms of the divisions that have evolved in this supposed 'global' movement, as representatives were often explicit about their perspectives on women's issues from their own national/ cultural perspective. The globalization of feminism was seen, by some, as a reflection of the agendas of middle-class feminists from Western countries. The result of this meeting, however, was the Platform for Action signed by 189 delegates. Thus culturally relativist arguments, which, for example, might allow or deem necessary acts such as polygamy, purdah and genital mutilation in some societies, were rejected in favour of universal values. What was especially notable about this conference also was the high representation from NGOs (over 4,000), far outnumbering the nation-state representatives. In 2011 a major step in the UN was taken when the UN Entity for

Box 5.4
continued

Gender Equality and Empowerment of Women was established and Michelle Bachalet, at the time former President of Chile (and at the time of writing President of Chile for a second term) was the inaugural director.

While globalization creates opportunities for the promotion of universalized women's rights, it also presents threats to the condition of women (Davids and Van Driel, 2009). Feminist geography is engaging increasingly with the nature and impact of globalization. At the broad level, globalization has a differential impact on women and men according to a host of contingencies in any given case. This work has included more traditional feminist analysis with respect to gender inequality and oppression of women. Christopherson (2002) argues that globalization has had the effect of differentiating women in new ways and that 'class has arguably increased in importance as the variable determining women's economic position' (p. 246). Further research has revealed the continued ignorance of the value-adding and social reproductive role of women in capitalism and the 'hidden ingredient' that is female labour in many new flexible industries and global commodity flows (see Barrientos et al., 1999). Christopherson (2002, pp. 236–237) argues, for example, that:

> [W]hile women continue to be disadvantaged relative to men with respect to wages, total work hours, and access to societal resources,

they are playing a critical role in the processes structuring a global economy. Their continued invisibility within these processes is a result of the measures we use to gauge economic and social contributions; of ideologies that marginalise women as either peripheral to, or victims of, economic change; and of the personal nature of many women's responses to exploitation.

Women's relative position is also greatly impacted by their location in regions that have been restructured in the neoliberal era. African and Latin American women have suffered greatly through the imposition of structural adjustment policies, especially in increasingly prevalent female-headed households (Chant, 1999, 2004).

There has also been a shift to adopt feminist *perspectives* (epistemologies and methods) on globalization that seek to complement positivistic empirical studies (Nagar et al., 2002). A more reflexive perspective built on feminist epistemology has been called for, as well as pleas for the mainstreaming of feminist perspectives on global change, which are 'missing a key explanatory variable' (Christopherson, 2002, p. 247). Partly because feminism is itself broad and diverse, there has been no single feminist response to globalization. The aim of building a global grass-roots feminism which delivers gender equality and strengthens civil society is a core aim of the movement, however (Pratt, 2009).

places, and civil society is reacting against the perceived loss of control over collective and individual destinies (Routledge, 1999). Others have linked the rise of transnational new social movements to the evolution of post-materialist society and associated dissatisfaction with the nation-state and its inability to cope with a multifaceted society. There is no single theory of new social movements, but an analysis of the literature suggests that they must contain the following to warrant that name (based on Cumbers, 2008; McCarthy, 2000; Mercier, 2003; Munck, 2006):

- a collectivity of people;
- a level of organization;
- a shared ideology;
- cross-class and cross-societal linkages;
- self-identification to that collectivity;
- an alternative or transformative social vision;
- a variety of conventional and unconventional tactics.

Whether any new social movements fulfil all of these characteristics is open to debate, and the distinction between 'old' and 'new' movements is deeply contested. These movements are considered *new* because they appeal to values which are cross-societal rather than focusing their struggle in the arena of the nation-state, as do traditional social movements such as organized labour (see Box 5.5). Furthermore, the techniques employed by such groups differ from those of the past and are focused away from traditional channels of representation such as established political parties. These are the forces that are seeking to regulate and resist the current mode of production from below (see above). People are losing faith, it seems, in formal politics.

The anti-globalization 'movement' – and it is debatable whether it should be termed a movement at all – is now a major phenomenon in world politics which has come to the fore over the past decade and peaked very recently. We explore the roots, evolution, nature and impact of this phenomenon below and ask whether it represents a new, or possibly even *newer* social movement.

The anti-globalization movement

The anti-globalization movement (AGM) is a nebulous term used to describe a wide range of protest, lobby and interest groups (McFarlane, 2009). Although the movement broadcasts its message through a variety of channels, it is the proliferation of street protests, often planned to take

Box 5.5

Globalization and the labour movement

There are two tendencies associated with globalization that are changing the conditions under which labour operates (Herod, 2002, 2009). First, the increased mobility of TNCs, given developments in transport and communications technology, means that any worker, or group of workers, has to compete with counterparts in other areas of the world. This has led to an increasing 'race to the bottom' as employers seek to push down wages to remain competitive at the global scale. Second, the end of the Cold War and the triumph of neoliberalism have led to a rewriting of the rules with respect to organized labour. In particular, the deregulation of labour markets across the world has seen labour unions undermined. In the West, regulatory changes that weigh against collective bargaining were brought into force with the surge to the political right in the 1980s. In Thatcher's Britain and Reagan's USA, for example, there were concerted efforts to 'break the unions' in the supposed search for efficiency. The EU is unique in the current era in that it has regulated to encourage international labour interaction. Under the Maastricht Treaty of 1993, each company employing over 1,000 people has to meet the costs of bringing labour representatives together on an annual basis where they can share information with counterparts. This grants rights to workers in around 1,500 TNCs across the continent (Dicken, 2011). In the Third World, labour is often organized weakly or regulated heavily by the state, and the shift to neoliberalism has compounded this trend. During the Latin American dictatorships of the 1970s and 1980s, for example, unions were outlawed in many countries, as they were seen as a source of economic inefficiency which did not sit well with the free-market model, and were considered hotbeds of radicalism in the context of the Cold War.

The rise of TNCs (see Chapter 4) presents special problems for the labour movement. Labour unions are traditionally nation-state based, although there were some examples of international solidarity movements in the nineteenth and early twentieth centuries (see Herod, 2002). TNCs by their very nature are not constrained by borders. Thus organized labour in host countries is often concerned that decisions made by TNCs are 'distant' and do not take into account the specificity of location. Although it is the case that the level of centralization of TNCs varies greatly, there has generally been a tendency to shift important decisions so that, in some cases, networks of unions within single large TNCs have evolved. Furthermore, international labour movements from the past, such as the various International Trade Secretariats (ITS) developed in the late nineteenth century (e.g. in mining, transport and metalwork) have been revitalized. The goal of labour solidarity in this case has been to prevent TNCs from playing off groups of workers in

Box 5.5
continued

one country against those in another, a process referred to a 'whipsawing'. Herod (2002) provides a number of examples of successful international cooperation of this nature. For instance, in 1990, metalworkers in a plant in Virginia were locked out in a dispute with the Swiss owners over work rules. The United Steelworkers of America networked with the International Metalworkers Federation (the industry's ITS) and evolved contacts with unions across the world where the same company had operations. Protests over the sacking of the Virginia workers took place in 28 countries, and they were rehired. A second transformation in organized labour has seen it foster the technologies of globalization to conduct its operations. Internet technology, for example, enables the diffusion of information about strikes and other protests at rapid speed between distant territories. Thus, among other things, unions have developed web pages, chatrooms and e-mail groups to divulge their message. The Association of University Staff of New Zealand receives regular e-mail bulletins from its counterpart in the UK, for example, allowing a sharing of information and potential responses to labour issues. The way unions structure their business has also altered due to technological change; paid 'organizers', whose job was previously to develop contacts, have been replaced by less formal 'networkers' whose goal it is to promote 'network activism'.

place at the same time as important capitalist summits or events, that has brought the AGM to the attention of the world. Actions have been staged in places as diverse as Seattle, Genoa, London, Hyderabad and Wellington. Some see this 'globalized resistance' as a unified reaction/resistance to transnational regulatory processes from above (Routledge, 2013; N. Smith, 2000; Spybey, 1996). For others, the range of the AGM is so broad and diffuse as to render the terms 'movement' and 'anti-globalization' inappropriate. The news media, which coined the term, jumped on the growing tide of protest against global capitalism and for the sake of simplicity gave it an all-embracing name (see Cartoon 5.1). In the *New York Times*, Friedman (1 December 1999) was particularly scathing with respect to its diversity, calling the WTO protests in Seattle 'a Noah's ark of flat-earth advocates, protectionist trade unions and yuppies looking for the 1960s fix' (quoted in Mercier, 2003, p. 35). It is certainly true that the protests which have taken place have involved the participation of a vast range of groups including environmentalists, peasant farmers, anarchists, feminists, consumers, unionists and workers – to name but a few.

Pieterse (2001) argues that the AGM crystallized around a number of linked concerns including the inherent instability of the post-Fordist capitalist system; the cycles of crises this has tended to create; increased

Cartoon 5.1 *Tell us your complex message in a 30-second sound bite*

Source: Kirk Anderson

global inequality since the dawn of the era of development; the perception that the USA is seeking to control the cultures and economies of the world; and the environmental degradation associated with the increased industrialization and modernization of the planet. These concerns, of course, existed before the dawn of the 'anti-globalization' movement, and indeed before the term 'globalization' itself appeared. However, globalization provides a 'catch-all' set of elements against which protest can be waged in a collaborative form. At its root, the AGM articulates a geographical concern – that the power of locality and individuals within localities has been usurped by forces which exist at 'higher scales' (Hertz, 2001; Cumbers *et al.*, 2008). At the core of the AGM is the conviction that neoliberal policies are further marginalizing the already impoverished and disempowered and that the globalization agenda is actually *designed* to exacerbate the concentration of wealth (Klein, 2001).

The evolution of the anti-globalization movement

As noted previously, resistance to neoliberal policies and corporate capitalism is hardly new (Munck, 2006). It has been continuing in various forms since the creation of industrial capitalism itself. More recently, across Latin America, in particular in the 1980s, there were widespread protests against the structural adjustment policies of the World Bank and IMF which imposed neoliberalism on the nation-states of the continent. But protest was not just evident in the Third World; in countries such as the USA, UK and New Zealand protests against Reaganomics, Thatcherism and Rogernomics respectively were mounted throughout the 1980s. Routledge (1999, 2013) argues that the first example of a 'global' protest against neoliberalism came in 1994, following the actions of the Zapatista National Liberation Army (*Ejército Zapatista Liberacion Nacional*) in Chiapas, Mexico (see Box 5.6). This group, then, is often credited as being the spark of the current AGM. In reality, there were a number of large-scale protests and other actions that took place in the years preceding the uprising; some of these are outlined in Box 5.7.

Box 5.6

The Zapatistas

On 1 January 1994, when the NAFTA came into force (see Chapter 4), about 3,000 lightly armed men and women took control of the main towns in Chiapas – a poor southern Mexican state. Most were peasants who had lost their land to large landowners who prospered in the restructured Latin American economy of the 1990s. Economic liberalization badly damaged local economies as protectionism for coffee, forestry and cattle were removed. People were killed in the ensuing struggle, a factor leading to widespread sympathy for the Zapatistas both at home and abroad. A ceasefire was announced in late January and negotiations over land reform and social change were initiated.

The Zapatistas were protesting for democratic reform to fight perceived domination by political elites and foreign TNCs. They explicitly pinpointed neoliberalism as the main adversary. In 1996 and 1997 they organized two 'encuentros' (meetings or conferences) 'for humanity against neoliberalism', bringing together a wide range of social movements from around the world – especially the Third World. The intention was to discuss the impacts of neoliberalism and how it might best be resisted. A major declaration was communicated as a result. Some credit this as the founding document of the anti-globalization movement. What is especially notable is how the Zapatistas used globalizing technology as a means of spreading their message across the globe, appealing to social networks with common interests (see www.ezln.org/).

Box 5.7

A selected chronology of the pre-Zapatista uprising 'anti-globalization' movement

1982 Eighth World Economic Summit (G7) in *Versailles*. Attack on the Paris representative of the World Bank and IMF as well as protests for peace and against Ronald Reagan.

1984 Tenth World Economic Summit (G7) in *London*. An alternative World Economic Summit was staged at the same time under the title of 'Global Challenge: Out of the Crisis, Out of Poverty'.

1985 Eleventh World Economic Summit (G7) in *Bonn*. Several protests, including one of 25,000 participants and a 'Tribunal against the World Economic Summit' held in the town hall of *St Godesberg*.

1986 Twelfth World Economic Summit in *Tokyo*. Critical flyers distributed and rockets fired over the guesthouse of the Japanese government exploded near the Canadian embassy. The rockets did not cause any damage.

1987 Thirteenth World Economic Summit in *Venice*. Several thousand demonstrators protested. Bombs exploded in front of the British and US Embassies in *Rome* on the second day of the negotiations. No one was hurt.

1988 Fourteenth World Economic Summit in *Toronto*. Around 500 representatives of the international Third World and economic groups participated in the 'Other Economic Summit'. Protests in several German cities

later that year (26–29 September) as the IMF and representatives of the World Bank met in *West Berlin*.

1990 The IMF and World Bank meet in *Washington DC*. Small demonstration and an alternative forum. March against child labour later that year in *New Delhi* (29–30 September) to coincide with the UN summit of children in *New York*. GATT World Trade talks in *Brussels* prompt a protest conference entitled 'GATTastrophe'. Demonstration by thousands of farmers in Europe as well as the USA, Japan and South Korea. Protests also took place in Argentina, Japan and Switzerland.

1991 World Economic Summit in *Munich*. Largely peaceful large-scale demonstration with about 17,000 participants. About 500 people were later arrested at the opening of the conference and 6,000 anti-globalization demonstrators protested, demanding the resignation of the Interior Minister and Munich Chief of Police.

1992 In *Bonn*, the autumn conference of the World Bank blockaded to protest the construction of dams in the Indian Narmada River, funded by the World Bank. On the occasion of the planned GATT Compromise between the EC and USA, 40,000 farmers demonstrated, arriving primarily from Germany and France but also from Japan, South Korea and Canada.

Box 5.7
continued

1993 Mass protests throughout the year against the GATT agreements, particularly of farmers: in *New Delhi* (particularly seed-patent provisions); in *Seoul* (against the opening of the South Korean rice market); in *Geneva* (including farmers from Switzerland, France, Spain, Japan, India, Canada and the USA); and throughout *Asia*. Leading Indian and Pakistani politicians complained that the USA was not willing to reduce its tax on textiles far enough.

After the second Zapatista *encuentro* (see Box 5.6), around 50 NGOs from across the world created a network called People's Global Action against free trade (PGA) with the intention of establishing coordinated protests. The first 'global days of action' occurred in late May 1998, to coincide with the WTO's second ministerial conference in Geneva. Protests took place in over 28 countries. In Hyderabad, India, 500,000 people took to the streets with the rallying cry: 'we the people of India declare the WTO our brutal enemy'. These groups also linked with movements from the West such as 'Reclaim the Streets' from the UK. By 1999, the movement had gained significant momentum and was beginning to identify a 'common enemy'. This concluded in a meeting at the 'Other' Davos by five large protest groups, running in parallel with the World Economic Forum in Davos, Switzerland. On 18 June 1999, the day of the G8 economic summit in Germany, PGA coordinated a day of protest against financial centres called 'Carnival against Capital' and action was taken in over 100 cities worldwide. On 19 August 1999, PGA met in Bangalore India to coordinate the next global day of action which would coincide with the opening of the 'Millennium Round' of WTO trade talks in Seattle (see Box 5.7).

The Battle of Seattle

The protests in Seattle in late 1999 are seen as a watershed in the evolution of the AGM and new social movements in general. Approximately 60,000 people descended on the city – the home of Microsoft and Starbucks – to take part in what has become known as the 'Battle of Seattle' between 29 November and 2 December. Simultaneously, protests in world cities around the globe, such as London, Paris and Bombay, took place. In Seattle itself, over 700 different groups took part. Steger (2002) argues that the truly transnational character of the participating groups set it apart from earlier anti-globalization protests. What was especially notable was that the protestors had come to the 'hearth' of global capitalism and corporate globalization, and a large contingent was from the USA itself.

Due to the protestors' efforts, the trade talks were postponed. Rioting took place and damage was relatively widespread – although the majority of participants used peaceful methods. However, early disagreements in the AGM with respect to tactics quickly became apparent, as some groups demanded more direct, and sometimes violent, action. Less radical contingents claimed that such tactics would be futile and counterproductive. The US military intervened and declared martial law. The media emphasized the anarchy and chaos of the protest, a trend that has continued in mainstream coverage of such events. Bill Clinton, then President of the USA, decried the method of protest – but acknowledged that the protesters had some worthwhile points to make.

Since Seattle, almost every major international meeting of this type has been targeted and the number of protestors has remained high. Although the exact focus of the protests might vary, the same range of groups would often take part. In 2000, for example, 250,000 people took part in a Paris anti-fascist demonstration in reaction to the rise of the right wing in France; in Sao Paulo, Brazil, 100,000 people demonstrated against neoliberalism in the home of anti-liberal protests; and an incredible 7,200,000 took part in a general strike to protest against structural adjustment programmes in Argentina. In 2001, the protests continued with, for example, 80,000 people in Quebec protesting the proposed Free Trade Agreement of the Americas. A notable development during this year was the establishment of the World Social Forum in Porto Alegre, Brazil, involving tens of thousands of attendees from across the political left, united, broadly speaking, by their opposition to globalization as currently practised. This conference – which ran in parallel with the meeting of economic leaders at the World Economic Forum held in Davos, Switzerland – was founded by a coalition of trade unions, NGOs and environmental groups from Brazil. Also in 2001, the 'Siege of Genoa' – involving an estimated 300,000 people – resulted in the first death of a protestor (See Plate 5.1). In many ways, the scale and nature of this particular protest represented the pinnacle of the AGM (Bygrave, 2002).

Plate 5.1 *Anti-globalization protests in Genoa, 2001*

Source: Reuters

9/11 – where did all the protestors go?

Following 9/11, the mainstream media suggested that the AGM would lose its head of steam. It was argued that the global War on Terror (see above) waged by the Bush Administration would seek to classify such protestors as 'terrorists' – raising the stakes for participants considerably. *The Guardian* newspaper from the UK asked: 'Where have all the protestors gone?' (Bygrave, 2002). The movement responded with its own question: 'Where have all the journalists gone?' (Taylor, 2002). It appeared that the new tactic of Western governments and allied media was to ignore the protests to the extent that this was possible, denying the global publicity given to the AGM up to that point. For example, the protest of 250,000 in the streets of Barcelona in March 2002 was covered far less extensively than were previous demonstrations on the news networks over which Western governments hold influence (Bygrave, 2002).

In reality, the attacks of 9/11 provided a renewed stimulus for some in the movement as the belief that they came about precisely because of the historical injustices created by the capitalist West became one of the central explanatory discourses adopted by the political left. At this point, however, the movement began to fractionate even more than it had hitherto. The invasion of Afghanistan and the rising possibility of war in Iraq provided new focal points for protest. The AGM partly metamorphasized into an anti-war movement, and this culminated in 350,000 protestors marching in London on 28 September 2002. Many protestors seemed to interpret the protest against a possible war as part of the anti-globalization ethic. Others did not see it this way and a questioning of the coherence of the AGM, which seemed able to reshape itself to any global issue at hand, permeated media discussion. The World Social Forum, which epitomizes the broad range of issues that unites the global left wing, has grown in importance and scale, with over 100,000 people attending the 2004 and 2005 meetings in Mumbai and Porto Alegre respectively. The issues dealt with at the 2005 forum – such as GM crops, the global environment, debt and the continued conflict in the Middle East – illustrated that it had become a general meeting of minds for alternative thinkers. Strictly speaking, however, it is more accurate to see the World Social Forum as part of the evolving alter-globalization movement (see discussion below) (see Worth and Buckley, 2009). The fervor surrounding the AGM of the early 2000s was not echoed until the early 2010s and for ten years dissipated somewhat. To claim dissolution, of course, implies a former unity – and this is controversial, as is discussed below.

Coherent movement or ineffective hybrid?

The term 'anti-globalization movement' was imposed externally, and there are a number of problems raised in using the term. First, the movement, given its inherent diversity in tactics and causes, has no one alternative model which it proposes. Second, the movement is itself facilitated and made possible by the processes of globalization, which is seemingly paradoxical. Third, leading from this last point, it is important to note that many within the movement are not opposed to globalization per se. Hardt and Negri (2001, p. 102) argue that 'the vast majority of the protestors are not against globalising currents and forces as such; they are not isolationist, separatist or even nationalist'.

Alternative titles for the movement have been suggested. Examples include the 'anti-capitalist movement', 'anti-imperialism' and 'anti-neoliberalism', all of which emphasize protest at the expansion of the capitalist nucleus (Petras and Vettmeyer, 2001). Other suggestions have included the 'movement against global corporatism' or the 'anti-corporate globalization movement', which emphasize the opposition to TNCs and supranational bodies (George, 2001; Starr, 2000). Further examples highlight the 'governance' aspects of the movement and include the 'Civil Society Movement', 'Citizens' Movement for World Democracy' (Bygrave, 2002), the 'New Democracy Movement' (Barlow and Clarke, 2001) and the 'Global Justice Movement' (Lefrancois, 2002, cited in Mercier, 2003). A further term that has been proposed is 'global justice network' (Cumbers *et al.*, 2008). Finally, and most nebulously, the term the 'Alternative Movement' has also been proposed. Naming is more than 'mere' semantics; the title chosen reflects the nature of the adversary, something of the politics of dealing with it, and helps sustain the identity required to operate as a movement (Routledge, 2009). The diversity of the movement as reflected in virtually all the suggested titles here is at once strength and weakness. It captures many of the pertinent issues in today's global society, while at the same time threatening to dissolve the concerns to the point of ineffectiveness.

The AGM, in a broad sense, claims a number of victories. Two of the largest and most publicized were the halting of the construction of the

World Bank-funded Narmada dam hydroelectric project in India and the blocking of the Multilateral Agreement on Investment (Brecher *et al.*, 2000). There have been others, but perhaps the main accomplishment has been symbolic and ideological – and thus very difficult to measure in the short term. Globalizing the issues of concern and bringing them to the attention of millions of people is a significant feat, although it could be argued that mounting large-scale media events have caused the institutions of global governance to retreat further into themselves, thereby lowering transparency further. However, despite the fact that much of the media tried, at first, to portray the movement as violent and anarchistic and later ignored it, the AGM has undoubtedly captured the public's imagination. There have even been a number of 'defections' by former insiders of neoliberal institutions (see Box 5.8).

Box 5.8

Defections from the inside

One factor which makes the anti-globalization movement so hard to pin down is that former 'insiders' have joined its ranks. While such people have not necessarily abandoned their adherence to market principles, they have criticized the manner in which prescriptions are made. Two notable examples are reported in Gwynne *et al.* (2003).

Jeffery Sachs was the World Bank's most prominent consultant in the 1990s. He has gone on to become a radical centrist author writing titles such as *The End of Poverty* (2011 [2005]), *Common Wealth* (2008) and *The Price of Civilization* (2012), all of which are critical of current globalization trends. Despite being part of the neoliberal apparatus, he reflected critically on the imposition of neoliberal reform on the Third World in 1999:

The Washington consensus listed 10 or 12 steps – the recipe for economic development. When you look at those, they're all pretty reasonable. But it's a kind of bland list of commandments, rather than a real blueprint of how to get from A to B. . . . It became a substitute for assistance, because the idea was 'you don't need us. You don't need any help. . . . You just have to follow the magic rules'. The actions of the IMF and World Bank became very stylised . . . at that level of simplicity, it just doesn't work.

(Sachs, 1999, in Gwynne *et al.*, 2003, p. 5)

Joseph Stiglitz was President Clinton's chief economic adviser between 1993 and 1997 and was the World Bank's chief economist

Box 5
contin

and vice-president from 1997 to 2000. He has since gone on to be a very influential author of books that are critical of the current globalization model such as *Making Globalization Work* (2006) and *Freefall* (2010). When he resigned in protest at IMF policies he wrote:

> When the IMF decides to assist a country, it dispatches a 'mission' of economists. These economists frequently lack extensive experience of the country; they are more likely to have firsthand knowledge of its five-star hotels than of the villages. . . . Critics accuse the institution of taking the cookie cutter approach to economics, and they're right. Country teams have been known to compose draft reports before visiting.
>
> (Stiglitz, 2002, in Gwynne *et al.*, 2003, p. 5)

Not surprisingly, the alternative or transformative vision of the AGM is not well defined. It has, at various times, included diverse suggestions such as: debt write-off; the end to unjust war; environmental sustainability; reform of global institutions; and a global welfare system (Mercier, 2003; Routledge, 2009). Through the work of authors such as Aguiton *et al.* (2001) in *The Other Davos*, and Bello (2002) in *Deglobalization*, we are seeing the formation of some concrete policy alternatives, which have at their core a global regulatory system that seeks to address inequality within a sustainable framework. In many ways, these suggestions bear much in common with earlier work on policies for 'sustainable development' (see Chapter 7) (see also Cumbers *et al.*, 2008; Hines, 2000; Houtart, 2001).

Alter-globalization from below?

From a theoretical standpoint the problem with the AGM, while powerful in its scope and force, is that it has tended to commit the error of *metanarrative* – ascribing all of the world's woes to the inexorable rise of globalization. In this sense the movement has become over-global and has lost sight of the localized forms of resistance from whence it was born. It is partly due to this that the movement has fragmented, as some seek to distance themselves from what they see as a relatively ineffective marriage of concerns. This has also involved the evolution of the *alter-globalization* movement, which is linked to the idea of *progressive globalization* (see Chapter 9). This perspective argues that it is not globalization per se which is the problem but the way it is currently practised. There is nothing inherently bad about the time–space compression of global society, but it does need to be regulated in order to prevent growing inequality and non-sustainability. These debates hinge crucially on the definition of the term 'globalization'. If we accept that globalization is synonymous with the corporatization of the world, then the AGM has a natural enemy (Hardt and Negri, 2000). If we accept, in contrast, that the corporatization outcome is but one possible outcome, then we can reconstruct the concept of globalization in a more progressive way.

This perspective has been especially influential in France where a large alter-globalization contingent gathered at the European Social Forum in September 2003. An editorial in *Le Monde* (*Guardian Weekly*, 20–26 November 2003, p. 33) argued that this group 'which is as heterogeneous as it is dynamic', had succeeded in creating a 'new political dynamic' in the country. In particular, in transcending accusations of being influenced by 'an antiquated form of socialism, they have now acquired a vigour which makes them capable of mobilising seasoned and inexperienced activists alike'. Furthermore, they have distanced themselves from the 'anti-globalizationists' who, according to the newspaper, 'want to barricade the nation-state and return to a policy of state control and protectionism'. However, the article points towards some of the ambiguities which still remain, the main one being the:

> lack of any convincing counter-project, while the alter-globalizationists can muster solid arguments against 'neoliberal globalization' . . . the alternatives they propose are lacking in clarity and coherence. . . . From the defence of 'free' software, the proposed Tobin Tax on capital movements and the defence of environment, to the denunciation of brand marketing, the struggle against poverty and the campaign against genetically modified foods, they peddle a host of totally uncoordinated notions.
>
> (*Guardian Weekly*, 20–26 November 2003, p. 33)

Another strand in the evolving alter-globalization movement, based on a more radical stance, is that of 'globalization from below'. Brecher *et al.* (2000, pp. 38–39) identify a number of common goals within this camp including: policies to strengthen the ability of national governments to counteract global economic forces; support for the empowerment of local people to take control of their natural resources; the organization of repressed groups in civil society; and the construction of a transnational social movement of organizations across borders (Munck, 2006). Mittelman (2000) argues that globalization from below should involve self-determination and economic self-governance by the majority. There are some groups that have been established specifically to promote grassroots globalization, as it is alternatively called, including the International Forum on Globalization, and People's Global Action. Undoubtedly, the crowning achievement of this part of the broader movement to date is seen as the establishment of the World Social Forum, which seeks concrete routes to world reform (Worth and Buckley, 2009).

Occupy and the global justice movement

For much of the 2000s the AGM had evolved, it seemed, into the anti-Iraq war movement, as noted above; this gave the protest a central focal point, although it took attention away from some of the broader points previously pursued through the AGM. The World Social Forum gathered pace timed with the World Economic Forum and by the end of the decade forums held in places as diverse as Porto Alegre, Brazil (2001–2003, 2005 and 2012), Nairobi, Kenya (2007), Mumbai (2005) and Dakar, Senegal (2011) were attracting tens of thousands of participants including increased representation from the Global South. An early criticism of the AGM was that it was overrepresented by interests and groups from the Global North.

Following the Global Financial Crisis (see Chapter 7), two notable public protest events on different scales took place that had a notable impact on subsequent global justice protests – the Arab Spring and the Chilean Winter. These reacted in part to the polarising impacts of the crisis and the responses, and grew out of general dissatisfaction and rising unemployment that added to a long list of other grievances. The first event was precipitated by political protest in Tunisia that led to the eventual downfall of the Mubarak regime in Egypt. Linked conflicts continue to this day with the current conflict in Syria aligned to this. The second event which was in turn inspired by the Arab Spring was what became known as the Chilean Winter of 2011 (see Box 1.2). This student-led protest broadened from calls for educational reform to general democratic and justice goals that echoed the anti-globalization movement of a decade before. One of the tactics of the movement in Chile – based on protests against the dictatorship many years prior to this was to occupy schools and universities. These two protest movements together with other events in places including Spain and France, influenced the rise of what became known as the Occupy movement (see Chomsky, 2012).

The first Occupy protest took place beginning on 17 September 2011 in and around Wall Street. Eventually thousands of people took control of Zuccurotti Park, being forcefully removed some months later. Marches were also organized around the Occupation. The protest was focused upon the slogan 'we are the 99%' which referred to what the movement claimed was the vast concentration in wealth distribution that had occurred through the 2000s. The movement was centred on Wall Street as it saw the banks and financial institutions located there as responsible for both the perpetuation of this inequality and the eventual downfall of the system in the GFC of 2007 onwards. The bailout responses of the Bush Administration and the perceived lack of strict regulation to prevent further accumulation were triggers for the event. The movement stressed the overly close links of large-scale institutions and policymakers in the USA and indeed across the world. President Barack Obama stressed that he understood the roots of the protest, although it is alleged that the protesters were monitored under the anti-terrorism act and they were eventually forcefully removed.

Within weeks, the Occupy movement had spread across the planet to over 80 countries and 200 cities worldwide. In New Zealand, for example, protests were seen in Auckland, Wellington, New Plymouth, Invercargill, Christchurch and Dunedin, all of them eventually being forcefully removed by early 2012 from occupied buildings and public spaces (see Plate 5.2). The goals of the Occupy movement were broad but more focused than its larger ascendant, the AGM; it focused on inequality and corporate influence over the political class. Large-scale protests in Brazil in 2013, Turkey in 2013 and Greece in 2011–2012 were linked in that similar issues were emphasized by the protesters – although the particular stimuli may have varied.

At the time of writing, then, it seemed that a global justice movement – that had risen from the earlier example of the AGM – was taking shape, focused on the perceived injustices of the GFC and the responses to it. However, this movement was different from protest movements of the past both in terms of it techniques and the composition of its adherents, media commentators started to point out. In *The Economist* (2013) with a cover showing a protester drinking a caffè latte and using a cell phone, the claim was made that protest has become a middle-class phenomenon. Writing in *The Guardian* in July 2013,

Plate 5.2 *The Occupy movement in Wellington, New Zealand*
Source: Authors

Beaumont argued that, unlike the protest movement of 1968 or Europe in 1989, these are movements with few or no discernible leaders and often conflicting ideologies. Beaumont makes a link between perceived protest and levels of trust among the public for their policymakers, a claim backed up in *New Global Revolutions* by Paul Mason (2013) – an author who argues that we are at the centre of 'revolution caused by the near collapse of global capitalism combined with an upswing in technical innovation' (quoted in Beaumont, 2013, p. 9).

New technology associated with globalization meant that these new networks were able to challenge traditional forms of leadership, and the more horizontal structures had potential for much greater breadth. It appeared that new networks had evolved that made the movement more flexible and able to react at great speed to events through technology such as Facebook and Twitter. Tali Hatuka, an urban geographer, echoes some of these points arguing that the new style of protest that involves multiple voices and horizontal structuring evolved from the anti-war protests beginning in 2003. She goes on to say: 'up to the 1990s protests tended to be organised around a pyramid structure . . . as much effort went into planning as the protest itself . . . now protest . . . is far more informal, the event often being immediate' (quoted in

Beaumont, 2013, p.10). This apparent newfound democratic flexibility was challenged by the uncovering of a number of cases of spying on civil society in 2013. When Edward Snowdon blew the whistle on the monitoring by the CIA of material from Facebook and other social media sites, it became clear that the freedom afforded by social media may well prove to be double-edged. Protests against the use of such data are likely to increase in the coming years and form a critical part of the evolving global justice movement.

Global Justice: movement of movements?

Callinicos has likened what he refers to as the 'anti-corporate' movement to the new social movements of the 1960s and argues: 'there has not been such a resurgence of activist activity since the Vietnam War' (2001, p. 387). As globalization comprises a complex set of processes, agendas and outcomes, anything defined in opposition to it will be similarly complicated. The common enemy is not as easy to define as for, say, the feminist movement or the anti-nuclear movement. It is perhaps best described as a 'movement of movements' – one which captures the globalization era and its dialectal processes.

Overall, it is uncertain to what extent the anti-globalization movement incorporates the elements thought necessary by commentators to define a new social movement (Routledge, 2009). Mercier's (2003) primary research on the anti-globalization movement in London illustrated a diverse, fractured and internally incoherent set of movements operating under a banner which many felt had been imposed upon them externally. It may be that current terminology is not able to capture what is happening. It is perhaps time to develop a theory which addresses the rise of 'newer social movements', 'global social networks', or some other term which captures globalized diversity within unity and the constellation of varied, shifting and dynamic interests which characterize the Global Justice Movement.

Conclusion – a multipolar world

Globalization is altering the way politics is practised. The nature of the nation-state is being transformed as a result, civil society is evolving rapidly, and governance practised at regional and global scales is flourishing. Political geography needs to move quickly to keep up with these changes. The argument presented in this chapter is that nation-states remain central players in global affairs. However, they will become less powerful as they proliferate due to the devolutionary pressures created by globalization and as their borders are transcended by cross-societal NGOs and NSMs that appeal to universal values. At the same time, global institutions will need to be reformed and strengthened in appropriate ways if they are to deal with the transnational flows of the contemporary world in an effective and equitable way. In this context, the reinvigoration of the United Nations is absolutely essential for global stability.

Geopolitics has evolved rapidly since the Second World War, moving through various 'world orders' around which global politics has been organized. Arguably, the Cold War was replaced by the neoliberal-democratic world order. More recently, and contentiously, the War on Terror might be seen as part of an evolving order that confirms the rise of the USA as the global hegemon. This chapter argued, however, that the War on Terror is best seen as an extension of the neoliberal-democratic world order, which – despite the overwhelming power of the USA – is in fact multipolar and multilayered, and involves the practice of governance at a range of scales. Underpinning the design of the successive 'world orders' and regimes of accumulation has been the capitalist cultural and economic mode of regulation. It is this imperative, and a redoubled research effort, that requires resistance to it if political geography is to contribute to a just and sustainable world.

Further reading

Brecher, Costello and Smith (2000): This is an excellent early review of the anti-globalization movement in all its complexity.

Chomsky (2012): This provides a very concise introduction to the aims and nature of the Occupy movement as well as suggestions for further action.

Gruffudd (2013): See this chapter for an excellent introduction to the concept of nationalism and how it has evolved.

Held and McGrew (2007a): This is a classic introduction to all things anti-globalization.

Khanna (2008): This book on world orders is a stimulating read.

McGrew (2008): In its discussion of 'politics beyond borders' this book sets the gold standard in terms of studies of political globalization and governance.

Munck (2006): A book on the rise of protest against globalization – radical and readable.

Ó Tuathail (2002): This chapter provides a solid introduction, still relevant today, to geopolitics and the concept of world orders, or 'geopolitical worlds', as the author terms them.

Rosenau (1990): This classic book presents some early arguments concerning the impacts of globalization on political governance and the state that remain salient.

Routledge (2013): This is a very good chapter on global resistance movements and survival.

Sidaway and Grundy-Warr (2012): An excellent introduction and geographical interpretation of the role and rise of the nation-state.

Sidaway and Mamadouh (2012): This chapter is a useful introduction to political geographies in the contemporary world.

Sparke (2013): The chapter in this book on governance is a very effective overview of global trends and tensions in the current era.

6 Globalizing cultural geographies

- Global cultural shifts and geography
- Defining cultures – landscapes, change and interaction
- Historical cultural globalization
- Contemporary cultural globalization – cultural imperialism?
- Criticisms of the homogeneous global culture arguments
- Cultural consumption – commodification and cosmopolitanism
- The global music industry case study
- Conclusion – towards progressive cultural spaces?

Advances in telecommunications technology in the twentieth century meant that cultural symbols – such as images, music or texts – flowed across the world as never before, and capitalism commodified these symbols and perpetuated their globalization. Cultures became less spatially bounded than they once were, leading to more intense cultural traffic. Some argue that this has given rise to a new 'global consumer society' based on the diffusion of Western culture, building around culturally imbued commodities marketed by companies such as MTV, McDonald's and Coca-Cola. While it is true that such symbols have become increasingly pervasive, the idea that their propagation is forging a homogeneous global culture has been rigorously debated, and this chapter pursues some of these arguments.

Although culture is seen as central in the *postmodern* epoch, in reality it has *always* been central to all forms of political, economic and social change. It is impossible to separate cultural, economic and political processes – especially in the context of globalization – but we do this in order to be able to deal with what would otherwise be an overwhelming complexity. There can be little doubt, however, that culture has become more *visible*. This new visibility is intimately linked to the evolution of globalization, as we will explore below. In this chapter, we consider the relationship between the process and agenda of globalization and how this relates to cultural change at various scales. The central question that we consider is whether globalization is leading to the homogenization of culture or, in contrast, a reassertion of the local. What we find is that this is indeterminate; cultural change is increasingly complex and differentiated and understanding it involves analysing the relationship between change at the global and local scales.

Global cultural shifts and geography

Over the past two decades, much of social science and human geography has been rewritten with a new emphasis on the way culture permeates other spheres. In the 1980s a cultural turn took place in the social sciences in general, influenced by developments within academia and society as a whole. In the academy, the rise of postmodernism, based partly on the humanist tradition of the 1970s, challenged 'grand' theories or metanarratives that sought to reduce social complexity to singular explanations (such as Marxism or World Systems Theory). This involved a resounding critique of positivism – an epistemology which views the world as existing independently of human perceptions of it and reducible to laws derived through empirical analysis. In contrast, postmodernism argues that 'reality' is a social construct and absolute truth is therefore impossible. This general critique gave rise to a range of perspectives on social change that is still unfolding today, including post-structuralism and postcolonialism. These critical perspectives often place 'culture' at the centre of their explanations and/or interpretations of the world. Postmodernism is not only a way of thinking; it may also be seen as an 'epoch' or time in world history and a set of associated material outcomes in which culture plays a more visible role than hitherto. Thus postmodernism has been theorized in the economic (e.g. flexible accumulation) and political spheres (e.g. post-materialist society) (see Chapters 4 and 5 respectively). The cultural turn was as much influenced by the changing world around the academy, and its shift to a postmodern condition, as by theorization from within.

In the late 1980s the postmodern 'cultural turn' in social sciences influenced human geography significantly (see Crang, 2009; Gibson and Waitt, 2009; Crang and Thrift, 2000; Jackson, 1989; Philo, 1991). This endeavour sought to elucidate the links between culture and space and emphasized the role of culture in social change. The shift involved the widespread use of *interpretive* techniques using qualitative methods, as opposed to empirical techniques (Johnston and Sidaway, 2014). At the core of the transformation lay a critique of the 'quantitative revolution' of the

1960s and 1970s – often by the very people who led that revolution. Geographical models and perspectives that sought to explain the world in terms of economic, political or environmental determinism were challenged – such as industrial location theory or core–periphery models. Geographers dealt with issues such as identity, race, gender, ethnicity, tradition, indigeneity, and the cultural roots of economic and political change in more sophisticated ways than previously (Johnston and Sidaway, 2014; McDowell, 1997). In contrast, traditional cultural geography had often described exotic 'mosaicked' (see Chapter 2) cultures from an anthropological viewpoint. Some new cultural geographers committed the same error of which they accused positivists – seeing culture as deterministic in social change. Notwithstanding this, a new vitality pervaded the discipline together with a range of new epistemological approaches and associated methods (Kindon and Pain, 2010). As is characteristic of a healthy discipline, the cultural turn in human geography has not been without its detractors, however (see Box 6.1).

Box 6.1

The 'cultural turn' and its relevance in human geography

The 'cultural turn', informed by postmodernism, sent shock waves through human geography that still resonate today (Crang, 2009). What are its implications for the relevance of human geography? This depends much on how one defines relevance. Two opposing views that were generated early in the debate are represented by Johnston (2000) and Pacione (1999). For Johnston, the turn helps human geography in its quest for relevance. He argues (2000, p. 696) that work in this area:

> is promoting awareness of how all knowledge is produced and situated in particular contexts and then used to promote privileged positions for some forms over others. . . . Such work is

highly relevant to the creation and maintenance of humane societies.

Pacione (1999, p. 4) is more critical:

> The failure to address real issues would seem to suggest that the advent of postmodernism in radical scholarship has done little to advance the cause of social justice. Discussion of relevant issues is abstracted into consideration of how particular discourses of power are produced and reproduced. Responsibility for bringing theory to bear on real-world circumstances is largely abdicated in favour of the intellectually sound but morally bankrupt premise that there is no such thing as reality.

Box 6.1
continued

To Pacione then, the cultural turn promoted conservatism and paralysing progressive collective action. New cultural geographers would argue that Pacione dismisses postmodernism because his conception of the term 'relevance' is too narrow, however. This debate remains of relevance in human geography today, although the initial fervour of the postmodern approach has made way for a more balanced perspective which seeks to draw on a range of epistemologies (ways of seeing the world). Notwithstanding, some journals are characterized by a relative focus on postmodern perspectives and associated approaches – such as post-structuralism and postcolonialism whereas others are more 'applied'. We argue that all perspectives are required a for healthy and thriving human geography that can claim to be global in scope. A significant challenge is encouraging dialogue across the unhelpful divide.

A central question in the study of cultural globalization from a human geographical viewpoint is: In what direction does the heightened cultural traffic referred to in the opening of this chapter flow? Does it move from the global to the local, the local to the global, or is it more complex and multiple? This takes us back to the discussion in Chapter 2 concerning the definition of, and interaction between, scales. The reality may be that the world is so complex that the traffic flows in all directions at the same time. Human geographers have engaged in the theorization of global cultural change at all levels. Global-scale work has focused on the movement of people, objects and images around the world through telecommunications, language, the media industries, radio and music, cinema, television and tourism (McEwan, 2012). This work has critically engaged with theses of homogenization, McDonaldization and Coca-Colonization, for example. At lower levels there has been particular interest in the reassertion of local cultures in the face of 'global' processes; how such processes are interpreted; and how they are resisted, regulated or reconstituted from below. A great deal of the work in this area has argued that global and local cultural processes interact in many different ways in different places to produce 'structurationized', 'hybridized' or 'glocalized' outcomes. For example, in the UK, CNN has presenters with UK accents and in Australia, TV actors with Australian accents overdub adverts made in the USA in order to appeal to the local market. Central to new cultural geography work in this area is how globalization is affecting cultural production and consumption, and how this in turn is affecting people's sense of identity, sense of place and cultural politics (Massey, 1994). This chapter seeks to throw light upon simultaneous global, local and glocal processes which make up cultural globalization. We can only understand how cultures change and interact in the current era by first defining what we mean by culture, however.

Defining cultures – landscapes, change and interaction

Defining 'culture' is a difficult task. There are many popular misconceptions of the term. McEwan argues that 'students of geography often imagine studying cultures to be either about looking at exotic peoples in other places, or about high culture, such as the arts' (2001, p. 155). These aspects are certainly part of studying culture, but do not present the full picture. Two definitions of culture are offered below:

> Cultures are maps of meaning through which the world is made intelligible.
>
> (Jackson, 1989, p. 1)

> Cultures are part of everyday life, and are systems of shared meanings that can exist on a number of different spatial scales (local, regional, national, global, among communities, groups or nations). They are embodied in the material and social world, and are dynamic rather than static.
>
> (McEwan and Daya, 2012, p. 273)

Culture provides direction, guidance, rules and limits concerning how humans interpret habitats and environments and make decisions on how to exploit resources, including each other. These elements are reproduced through formal and informal means. Cultures are thus constructed by society and are subject to the operation of power relations. There is often tension between different cultures in any one given space or place. Often, a dominant culture may seek to impose meaning of what 'culture' is or should be, and subcultures – some of which arise in resistance to dominant culture – may challenge this. An important contribution in recent years to thinking about culture is its conceptualization as a *process* rather than as a fixed entity. Culture involves the practising of identities – be they political, sexual, ethnic or national – around sets of shared beliefs and values (see Barnett, 2009).

Cultural traits are individual components of a cultural complex. A group of people that is distinguishable by its possession of a unique assortment of traits may be called a cultural group. Traits are the building blocks of culture and may be divided into three categories:

- sociological – interaction forms, laws, regulations;
- ideological – religion, politics and non-secular belief systems;

- technological – technical systems and abilities resulting in infrastructures.

In tracing the geographies of the globalization of culture we need to examine the stretching of these traits across space and time through social networks and the resultant impacts on cultural 'landscapes'.

Cultural landscapes

Cultural landscapes were one of the central components of 'traditional' cultural geography, developed largely by Carl Sauer (1925). He argued that 'landscape' – and this meant far more than the 'natural' landscape – reflects the traits of the past and/or currently occupant culture. It was argued that humans as an agent of change are constrained by their environment and culture. Landscapes thus differ greatly according to local geographical conditions, defined in the broadest sense. Material culture consists of the physical objects created by cultures as they carry out their way of life. For example, urban-industrial society creates landscapes of concrete and steel. Material culture reflects the non-material aspects of culture such as religion, and non-secular values. An obvious example would be the building of Christian churches or Muslim mosques; a less obvious example might be the clear felling of forest to make way for agro-industrial development.

Ideas concerning cultural landscapes have shifted of late as cultural globalization has made cultural landscapes much more difficult to discern, and bounded cultural spaces have 'melted into air' (Harvey, 1989). The reality is that, even in the supposedly 'mosaicked' past, few entirely isolated cultures existed and the majority have been subject to outside influence of some form. This disrupts notions of 'indigeneity', 'tradition' and 'authenticity'. Isolation is a relative concept of course. Many still expect cultures, such as those in the islands of the South Pacific, to be relatively untouched, or 'authentic', by virtue of the large absolute distance between populated land. There is no necessary correlation between absolute distance and cultural isolation, however. One of the most physically remote islands on the planet, Tahiti, as part of French Polynesia, is far more culturally connected to the West, and exhibits a cultural landscape constructed accordingly, than many other Pacific Island nations at closer proximity to Western

Plate 6.1 *Cultural landscapes – Kuching, Malaysian Borneo*

countries such as the Solomon Islands or Papua New Guinea. This arises because of the territory's intimate connections to the colonial power France. Conversely, there are large cities in densely populated regions such as Tehran (Iran), Baghdad and Kinshasa (Democratic Republic of Congo) that, for historical reasons, are largely bypassed by globalized cultural flows becoming what Rennie-Short (2004)
refers to as 'black holes'. In short, the influence of globalization on cultural landscapes is a function of the connectedness of that landscape to global networks together with the geographical and historical contingencies which have hitherto shaped that place. All cultural landscapes remain unique, but they are more complex and hybridized than they once were. It is these new globalized cultural landscapes, and to a lesser extent the 'black holes' in the new networks, that form the basis of much investigation in contemporary cultural geography (see Cosgrove, 2009).

Cultural change and interaction

Before moving on to look at the arguments surrounding the 'globalization of culture' in more detail, we need to understand a little more about concepts of cultural change. Cultures change at different speeds across space and time. Both 'internal' and 'external' forces influence them – and there is a range of different 'modes of interaction', as we will see below. Internally, innovations such as the development of new technology or new philosophical understanding can alter cultures. External influences may include colonialism, global marketing and tourism. In reality, there is a blurring between what comes from 'outside' and what comes from 'inside'. The problem with understanding cultural change today is that distinguishing between the 'internal' and the 'external' has become increasingly problematic given the processes of deterritorialization.

Human history may be traced in terms of a series of *cultural revolutions* (e.g. agricultural, industrial, democratic, technological). A cultural revolution occurs when a culture changes dramatically and the roots of this change may be multiple. It may occur subsequent to the invention of a new groundbreaking technology such as the wheel, settled agriculture, the combustion engine or the internet, but deeper historical forces often drive technological innovation. For example, the industrial revolution was inspired in a technological sense by the invention of the combustion engine which resulted in mechanization and possibilities in terms of the division of labour and economies of scale, all of which facilitated capitalist accumulation. However, the industrial revolution would not have occurred or been sustained without the earlier agricultural revolution and, later, industrial colonialism. Although cultural revolutions, by definition, have shattering consequences at the time, a more important process over the long run is cultural diffusion as was witnessed, for example, during the Maoist cultural revolution in China which sought to explicitly reverse what was seen as the creeping influence of Westernization and in particular Americanization.

Cultural diffusion is the process of the spreading of cultural traits from one place to another. A locality or region that has been a rich source of traits is known as a cultural hearth. In the past, hearths were bounded spaces; for example, Mesopotamia (the area roughly corresponding to Iraq today) became a cultural hearth from which agriculture diffused. Diffusion arises through various processes of personal and impersonal interaction. In the past, migrations and empires were the most important diffusing conduits, whereas today shared meanings, ideas and cultural norms are diffused through the broader spaces of flows. This leads to the possibility of deterritorialized hearths. Arguably, neoliberal globalization's cultural hearth is the world city network, dominated by the USA and Western Europe.

As cultures move and/or expand, either in real space or cyberspace, they will inevitably interact with others. *Cultural interaction* may be conceptualized in three main ways, which exist on a continuum and are subject to overlap:

- *Assimilation* occurs when a culture changes completely and loses its original traits due to subsumption by a dominant one. The geographical outcome of this process, if it is extended globally, is universalization and spatial homogeneity.
- *Acculturation* takes place when one group assumes the traits of others while retaining some of its own. The geographical outcome of this is glocalization or hybridity, characterized by heterogeneous patterns.
- *Autarkism* occurs where a culture reasserts its authenticity and specificity in the face of a perceived or actual threat from another culture. The geographical outcome of this is localization and 'mosaicked' heterogeneity.

As in every other sphere, commentators are divided on the extent and nature of cultural globalization. Broadly speaking, opinions may be divided into the three camps as outlined in Chapter 2. The arguments of the three schools correspond, in part, to the models of cultural interaction outlined above.

- *Hyperglobalists* – argue that globalization is leading to the homogenization of world culture, towards a Western – especially US – form. Radical hyperglobalists (i.e. from the left wing) agree and see it as cultural imperialism. This corresponds to the *assimilation* model that leads to universalization.
- *Sceptics* – argue that hyperglobalists' claims are greatly exaggerated and that national culture in particular continues to exert the overwhelming influence on cultural change. This corresponds, in part, to the *autarkic* model, although less radical interpretations might favour a mild *acculturation* model. In both cases, heterogeneity is the outcome although this may be localized or glocalized in form, depending on the exact nature of cultural interaction that is envisaged.
- *Transformationalists* – argue that the processes of globalization are leading to an intermingling of cultures creating new hybrids and

networks. This corresponds to the *acculturation* model that leads to glocalized heterogeneity.

Historical cultural globalization

Held *et al.* (1999) argue that, to date, much of the discussion in this area has focused exclusively on the impacts of contemporary processes and that the enduring legacies of earlier historical periods on current patterns of cultural globalization have been underestimated. Cultural globalization, arguably, pre-dates significant globalization in other spheres. Global migrations and the populating of the Earth, the diffusion of language families and the initiation of trade routes were the first manifestations of this. Following this, arguably, the first true examples of global networks evolved in the form of *world religions*. There are six religions that are conventionally described as world religions – Christianity, Islam, Confucianism, Hinduism, Judaism and Buddhism. The latter four, although significant in numbers both in the past and today, are essentially regionalized (see Map 6.1). It is the first two, however, which can lay claim to being global in their extent. In the case of Islam and, especially, Christianity, political networks and military force were combined to propagate the respective religions, which were used as both justification and means of conquering territory. Christianity was especially efficient in this respect and during the sixteenth and seventeenth centuries began a major diffusion across the globe linked to the expansion of the European empires. This has left an enormous legacy in terms of the commonality in belief systems across Latin America, North America, much of Africa and the Pacific Islands. It should be noted that Christianity was often acculturated according to existing belief structures (see Plate 6.3) and continues to form a major part of cultural identity in places across the Pacific such as in the example of the *fa'a Samoa* (Samoan Way).

Linked to the rise of religion was the rise of *global empires*, which have left major legacies. The most successful empires, such as the Roman and British Empires, are the ones where 'the extensive reach of military and political power has been reinforced by the reach of cultural power' (Held *et al.*, 1999, p. 334). In this way, global empires were able to overcome the inherent dialectic of imperialism whereby pressure is placed on centralized control by the inherent decentralization of political networks. Thus, in the case of the British Empire, for example, extensive

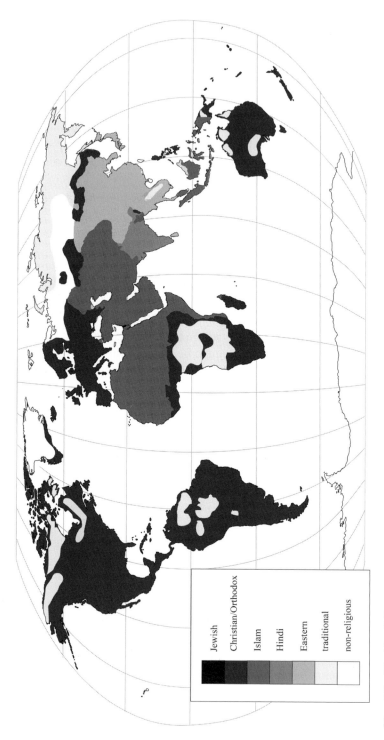

Map 6.1 World religions

Jewish
Christian/Orthodox
Islam
Hindi
Eastern
traditional
non-religious

Plate 6.2 *Tongan cultural landscape – a graveyard*
Christian graves in Tonga are often adorned with tapestries with pre-Christian motifs.

work in communications infrastructure, including the development of the telegraph and undersea cable system, gave the British a distinct advantage. This allowed the transmission of cultural information intended to create a sense of shared identity. Education policy was also instrumental in holding the empire together through the production of standard textbooks, curricula and the Oxbridge education of colonial elites.

In the above way, European imperialism laid the cultural infrastructure for more recent cultural globalization. Throughout the nineteenth and twentieth centuries, the emphasis shifted to two areas which remain salient today – the rise of *national culture*, and *secular ideologies* of liberalism, socialism and science. As described in Chapter 5, nation-states came to prominence in the nineteenth century. An essential part of this rise was the construction of nationalism and national cultures, which were only sometimes based on pre-existing cultural commonalities. The concept of the nation-state was a Western one, and this was diffused across the world largely through the cultural infrastructures of colonialism. Crucial to the establishment of the nation-state system was the forging of identities that cut across class. Nation-building involved a variety of tactics including the pursuit of a common language, an education system, national communication networks, national symbols (e.g. flags) and a standing military. Paradoxically, empires sowed their own seeds of destruction through the diffusion of national self-determination.

Alongside the development of national cultures, various secular ideologies or discourses were also evolved and dispersed. *Socialism*, based on early work by Marx subsequent to the industrial revolution in Britain, was one example of this. *Liberalism*, which remains of central importance in the current era, also arose during this period. The role of *science* as a rationalizing element in society was also propagated at this time. It was through the nation-state system that the culture of *capitalism* was diffused across the world, and this remains the dominant force in global society and culture today. On this basis, it could be argued that cultural globalization reached a climax during the colonial wave. It is certainly the case that the global networks of contemporary culture were laid over many centuries. World religions, capitalism and nationalism are arguably more important today than they have ever been, with the latter making a recent comeback in the form of the reassertion of local identity in the face of globalizing processes.

Contemporary cultural globalization – cultural imperialism?

Current cultural globalization departs from that of the past in terms of six trends and processes (based on Held *et al.*, 1999):

- new global cultural infrastructures of technology operate at unprecedented scale and efficiency;
- the consequent rise in the velocity of cultural exchanges across borders and between societies;
- the unparalleled rise of Western culture as the core marker of global cultural interaction;
- the rise of TNCs in the culture industries which create and perpetuate the infrastructure necessary for heightened diffusion;
- the rise of business culture as the main driver of cultural exchange, linked to the preceding point;
- the shift in the 'geography' of cultural interaction compared to the pre-Second World War world.

In the case of the latter, Held *et al.* argue that, although flows from the West to the rest of the world are far greater in an absolute sense, we are seeing a partial reversal of this through increasingly complex patterns of migration (see Box 6.2), the evolution of mass tourism, and the rise of new sectors such as the world music industry (see case study on global

---

Box 6.2

The 'age of migration' and cultural change

International migration is playing a more central role in demographic and cultural change than ever. Castles and Miller (1993) referred to the late 1900s as the 'Age of Migration', and this movement of people has increased since that time significantly (Munck, 2008; Ritzer, 2011). There are approximately 200 million international migrants per annum at the current time. The United Nations has predicted that this might rise to 400 million by 2050. We can observe four key trends in international flows:

1 Migration is globalizing in the sense that the range of origin and destination countries has greatly diversified over the recent past.
2 The volume of migratory flows is increasing in all regions.
3 The nature of migrants, and their reasons for migration, is becoming increasingly differentiated. Today, no one type of movement dominates with flows characterized by combinations of economic, environmental, political and social migrants.
4 Migration is being feminized as women move increasingly for economic reasons and dominate refugee flows.

Migration, of course, has always been important in demographic evolution, and it has played a major role in historical cultural globalization. Flows of note include the Atlantic slave trade and the mass migrations from Europe to the 'New World' between 1850 and 1950. The latter example saw 55 million people relocate; 30 million people migrated into the USA between 1860 and 1920 alone. However, Champion (2001) argues that migration has become more important in contemporary demographic transformations because of the decline in natural increase across the world, but especially in the West. Thus migration accounts for an increasing proportion of population growth, and this is especially the case in richer countries. Indeed, in Europe over the past 30 years, migration has offset natural decrease and usually has been the *only* source of population growth. Overall, migration accounts for three-fifths of population growth in Western countries as whole. In the Third World, where natural increase remains relatively high, migratory movements are less important in demographic change.

At the global scale, international migration over the past 50 years has exhibited some

Box 6.2
continued

clear spatial patterns. There has been a general shift from the Third World to the West, with Europe and North America receiving the bulk of migrants. This has led to the establishment of significant diaspora populations in the West and also increased cultural hybridity. In Europe this has been fuelled particularly by the end of the Cold War, the Balkan crises of the 1990s and the widening membership of the EU all of which led to large-scale movements from the east to the west of the Continent. In North America, the two major source regions of recent migration have been Central America (especially Mexico) and East and Southeast Asia (especially the Philippines, Vietnam and China). The USA has also received a steady flow of Cuban migrants since the revolution of 1959. Mexican immigration dominates, however, and the 1990s saw the legal shift of over two million people across the border into the USA, most of these in search of better economic opportunity. As a consequence of Hispanic immigration, the culture of the USA is being transformed; for example, linguistically there has been a major shift to Spanish over the past 20 years.

Three groups of migrants dominate the flows that are currently taking place. The first is 'professional' migrants who move for career prospects. The rise of TNCs has seen an increase in postings abroad, and the shortage of skilled labour in some Western economies has led to a 'brain drain' from some poorer countries. The second major group is unskilled labour, which often moves without specific offers of work. This has led primarily to increased flows from the Third World to the West as employers search for workers who are willing to labour for lower wages. A third major group is refugees and asylum seekers which was larger in magnitude in the recent past than at any other time in history, fleeing conflict and persecution in countries including Syria, Iraq, Afghanistan and Somalia. By the early 2010s it signified a global crisis arriving at over 15 million persons, according to the United Nations. This does not count the large numbers of people 'displaced' through conflict inside their own countries which was estimated in 2013 to be at approximately 30 million. Although the refugee 'crisis' is highly visible in the public sphere in Western countries – it has dominated recent election politics in Australia, Germany and the UK, for example – in absolute terms, the number of refugees entering Third World countries is far higher. At the time of writing there was a crisis in the Middle East as over 2 million refugees from Syria poured into neighbouring countries. Because immigrants into Western countries have a significant impact on population composition, migration has become a highly politicized issue. This has led to the revival in some European states in the 2000s, for example, of nationalistic political groups who bemoan the economic and cultural impacts of immigration. In the USA and Australia such matters also continue to cause significant controversy. Migrant groups are often painted as undermining the position of 'indigenous' non-skilled labour, and, in the case of the latter, the arrival of people with different traditions, religions and languages is seen by the right wing as undermining the integrity of the national identity and the state (see the academic journal *Migration Studies* for the latest studies in this area).

music below). Some have argued that globalization inexorably leads to cultural homogenization, however. James Petras (1993), together with Ritzer (2008), for example, claims that a global culture is emerging which is directly linked to the USA. This has both political and economic goals and is part of the geopolitics of capitalism. We consider the arguments for homogeneous global culture below.

Cultural imperialism

Why is it that some aboriginal people in Australia send each other Christmas cards featuring snow? Why do most maps of the world show Britain in the centre and at the top? Why is the Middle East called so? The answers to these questions are found in the history of imperialism and spread of Westernization. In this context, Pieterse (1995, p. 45) argues:

> The world is becoming more uniform and standardised, through a technological, commercial and cultural synchronisation emanating from the West, and globalization is tied up with modernity.

Proponents of the cultural imperialism thesis date its inception to the industrial colonialism phase. As discussed in Chapter 3, it was during this phase that colonialism reached its zenith, peaking just prior to the First World War when the British Empire reached its maximum territorial extent. Christian missions played a significant role in preparing the ground for the diffusion of capitalism and imperialism – acting as both a vanguard of and a rationale for expansion as it had done centuries ago in the case of Catholic missions in Latin America. Through colonialism, Western culture was diffused, driven by and driving the penetration of capitalism. The end of formal colonialism in the second half of the twentieth century did not spell the end of cultural imperialism. According to this argument, cultural imperialism is no longer played out explicitly by nation-states, and it has become an economic process as well as a political one. It is constructed by TNCs that represent the interests of the elite, especially those of the USA. Powerful Western governments facilitate cultural imperialism, and it is bolstered further by various transnational institutions (IMF, World Bank and WTO especially) that are designed to serve capitalist needs. The powerful companies, institutions and countries that have the greatest say in the governance of the contemporary system have thus built directly on the spoils of colonialism. As the most powerful country with the largest corporations, the USA

Plate 6.3 *Cultural imperialism – Tahitian Santa Claus*

plays the major role in the diffusion of Westernization and thus has great influence over its nature.

In 1993, Ritzer wrote an influential book concerning the McDonald's franchise. He defined *McDonaldization* as the process through which the organizational, productive and representational principles of McDonald's were redefining globalization. At the core of the new model is the efficiency and standardization of the McDonald's approach; 'Drive-thru' windows and assembly line production impose a new rationality on the process of production and consumption. A key point is the predictability of the McDonald's menu, virtually standardized across the world, imposing uniformity on the customer and the workforce alike. Authors such as John F. Love (1995) celebrate McDonald's, and see it as playing a positive role in 'exporting Americana'. It is notable in this supposed age of flexibility and niche marketing that standardized production epitomized by McDonald's has become so prominent. This thesis has been criticized, however, by those who argue that McDonald's products often reflect the locality in which they are marketed (see Plate 6.4). What is clear is that 'culture' helps sell products as evident in the global reach and perpetuation of Santa Claus, Halloween and Father's Day for example.

Plate 6.4   *McChurrasco – a Chileanized McDonald's product*

Those who subscribe to the cultural imperialism thesis worry that the world's cultural diversity will be threatened by the diffusion of Westernization/Americanization (Waters, 2001). There are around 300 million people who identify as members of an indigenous culture, and there are many examples of threatened survival and/or erosion of these in regions such as the Amazon Basin, North America and Australia. Aside from the immorality of culture death, with the decline of such entities much 'traditional' wisdom is lost, rather as reduced biodiversity leaves a depleted ecosystem. Given the rise of universal human rights, however, these transitions have taken on a political form.

Box 6.3 presents some arguments that support the thesis of cultural imperialism. Evidence in five areas is discussed: language, tourism,

---

Box 6.3

Global cultural imperialism?

Language – There are approximately 6,000 languages in the world and this may drop to 3,000 by 2100 (Crystal, 2010). Approximately 60 per cent of these languages have less than 10,000 speakers; a quarter have less than 1,000. English is becoming *the* world language.

Although Mandarin is more widely spoken as a first language and second language, if second-language speakers are taken into account, the total number of English speakers is close to one billion. This number increases significantly if those who have a working knowledge of the language are included. English is the medium of communication in many important fields including air travel, finance and the internet. Two-thirds of all scientists write in English; 80 per cent of the information stored in electronic retrieval systems is in English; 120 countries receive radio programmes in English; and at any given time over 200 million students are studying English as an additional language. It is an official language in much of Africa, the Pacific, and South and Southeast Asia. This dominance may be explained by the cultural imperialism of the British and the current hegemonic role of the USA, and new technologies are perpetuating this trend (Crystal, 2010).

Tourism – Tourism is now the world's largest industry, with revenues of over US$1 trillion in 2011 (1,000 billion). Mass tourism has often been compared to Fordist methods of production. The standard package holiday seeking sea and sun, often involving accommodation in a global hotel chain such as the Holiday Inn or the Sheraton, has proliferated as a desirable form of relaxation. The journey of many British to the Costa Del Sol, Spain, where they practise cultural traits such as drinking beer

Box 6.3
continued

and eating fish and chips while lying on crowded beaches surrounded by tall buildings, is a stereotype which captures the essence of this type of standardization.

Global brands – The growth in the influence of TNCs is the rise of global consumer culture built around world brands. McDonald's, for example, operates in close to 35,000 outlets in over 120 countries. Coca-Cola is sold in nearly every country. It is a transcultural item yet it is very much linked with US culture. This type of standardized consumption is mirrored across many sectors and is perpetuated by the capitalist system it supports and is borne out of.

Media – National media systems are being superseded by global media complexes. With global deregulation during the neoliberal era, and WTO regulations since 1997, the media have become more expansionist and consumerist, rather than aiming coverage at citizens of a particular nation-state. The industry has become increasingly concentrated, privatized, vertically and horizontally integrated, and predatory. Around 20 to 30 large TNCs dominate the global entertainment and media industry, all of which are from the West and most of which are from the USA; and this concentration has increased of late. These include giants such as Time-Warner, Disney, New Corporation/Fox, NBC/ Universal and Sony. In terms of the news media there has been a standardization of presentation, with increased focus on 'human interest' stories, and a generally unquestioned bias towards the West and its allies – powerfully illustrated during the CNN, BBC and Fox news coverage of the war in Iraq (2003–2011), the Syrian Civil War (2011–), and the ongoing conflict in Palestine. Herman and Chomksy (1988) predicted a powerful intentionality at work in this trend which they termed 'manufacturing consent' almost 30 years ago.

Democracy – As outlined in Chapter 5, the spread of liberal democracy has been profound and is now practised in the vast majority of nation-states across the planet. Forms of democracy vary but over 120 nation-states practise what might be reasonably termed as such. Underlying this diffusion is the Western Enlightenment belief that it is the most desirable form of governance, and this is evidenced by its defeat of socialism. Parliamentary systems have spread to all corners of the planet, diffused through Western empires and successor states and more recently through the conditionality of loans from the IMF to poorer countries. This is eclipsing the diversity of governance methods that other cultures exhibit, which in some cases have deep historical roots.

consumerism, media and democracy. The global music industry is considered separately in a subsequent case study.

A variation on the cultural imperialism argument sees the creation of a universalized hybrid culture. This type of culture is homogeneous but not entirely Western in nature. Thus we are still left with the impact of

Box 6.4

Alcohol and globalization

Although banned in some parts of the world and uneven in terms of rates of consumption, alcohol in its different forms may be regarded as a global product. Alcoholic beverages though usually were developed very much in local settings, using processes such as fermentation or distillation to produce alcohol from surplus agricultural products, such as grain or grapes. Beer remains a local product because of its relative perishability but wine and spirits (such as whisky and vodka) keep well and can be transported over long distances.

The alcoholic beverage industry is now highly globalized and characterized by well-recognized products and brands. Large global beverage corporations, such as Constellation Brands, Pernod Ricard, Diageo, Heineken and SAB Millers, dominate the world's supply, often by holding large and diverse portfolios of brands spanning beer, wine and spirits. They are responsible for shipping large quantities of products to markets around the globe or, in the case of beer, establishing brewing operations in many countries that manufacture a local version of global brands such as Carlsberg, Fosters, Guinness, Heineken or Stella Artois. In this respect, beer is similar to Coke or Pepsi – consumers seek a global brand that will be basically the same wherever it is purchased (or made) throughout the world and it will be promoted by global marketing campaigns and receive global recognition.

Alcoholic beverages are widely sought after and span a huge range of price points. At one end, it might be cheap, mass-produced drinks such as beer, bulk wine or 'ready-to-drink' spirits. At the other end, we might find very expensive wine from Champagne, Bordeaux or Burgundy, aged single malt whisky or fine vintage cognac. The consumption of alcohol is usually associated with social interaction and a lessening of social inhibitions. There are cultures of alcohol consumption that revolve around maximizing intake and 'getting pissed' – these seek low price alcoholic products. Yet there are other cultures of consumption which associate moderate consumption with everyday dining or sometimes combining fine dining with expensive and rare wines. In addition, in many parts of the world, consumption of some beverages, particularly expensive wines, is seen as a mark of prestige – a conspicuous show of consumption to indicate to others that the consumer supposedly has taste, sophistication and wealth (Overton and Murray, 2011). These consumers will want more differentiated products, ones characterized by unique attributes relating to place of origin (a Scotch whisky or a Bordeaux wine), age and price/scarcity. In addition, knowledge of the geography of the product (knowing the difference between an Australian Shiraz and a French Hermitage) can further reinforce the status supposedly derived from the conspicuous consumption of the product.

Box 6.4
continued

Plate 6.5 *A liquor shop in Honiara, Solomon Islands*

Box 6.4
continued

The geography of the bottle thus reinforces its value to the producer (and consumer). As a result of these differing demands, beverage companies provide a wide range of products: some may be low-priced, high-volume cheap drinks with no defined geographical origin; others may be expensive and rare and closely linked to a certain place (whether a small distillery on the island of Islay in Scotland or a certain vineyard in Burgundy in France).

Thus these TNCs embrace both placeless mass products and highly localized luxury ones.

Yet whilst we can see in these ways that the geography of a beverage's origin has been globalized through marketing strategies and the way large corporations acquire brands with distinct local identities, there is another side to the globalization of alcohol. Many beer drinkers in Britain in the 1970s and 1980s grew tired of the mass-produced beer they were being offered by an increasingly small number of beer-making companies there. The Campaign for Real Ale (CAMRA) was formed in 1971 and it mobilized consumer support for traditional beer styles and small local breweries in opposition to the large companies. The success of this campaign led to the revival of many small breweries in the UK and it greatly diversified the products on offer. We have seen similar movements in other parts of the world,

including USA, Canada and Australia and the rapid growth of craft beer making. Furthermore, the movement to revive traditional beer making and small local breweries has been paralleled by consumer demand for 'artisanal' wines: low-volume, locally produced and distinctive wines that vary in style and quality from place to place and year to year. It is a markedly different model of production compared to the 'Fordist' manufacturing of beverages that are high volume, low cost and consistent in style over time and space.

Thus we see with alcoholic beverages some complex and sometimes contradictory processes of globalization. There are classic trends towards homogeneity, mass production and global reach; yet there are also strong local resistances to this in the form of revivalist movements that promote small-scale place-based production of distinctive wines and beers often only available within small spatial bounds. Furthermore, forms of hybridization are also apparent as global corporations seek to tap into the demand for local, distinctive and traditional products by either mimicking local craft-type beers or adding prestigious place-based wines (a Champagne house or a Bordeaux chateaux) to their wide portfolios of beverage brands (see Overton et al., 2012).

Westernization as something that is very powerful – but it is not unidirectional. The influence of non-Western cultures on dominant Western culture has been marked. For example, the British drink tea because of the British imperial connection with India, and a number of words in the English language, such as *bungalow*, *shampoo*, *thug* and *pyjamas*, are borrowed from languages of the subcontinent. These traits are acculturated into the dominant culture and form part of the subsequent round of Westernization as they are disseminated through neo-imperial networks. The influence of Black American and Hispanic dialects on rap, the most popular music globally at present, and the fact that football (soccer), which diffused through the British Empire, is thought to have been invented in China are further examples of universalized hybrids.

Criticisms of the homogeneous global culture arguments

So what are we to make of the two main global culture theses? Both the idea of a cultural imperialism and a universalized hybrid culture have been criticized. This is particularly the case in the context of the former. It has been argued that the concept of cultural imperialism ascribes globalization with too much determining power and is too broad a generalization of the processes that are occuring on the ground (Tomlinson, 1999). The power of locality, and of local culture, is thus overlooked. The result is a metanarrative or grand theory – the kind that geographers have been rallying against since the mid-1980s at least (Johnston and Sidaway, 2014). In particular, Tomlinson has argued that grand theories of homogenization have misrepresented, even patronized, the supposed recipients of dominant cultural signals and symbols emanating from the West (Tomlinson, 1991). It is naively assumed that masses of the population of the culturally marginalized will passively accept these transformations. This is Eurocentric and invokes a one-way flow of cultural symbols and meanings without recognizing the countervailing possibilities of localized hybridization or localization. Furthermore, the rise of 'cosmopolitan consumption' in the West casts further doubt on the homogenization thesis (see below). In sum, McEwan and Daya (2012, p. 281) argue that:

> in reality both of these processes are flawed explanations for what is happening today. If global culture exists, it is far from a product of unidirectional 'Westernisation'. However, alternative ideas about cultures mixing to produce a universal global culture are also problematic. Cultures are mixing, but this mix does not necessarily mean we are all becoming the same.

Samir Amin (1997) argues that globalization has actually led to a reassertion of ethnic and cultural difference (see also Barnett, 2009 on

Box 6.5

A critical view of the 'global gay identity' thesis

Bell and Binnie (2000) are highly critical of what they see as 'the export of western definitions of sexual practices, identities and cultures around the world' (p. 116). This has led to the proposition, in some circles, that there exists a 'global gay identity', with critics such as Herdt (1997) associated particularly with this idea. In the book *Same Sex, Different Cultures* the latter argues for a model of gay behaviour which is diffusing from the West to marginalized regions and localities. Given the recent wave of same-sex marriage bills across the world in places as diverse as Argentina and New Zealand and the reactions to this, such as harsher anti-homosexual laws in Russia and the introduction of the death penalty in Uganda for homosexuality, one might be forgiven for thinking that such an expanding nucleus of cultural change exists. An integral part of this evolution is the development of what Sinfield calls a 'queer diaspora' which cuts across traditional boundaries of state and citizenship. Watney (1995, quoted in Bell and Binnie, 2000, p. 117) argues that what defines this:

> is the sense of relief and safety which a gay man or lesbian finds in a gay bar or a dyke bar in a strange city in a foreign country. Even if one cannot speak the local language, we feel a sense of identification.

Bell and Binnie's critique of the globalization of such identities is multiple. They argue that the global gay culture thesis presupposes a common form of homophobia across the world – rather as the globalized feminist movement assumed a single form of patriarchy. In reality, homophobia is constructed in vastly different ways in different local and national contexts. Second, the construction of what Cant calls a 'McPink' economy does not take into account those who are marginalized from the global gay capitalist circuit. As such, 'there is a danger in exporting a mode of "being gay" from late-modern westernised liberal democracies that does in fact "McDonaldise sexual minorities"' (Bell and Binnie, 2000, pp. 116–117). In criticizing this point, they echo Cant (1996, in Bell and Binnie, 2000, p. 117), who argues that globally speaking:

> The romantic myth of homosexual identity cutting across class, race, and so on doesn't work in practice any more than it does in the West. The experience of sexuality in everyday life is shaped by such variables as the gap between city and country; ethnic and religious differences; and hierarchies of wealth, education and age.

Overall, the globalized model privileges Western sexual identities and emphasizes 'sameness over difference' between and among lesbians and gay men. The overemphasis on the role of the affluent, middle-class, jet-set gay subject masks 'a

Box 6.5
continued

multitude of local differences' (p. 119) whose reality is far more embedded in local practices and regulatory policies at the level of the nation-state. In general, Bell and Binnie argue that a postcolonial analysis is required in order to more adequately capture the diversity of experience, claiming that (p. 122):

The complexities of national, transnational and global identifications and disidentifications thus reproduce their own disjunctive flows around the world, mapping out the complex implications that globalization has upon sexual citizenship.

this point), and that this has come about as a form of resistance to homogenizing tendencies. Another term for this is 'reterritorialization', as used in post-structural political geographies. Paradoxically, globalization makes space for such assertions – which mirrors change in the political sphere. In the field of language, for example, there has been a revival in smaller and hitherto oppressed European languages (such as Welsh, Irish, Scots, Catalan and Basque) over recent years (see Plate 6.6). There has been a similar revival in New Zealand where Māori is an official language, and similar policies are coming to the fore in places across Latin America.

(a)

(b)

Plate 6.6 *The reassertion of language – road signs in (a) Ireland and (b) Wales*

Source: Photo by Jo Heitger and authors

Box 6.6

The globalizing internet: search engines and social media

It is easy to argue that the internet has become a powerful force for globalization. Search engines, such as Google, allow users to access the huge volume of material that is available on the internet. In a sense they are portals to the global bank of knowledge. This virtual democratization of knowledge is important for it breaks down many former barriers to accessing knowledge hitherto held in libraries, archives and individual minds. There are, of course, still important barriers to access: without access to computers or internet access, or without a knowledge of English (or the other languages used), many millions of people are still excluded.

Similarly, social media platforms (Facebook and Twitter being the two most popular currently) allow people to communicate across the globe and both maintain and build interpersonal contacts and relationships. Thus, people can easily and cheaply keep links with friends and family living in other countries but also build wider networks of acquaintances and more closely follow the activities and views of others, including celebrities and politicians who are now realizing the value of these media. Social networking allows people to bypass traditional media (television, radio, newspapers and magazines) and be more selective and direct in the way they obtain information about others. Global social networks are being developed as never before.

To a large extent, we can see social media and internet search engines as forces for homogenization. We have noted how the internet is dominated by the English language, and the major companies that dominate internet search engines and social media (Google, Yahoo, Facebook, Twitter, etc.) are American-owned. Early versions of search engines were based on the English language only, and the way they accessed information was dependent on the information available (largely Western in origin early on). Similarly, social media sites often tend to encourage top-down modes of information flow (global celebrities will use their Twitter accounts to keep fans in touch worldwide, and companies use Facebook sites to help sell products.

However, as the internet has evolved and become both much larger and more sophisticated, the potential of it as a force for promoting diversity and bottom-up communication has been partly realized. Search engines now allow for many different languages to be used (Google list 87 languages that are supported by its input tools) and these allow users to access a rapidly increasing volume of material in these languages. Furthermore, a translation feature in Google (currently supporting 71 languages) allows users to gain access to, and use, data in languages they might not understand at first. Such features can be seen as ways to support and strengthen local and sometimes endangered languages

Box 6.6
continued

and promote some multilingualism – though it notable that a digital divide still exists: whilst Google does well to support minority languages in Europe (such as Welsh or Catalan), not one of the 800 indigenous languages of Papua New Guinea features or is likely to! Another interesting feature of search engines is the way they provide an unseen personalization of searches. Recording a user's previous searches and locality helps Google tailor the search results so that typing in a search for 'best hotels' for example will lead to a list in cities close to the user (and provide links to hotel websites on the right of the screen in a subtle form of advertising). Thus whilst users may think that a search engine will give them direct and instant access to a global body of knowledge, in reality that access is much more narrow, localized and personalized than they may realize.

Similarly social media in practice have become heavily localized. The success of Facebook, to a large extent, is due to its ability to allow users to start from real existing or past social relationships and help reinforce these. Twitter, too, although it may involve 'celebrity chasing', does base itself on providing a new means for people in existing relationships to communicate. Furthermore, although the companies which own the sites may be Western, the means

of communication, like a telephone, is multilingual. Thus, during the Arab Spring of 2010–2011, people in North Africa and the Middle East were able to communicate instantly and without interference about political events in particular localities and both mobilize support and disseminate news to the outside world. The supposedly homogenizing, top-down instrument had become a means for promoting local-level change, spreading it outwards and successfully challenging well-entrenched power structures.

How the internet evolves in future will likely be complex and contested. Commercial and political interests will seek to both control and manipulate it in order to pursue their own agendas. These underlying forces will undoubtedly wield strong influence – the floating of Facebook on the stock exchange in 2012 will lead to demands for financial return to investors through more advertising whilst repressive regimes may seek to block internet access and social media. Yet an ever-larger and more complex internet, and no doubt future social media and search engine software, may well provide even better tools for the local to access the global and build upon, preserve and strengthen not only local relationships, knowledge, language and resources but also human and political rights.

In what has been redefined as the 'plurination' of Bolivia, as well as in Ecuador and Venezuela, dozens of indigenous languages have been given official status. Potentially, communications technologies can play an important role, informing us of other cultures and providing the impetus and infrastructure necessary for the protection of cultural diversity.

Localized hybridity

The idea of localized hybridity allows us to move beyond what McEwan terms 'mosaic' (multiculturalism) and 'melting pot' (assimilation) concepts of cultural interaction. Multiculturalism is a term which is often considered 'progressive' but which can ghettoize and segregate cultures. For example, in a migratory setting, arrival cultures may be accepted and tolerated but aspects of identity (be they racial, social and so on) are essentialized and used as a means of differentiation. Localized hybridity refers to a situation where cultures mix to produce unique outcomes. There are many instances of this (see Box 6.5 and global music case study). For example, 'Australian cuisine' is a mixture of traditional British cooking and exclusively Australian ingredients (such as 'Moreton Bay bugs' and barramundi), together with a strong Asian influence and the fare of the many waves of migrants that have arrived in the country including Greeks, Italians and Spanish. The essential point is that these forms do not just exist in a multi-culinary form – they are often combined in new, and tasty, ways. Indeed, Australia is becoming known for what is sometimes referred to light-heartedly as *con*fusion cooking. A second example also relates to cuisine. Curry is the most popular food in Britain and central to the identity of millions of Brits. British curry, and derivatives such as balti and butter chicken, are actually constructed foods based on Indian cuisine, which Indians from India would probably not recognize if they dined at a restaurant in Small Heath, Birmingham – the self-proclaimed 'balti capital of the world'. Another example of localized hybridity comes from the field of language. There are many words in Chilean Spanish that differ from Castillian Spanish because of the influence of Mapudungun, the language of the Mapuche (Southern Chilean indigenous). Indeed, as is argued in Box 6.5, there are many local varieties of 'global' languages that have developed based on contingent factors.

In the context of the above, the concept of *glocalization* is particularly important – whereby neither the global nor the local is given ascendency in terms of determining power and outcomes turn on mutually conditioning global local relations (this is similar to the idea of structuration as suggested by Giddens and posited for human geography by Johnston) (Swyngedouw, 2006). In cultural geography, this means that processes at different 'scales' interact to produce hybridized outcomes (Sparke, 2009b). As such, local agents are no longer seen as victims of global cultural shifts but as interpreters of such processes. When we accept this proposition, we are able to more accurately analyse the influence of non-Western cultures on other Western cultures, the influence of Western cultures on Western cultures, and the influence of non-Western cultures on other non-Western cultures. In other words, this perspective gives us more subtle tools for interpreting complex cultural change. Lash and Urry (1994) take this argument a stage further, arguing that diversity and uniqueness become part of cultural capital and may be used in some cases in order to promote livelihoods in the new postmodern economy (see section on cultural commodification below). An extension of this idea leads to the concept of renewed localism (Levitt, 2001; Rycroft, 2009). This process might entail the rise of ethnic fundamentalism – where a group attempts to recreate and rediscover its fundamental and radical roots, for example. Although the evidence points towards increased hybridity, there can be little doubt that there are some very real examples of renewed localism in all spheres (for example, in the form of delinked local economic trading schemes (LETS) or devolution movements across Europe).

In some ways cultures have always been hybridized, since no culture is static and cultural interaction is constant. But there can be little doubt that the speed of cultural 'traffic' in the contemporary world is such that interactions – or processes of transculturation, as McEwan and Daya (2012) call them – are speeding up. To many commentators, localized hybridity is liberating and exciting; it provides a means of resisting and subverting dominant and hegemonic cultures and disrupting homogenizing processes. It can create a richer and more 'progressive' sense of place (Massey, 1991).

Diaspora

The fixed relationship between place and culture is disrupted by the concept of diaspora. A diaspora refers to the scattering of a culture from its 'homeland' and was originally used to refer to the Jewish peoples. With the high contemporary incidence of migratory movements (see Box 6.2) such 'communities' have become increasingly common. Cyberspace provides particularly fertile space for diasporic cultural interactions (see Box 6.6). The Indian diaspora, for example, stretches across all continents with significant 'Indian' populations found in places as diverse as South Africa, Brazil, the UK and Fiji. The diasporas of the Pacific Island nations are, relative to the size of 'hearth' populations, among the largest in the world (see Box 6.7 on the Niuean diaspora). McEwan and Daya (2012, p. 271) argue that:

> Diasporic identities are important because they are at once local and global, and are based on transnational identifications encompassing both 'imagined' and 'encountered' communities.

Within 'diasporic space', processes of hybridization are paramount as migrants interact with 'host' cultures to create new forms (Cohen, 2008). Britishness, for example, can take on many forms – African-Caribbean, Indian, Jewish and so forth – and these diasporic identities recondition the original concept of 'Britishness'. Diasporas stretch across traditional boundaries, forging new connections between places, creating cultural forms that challenge accepted hierarchies and structures. This challenges many accepted geographical notions – such as 'home' and 'belonging' (Dwyer, 2013; Gamlen, 2011, 2012). As such, diaspora are profoundly transforming the spatiality of nation-state cultures.

Overall, it is easy to run away with the idea that globalization is destroying cultural diversity. It is certainly altering the cultural map of the world profoundly. However, as illustrated above, there are signs that the process is not always characterized by one-way cultural traffic. To

---

Box 6.7

The Niuean diaspora

Niue is a country located in the Western South Pacific Ocean, occupying one island of approximately 250 square kilometres with a population of approximately 1,500. It is one of the smallest independent territories on Earth and is isolated at over 200 km from its nearest neighbour, Tonga, with only one wharf and a small airport (see Map 8.5). Populated by Polynesians beginning some 1,000 years ago, it was colonized first by Great Britain in 1901 and handed over to New Zealand jurisdiction from 1904 onwards. It was not a colony settled by Europeans and has retained a strong indigenous identity. In 1974 it gained independence as a 'territory in free association' with New Zealand, with the latter responsible for its defence and foreign policy. Niue receives substantial amounts of aid, sometimes totalling 50 per cent of GNP, the rest being made up by a mixture of remittances, tourism receipts and agro-exports (Overton and Murray, 2013; Murray and Terry, 2004).

Since independence, Niue has been losing its population rapidly, falling from 5,000 in 1966 to the current level (see Plate 6.7). Today, there are far more Niueans living in Pacific Rim countries than on the island itself. By the 2000s there were over 20,000 living in New Zealand and thousands more in Australia and elsewhere. The Niuean diaspora has close ties with 'home' – extensive remittances in cash and kind are sent by first-, second- and third-generation migrants. Cultural practices are relatively well maintained in Niuean population centres, such as Auckland,

Box 6.7
continued

Plate 6.7 *Outmigration from Niue*
Air travel has made outmigration from Niue easier and cheaper. A result of this is
abandoned housing, now a common sight in the country.

Box 6.7
continued

where earpiercing and other rituals regularly take place. Since independence, Niueans have had the option of dual New Zealand/Niue citizenship which has fuelled outmigration. Friction can be experienced when expatriate Niueans return home, as some of their cultural traits have been transformed. The adoption by young Polynesians of New Zealand-mediated US hip-hop culture is widespread, for example, and this is alien to those who have stayed on the island. Cultural values are thus changing rapidly, and this can lead to conflict. One salient example is the issue of land. All male Niueans are entitled to land as part of their *magafoa* (family group). As most Niueans are away from the island, this has left much land idle. Those who still remain have become increasingly frustrated with the desires of the diaspora population to retain their titles. The latter group cite the right to land as part of the *Moie Faka Niue*, or Niuean Way, however (McIntyre and Soulsby, 2004).

From isolation comes the need to develop links, however; and Niue is the only country in the world where the population is provided with free access to the internet through wi-fi services in villages. As a consequence, we are seeing the emergence of a 'cyberdiaspora' as Niueans communicate over the internet through e-mail, a dedicated chat room, and numerous Niue-oriented sites (Gibson, 2004). The realities of isolation, and vulnerability to climate, were highlighted in January 2004, however, when Hurricane Heta destroyed much of the island's infrastructure and agricultural crops. It is predicted that this, together with pressures on the country's economy, will lead to further depopulation. Economic problems were heightened in the 1990s by the insistence of development 'trustees', such as the New Zealand aid programme, that Niue undertake neoliberal-like reform. The resultant restructuring led to a collapse of the state employment sector in the mid-1990s (Murray, 2002b; Overton and Murray, 2013). However, in the past decade, aid funding has returned mainly because there is a realization that the country and its welfare would simply not be sustainable without significant amounts of aid and the maintenance of basic government services. Currently, aid levels are very high by international standards, equivalent to over $US 10,000 per resident per year. Is it possible that we have witnessed, through a mixture of voluntary and involuntary outmigration, a relocation of the cultural hearth of this society? Or is the culture so changed as to be unrecognizable? These are the kinds of political questions that cultural geographers often grapple with in the context of marginalized peoples.

most geographers, the conditioning or deterministic impacts of cultural globalization have been greatly overemphasized. Five examples, corresponding to those explored in the context of the global cultural homogenization thesis in Box 6.3, are presented in Box 6.8.

Box 6.8

Global cultural heterogeneity

Language – Although English is becoming the dominant world language, enormous variety remains. Important world languages include Mandarin, Spanish, Hindi, Arabic, Portuguese, Bengali and Russian – all of which have over 100 million first-language speakers. There is no reason to suppose that these numbers will decline. It is likely, however, that an increasing number of people will learn English as a second language, leading to increased bilingualism. In parts of the USA, such as southern Florida, given the high Latino population, English and Spanish are equally visible. *Dora the Explorer*, a bilingual US children's television character with Latin American cultural roots, embodies the shifting linguistic landscape of the USA. Within English itself, there are many existing and evolving dialects and communalects that contain elements – including grammar, intonation and vocabulary – of both the indigenous language and received English. Jamaican English, Singapore English (Singlish) and Fijian English are unintelligible to some English speakers from England, for example. In history, the rise of pidgin and Swahili as intermediary languages that evolve to make trade possible point to the differentiating impacts of the diffusion of a dominant language culture (see also Box 6.6 on language and the internet).

Tourism – For a growing number of people, 'otherness' is what makes a destination attractive. It is the very process of the globalization of the media which has made people more aware of otherness and increased the demand for it, leading to – among some – a conscious rejection of mass tourism and the search for the 'authentic' and different (Palomino-Schalscha, 2013). Of course, the concept of authenticity in this regard is open for debate. It has been argued that some places are incorporated into global capitalism in ways which reinforce rather than erode their distinctiveness. The rise of ecotourism, and even poverty tourism, seems to point to a shift away from the 'Fordist' tourism outlined in Box 6.3 (Scheyvens, 2002, 2011).

Global brands – Despite the undisputable rise of global brands, products are often adapted to local conditions. For example, mutton is used in Indian Big Macs and vegetable oil is used for frying. MTV, one of the archetypal examples of cultural pop globalization, is adapted for audiences across the world, and MTV Asia and MTV Latin America play a mixture of Western and regional videos. The Barbie doll comes in 30 different national varieties.

Media – Although the global media is increasingly concentrated and standardized both within and beyond major corporations, product differentiation is actively pursued. Local newspapers, radio stations and television stations remain important information providers and offer their own particular slant on the news, although many

Box 6.8
continued

of them are ultimately owned by large media conglomerates. During the Afghanistan and Iraq wars the Qatar-based Arabic news network *Al Jazeera* (still owned by the Government of Qatar) came to the fore as an alternative voice, which was often highly critical of US and UK actions in the Middle East. Small entertainment operations exist, such as independent film-makers, although it is increasingly difficult for them to find distribution channels.

Democracy – Just as there are alternative capitalisms, there are alternative democratic structures. Democracy in many Pacific Island nations is grafted on top of traditional and other forms of governance. In New Zealand, for example, seats are reserved for ethnic Maori in Parliament. Outside of the democratic hegemon, there exist many other formations, which account for a significant proportion of the global population. The transition in China, for example, is highly nonconformist since it attempts to combine liberalized economics with authoritarian politics. Socialism persists in a range of other countries such as North Korea, Cuba and Vietnam – albeit in very different ways with very different outcomes.

Cultural consumption – commodification and cosmopolitanism

Recent work by geographers has dealt with the evolving nature of consumption in the face of cultural globalization (see Crang, 2008; Mansvelt, 2005). There has been considerable debate concerning the rise of 'consumer culture' over the past three decades in particular. Consumption has come to be understood as far more than a material transaction process. Items are constituted of symbolic as well as material value and their consumption is tied up with the creation and expression of identities. In particular, 'brands' have become increasingly important markers of identity. Adidas shoes and clothing, for example, are linked with the hip-hop culture of the USA, and many purchase the clothing in order to associate themselves with this particular youth movement. In this sense then, consumption has become a key source of social and cultural differentiation. This is linked to globalization in a number of ways – not least as such 'signifiers' are rapidly diffused across the planet, constructing global cultures that transcend borders. Some have argued that this emergent culture is now beyond the control of any one group (Baudrillard, 1988). That such cultures are evolving is beyond question – whether they are creating one global culture is another matter, however, as the rise of the 'cosmopolitan consumer' discussed later in this section illustrates (see also previous discussion).

When the objects, ideas and traits of a culture become part of the capitalist system of exchange and are bought and sold, a process of cultural commodification is said to be occurring. Linked to the rise of the cosmopolitan consumer has been the increased demand for and supply of 'authentic' and 'exotic' cultural experiences. This is most often associated with the tourist sector where, as previously noted, the dominant mode of tourism provision has shifted from the Fordist/mass market 'sun and sea' model to a more differentiated sector based around 'new' cultural experiences (see Box 6.8). Cultural commodification is an increasing feature of global society. In supermarkets across the West, for example, new rows of produce from non-Western countries are conflated under the label 'ethnic' foods – a process not unlike the evolution of 'world' music (see case study below and Box 6.4 above). In urban areas, especially world cities, a prominent marketing strategy of local administrations of late has been to recreate and promote 'ethnic' districts such as Chinatown in Soho, London, or the Arab Quarter in Singapore (Chang, 2005). Ironically, in the past such areas may well have been the 'ethnic ghettos' that tourists were encouraged to avoid. There are also frequent attempts to create reconstructions of imagined rural idyllic lifestyles (see Plate 6.8). Cultural commodification works both ways, involving flows from the Western to the non-Western world. Indeed, in many respects this direction of flow pre-dates the flow from poorer to richer countries. Symbols such as Coca-Cola and McDonald's are marketed as Western ideals, implying that consumption confers status. In both the Western and the non-Western world, experiencing the 'other' can confer status and 'cosmopolitanism' on the consumer.

Bell and Valentine (1994) define 'cosmopolitan consumers' as a consumerist elite which seeks cultural signifiers that are not touched by homogenizing forces and actively explores transcultural diversity (Cook, 2012). This group accumulates cultural capital – such as music, clothing, food, beverages or furnishings – as a means of disidentifying themselves from the mainstream (Shurmer-Smith and Hannam, 1994). Increasingly, such consumers do not have to leave their localities in order to achieve their goals, and thus heightened cultural flows have the dialectical impact of escalating the potential for heterogeneity. Cosmopolitan consumers are not only found in richer countries, although they are relatively more common in such contexts since practising such accumulation requires financial capability and access to global networks. In poorer countries such elites do exist, and they are often highly visible. In Chile, for example, a cosmopolitan consumer might take his or her holidays in

(a)

(b)

Plate 6.8 *Cultural commodification in (a) Denmark and (b) Malaysia*
Cultural centres are common in places as diverse as Belgium, New Zealand, South
Africa and Thailand. They often present an idealized picture of a rural past. This
Hjerl Hede cultural museum in Denmark was founded in 1930 and features
reconstructions of a village and demonstrations of 'ancient traditions'. In Sarawak,
Malaysian Borneo, a cultural centre near Kuching advertises traditional practices
such as shooting blowpipes.

Source: Photos by Jo Heitger and authors

Miami, drink Chivas Regal whiskey and listen to rock music in English. The fact that the same markers of identity would not confer cosmopolitanism on consumers in the USA, for example, highlights the fact that cultural symbols are interpreted in very different ways across space according to local contingencies.

Baudrillard (1988) is a post-structuralist who has written extensively on the process of cultural commodification and has influenced a number of human geographers in this area. He argues that the 'postmodern' culture is one of signs and symbols. Today's world is characterized by the proliferation of simulacra where the symbols or signifiers of an event or item replace direct experience. As such, 'reality' is experienced chiefly through representations of it – meaning that distinguishing between the concrete and the imaginary becomes increasingly difficult and redundant. This creates a social environment of 'hyperreality' which disembeds cultures and their component traits, and creates new spaces where the former contingencies of history become less important. Geographers and others have extended these ideas to the analysis of capitalism and economics and services such as tourism.

It may be argued that cultural commodification, if managed appropriately and sensitively, offers the prospect of improved livelihoods for marginalized cultures as well as helping to preserve some cultural 'traditions' (see Box 6.9). In contrast, some see such commodification as

---

Box 6.9

Tourism and the invention of traditional culture in Fiji

Culture is used to sell things, and tourism – the largest single economic sector in the global economy – is one of the best examples. During the late 1980s, the mass consumption beach and sun ideal was challenged by the growth in tourism to more 'exotic' locations. In the South Pacific, for example, ecotourism and cultural tourism have become increasingly important within the general context of tourist industry growth in the region. In Fiji, in relatively mainstream resort holidays and cruises, the

'Fijian experience' is sold to attract visitors. This has seen the reproduction and adaptation of certain cultural practices, such as the *kava* drinking welcome ceremony, and the invention of others such as new forms of *meke* (dance) to cater for those seeking an 'authentic' experience (see Plate 6.8) that emphasizes the friendly and welcoming nature of Fijians which serves as an important antidote to the recent history of military coups and violence there. The Fiji Visitors Bureau website

Box 6.9
continued

dedicates considerable space to explaining traditional Fijian social structures, beliefs and ceremonies. In the world outside of the tourist resorts, such 'traditions' are rapidly evolving. Most Fijians now live in urban areas and are increasingly Westernized in both their tastes and attitudes; their dress, music and other aspects of contemporary culture are a long way from what is simulated in resorts, and the strong Christian elements of everyday life are not extenuated in the resorts. 'Traditional' practices still exist in some interior and coastal villages, but they are largely the lot of beachside resorts, which have become commodified cultural enclaves. Tourism, trading on both invented and authentic Fijian traditions mixed with modern resort-based provision, is a truly glocal product. Some have argued that globalized tourism in places such as Fiji relies on a damaging 'orientalism' of the Pacific (see Nicole, 2000). Others have argued that commodification of this nature actually helps preserve culture and ethnicity in ways that would be otherwise challenged.

Plate 6.9 *Fijian tourism – villagers perform a 'traditional' dance for tourists*

tantamount to the 'prostitution' of culture (Britton and Clarke, 1987; Scheyvens, 2011). In this latter argument, cultures move out of the control of those that inhabit them and into the hands of market forces. Such forces are characteristically fickle and cyclical – and thus actually threaten sustainable livelihoods. Commodification can also compound stereotypes of what 'authentic' and 'indigenous' mean. In this way, creating marketable cultures may 'straitjacket' cultures so that they do not evolve as they would have under less 'exploited' circumstances. It could be argued that this last perspective patronizes and simplifies the marginalized who are not necessarily passive receivers of market signals. The new cultural politics is a highly contested terrain.

The global music industry case study

It is a well-worn cliché that music is a universal language. Although transferral between cultures is not unproblematic – due to vast variations in terms of both musical style and structure and lyrical subject matter – it does not depend on written or spoken communication explicitly for its major impact. It is therefore a prime candidate for cultural interpenetration. The geography of the production and consumption of music is highly uneven however, with the bulk of commercial musical flows moving from the West to elsewhere, reflecting economic and political relations of power that are rooted in colonial history. The industry is dominated by a few very large TNCs that have been instrumental in diffusing Westernized forms of pop and rock across the planet. Some of the results – MTV, Michael Jackson, Elvis Presley, Madonna, Adele, Robbie Williams and Lady Gaga – in some ways epitomize the homogenization of global culture thesis.

In the past, the diffusion of music was limited by the absence of technology required to record and broadcast it. Thus the site of production was the live performance. Musical forms diffused through early globalized networks including world religions, empires and associated migratory patterns. Some flows, such as the slave trade from Africa to the Americas, left an especially lasting legacy in terms of musical hybrids, leading to blues, gospel and, indirectly, to rock and roll, soul music, RnB, and hip hop/rap. The conquest of the Americas, for example, by the Spanish and Portuguese led to the formation of new musical forms combining flamenco, Western classical music and Andean music which can be heard clearly today in the music of Latin America. Behind each of these forms lies a rich geographical history involving migration, conflict, conquest, coexistence and other forms of cultural diffusion and interaction. Flamenco, which emanates from Andalucía in Southern Spain, is a musical form that combines Romani rhythms and themes (referred to as *Gitano* in Spain) with Moorish harmonies and structures – reflecting the interaction of those two cultures in Spain until the start of the 1600s – that subsequently evolved into what we recognize today as traditional flamenco and even further into flamenco fusion as epitomized by groups such as Ojos de Brujo from Barcelona. The globalization of music over the past 100 years illustrates the important facilitating role of technology as distinct and increasingly intense waves of globalization were linked to new infrastructures. Storage and reproduction of music has been revolutionized, moving chronologically through vinyl record, cassette, compact disc, minidisc, the MP3 (digital storage) and, most recently, through streaming digital form, becoming increasingly 'footloose' with each successive invention. The rise of infrastructures of transmission associated with broader cultural globalization such as the radio, television and most recently the internet have lubricated this increased flow. Large cultural-production TNCs have been instrumental in the development of this technology and have taken advantage of the opportunities afforded by it. As a consequence of these trends, music has globalized in three major ways: (1) music TNCs have evolved to foster global flows; (2) national markets have been increasingly penetrated by foreign performers; (3) Western styles and images have been diffused through the music industry.

The industry has long been dominated by a handful of very large TNCs. In 2012 the total revenue generated by industry sales was approximately US$16.5 billion (IFPI, 2013). This, however, had fallen from US$32 billion in 2003 with unit sales of 2.9 billion (IFPI, 2004). As this reduction in sales revenue has continued due to technological change (see later), the industry has become increasingly concentrated. In the 1990s six companies controlled the vast majority of industry sales: Universal, Sony, Warner, EMI, Polygram and BMG. Competitive pressures during the 2000s led to horizontal integration through mergers and acquisitions, and by 2012 only three large companies dominated approximately 90 per cent of all global trade in commercial music: Universal (which had acquired EMI) accounting for over a third of all trade, Sony just under another third, and Warner Music just under one-fifth. Independent labels accounted for less than one-tenth of all global market share at this time (IFPI, 2013) Such TNCs, then, are often vertically as well as horizontally integrated and have been increasingly aggressive in terms of buying smaller, formerly independent, labels together with recording, supply and distribution brands and facilities. These giant firms, which operate a host of linked labels beneath them, have their sales concentrated in the USA and, to a lesser extent, the UK, and they are generally anglophone in orientation. As noted above, the

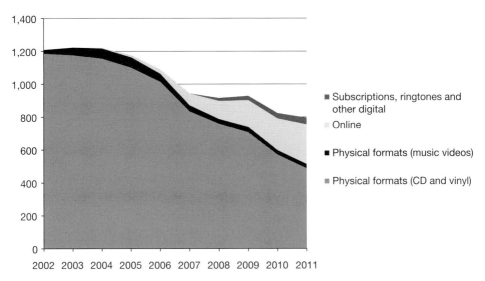

Figure 6.1 *British music industry income 2002–2011 (£millions)*

Source: www.bpi.co.uk/assets/files/Industry%20Income_2012%20Yearbook.pdf accessed 4 October 2013

globalization of pop and rock has largely involved the diffusion of cultural forms from the USA and the UK to the rest of the world. In the aftermath of the Second World War, rock and roll, blues and soul were the major export styles. Important forms in approximate chronological order that have subsequently globalized include country and western, urban folk, heavy metal, punk, new romantic, gothic, acid jazz, house/rave, grunge, Britpop, rap, hip hop, boy/girl bands, new RnB and, more recently, new soul and new folk styles.

Few forms have captured the global imagination as has rap/hip hop however. Emanating from the ghettos of major US cities in the early 1980s after pioneering innovation by artists such as DJ Herc in the 1970s, it celebrated its fortieth anniversary in 2013. This form has risen to global predominance in the singles charts of virtually every country and is emulated, both as a musical form and lifestyle, by youth on all continents. Rap culture is diverse and complex, but it often goes hand in hand with other markers of neoliberal cultural globalization such as a global brand culture and ostentatious materialism. The coopting of rap/hip hop by capitalist circuits was not necessarily intended or even predictable at its inception. The pioneering work of groups such as the Sugarhill Gang, and Grandmaster Flash and the Furious Five formed part of a political protest movement against 'mainstream' US culture. It could be argued that many eventually globalized musical forms such as blues, punk and grunge started life as countercultural movements only to be appropriated by cultural TNCs.

No countries have evolved globalized popular music to the same extent as the USA and the UK, and these markets have historically been relatively impenetrable to artists from other markets. The Guinness British Hit Singles (Roberts, 2001) top 40 groups of 1952 to 2000, based on total weeks on the sales chart, featured 20 UK acts, 19 US acts and 1 Australian act (the Bee Gees, who were actually born in the UK). In other anglophone and part-anglophone countries such as Australia, Canada and New Zealand, dominance by US and UK artists remains pronounced despite government-led efforts in each case to promote national content on radio. The first two countries have produced a number of global 'stars' including INXS, AC/DC, the Bee Gees and Kylie Minogue in the case of the former, and Celine Dion, Bryan Adams, Neil Young and Rush in the case of the latter. New Zealand has found it especially difficult to penetrate global markets with only Crowded House achieving anything approaching global status. The New Zealand national market is particularly saturated by overseas acts. Of the Top 40 acts in 1966 to 1996, based on weeks on the singles charts, only 1 (Dance Exponents) was from New Zealand, while 19 were from the UK, 15 from the USA, 3 from Australia (Bee Gees, Olivia Newton John, Jimmy Barnes), and 1 each from Ireland (U2) and Sweden (Abba) (authors' calculations).

Within Europe, France and Germany have strong domestic industries but, aside from a small number of 'Europop' hit groups, such as Kraftwerk, Trio and Nena, they have not captured global markets. Scandinavian countries have enjoyed a little more success, producing Abba and, more recently, A-ha and Roxette. Spain has produced few global names in popular music, with notable exceptions being father and son Julio and Enrique Iglesias. Spanish artists are very popular in Latin America, however, in part because of language commonalities and cultural affinities due to the nature of colonization and subsequent interaction but also due to the growing role of Spanish TNCs in the continent in the 2000s. The diffusion of non-UK and non-US music through established networks is hindered by the general trend towards English global pop and by the paucity of distributing companies located outside of these two countries. Consequently, European charts are heavily populated by US and UK acts. In Latin America, Spanish and US acts are overwhelmingly popular, and national acts that make inroads to commercial success – such as Maná (Mexico), Los Tres (Chile), Fito Paez (Argentina) and Charly Garcia (Argentina) – are relatively rare. In Asia, similar patterns are observed, although the rise of Bollywood as a cultural force in South and Southeast Asia is partially offsetting this trend. There has been a rise in commerical pop from China and Asia in general over the last decade – most notably in the form of K-Pop from South Korea as epitomized in 2012 with the PSY hit 'Gangnam Style'.

One country that apparently bucks the above trend is Jamaica, which has been hugely influential in global markets through the reggae (Bob Marley) and, more recently, ragga (Shaggy) forms. These forms are highly commercially successful in Western countries and have produced a number of emulative groups such as UB40. Reggae has become hugely popular across Africa and the Pacific Islands, and it is often seen as a form of resistance to globalized music, encapsulated in the emancipatory messages of Bob Marley. Pacific Reggae has evolved a unique form, marrying the ska and off-beat of the Caribbean style with the melody, chord progressions and instrumentations of the Pacific. Another non-Western form which is achieving considerable global impact is the style broadly known as 'tropical', which incorporates salsa, merengue, cumbia and so on. More traditional tropical rhythms and melodies were adapted and evolved by Puerto Ricans and Cubans in New York in the 1970s, including breakthrough artists such as Ruben Blades. These were popularized into the mainstream through the 'pop tropical' sounds of Gloria Estefan who sings in both Spanish and English. Carlos Santana, a member of the early 1970s psychedelic rock movement, has found a revitalized market for his particular brand of 'tropical rock' in the mainstream West and among the Hispanic diaspora in the USA. Across Latin America, 'tropical' artists such as Juan Luis Guerra, Willie Colón, Joe Arroyo and Celia Cruz challenge Western rock and pop styles to a certain extent. Reggaeton, which is popular across the world but especially in Latin America and Hispanic USA, is also derived in part from Jamaican styles mixed with hip hop and samba and can be considered the latest in this wave of musical forms that do not emanate from the 'core' of the global musical economy.

Homogenization is not necessarily the outcome of the globalization of Western forms of music (Hudson, 2006). One of the most enduring impacts of the diffusion of US and UK forms of music is the creation of localized hybrids. A few examples only can be provided here. Latin American rock, beyond being sung principally in Spanish, does have a quality which sets it apart from Western rock often involving the incorporation of 'latino' rhythms and flamenco-like guitar breaks in terms of both scales and timing (as in Carlos Santana); New Zealand hip hop fuses US and Polynesian rhythms, tones and subject matter in a way which is unique (as in Che Fu); African reggae which mixes Caribbean sounds with the soukous style is unique; British Bhangra-Beat epitomized by Apache Indian from Birmingham in the early 1990s welded hip hop, reggae, ska and bhangra. What is undeniable however is that these forms rarely flow back to where they came from and recondition the forms from which they were derived. That was, arguably, until the recent rise of so-called 'World music'.

Of course, outside of commercial circuits of capital, infinitely rich and diverse musical forms exist that are reproduced daily. In the early 1980s, frustrated by the seeming impossibility of classifying such diverse styles, a small record store in London coined the phrase 'World music' (see Plate 6.10). Despite the neocolonial overtones of conflating the diverse musical heritage of over 90 per cent of the world's population, this new

Plate 6.10 *The Vinyl Countdown record store, New Plymouth, New Zealand*

descriptor created a space for otherwise 'invisible' music and provided a model for larger record stores. Rising demand for World music led to increased airplay and the development of dedicated festivals such as WOMAD (World Music and Dance). A new industry, populated by relatively small-scale firms, has been stimulated, and there are now over 3,000 World music labels. There has also been a flowering of devoted internet sites (somewhere in the region of 10,000) and specialist magazines such as *Global Rhythm*. (See also Miller and Shahriari, 2013.)

Despite the predominantly one-way traffic, then, it is now possible to hear alternative music in Western countries more than ever before. This music has been marketed in its 'authentic' form, and has also become acculturated into Western pop music through the work of artists such as Paul Simon (Southern African), Damon Albarn (African) and Ry Cooder (Tropical Caribbean) to mention but a few. This trend builds on earlier experiments with explicitly non-Western form by groups such as The Beatles, Led Zeppelin, The Jimi Hendrix Experience and The Doors in the late 1960s and early 1970s. Some might argue that it was George Harrison that was the true pioneer in terms of seeking to integrate and diffuse non-Western styles to Western audiences which involved working with one of the great sitar players – Ravi Shankar. There can be little doubt that this shift is correlated with increased travel and the facilitating role of communications technology. Indeed, critics would argue that the absorption of non-Western forms and its morphing into more marketable forms for commercial purposes has been central in the history of the global music industry – the foundation of contemporary music blues and jazz, after all, was rooted in Africa. The celebration of the rise of resistant musical voices may be premature however. World music accounts for a tiny proportion of total sales, standing at less than 1 per cent in the UK.

In the decade of the 2000s, the global music industry had undergone a revolution related to the technologies that facilitate globalization. The increased role of the internet as a form of distributing music had, to an extent, undermined the hegemonic role of the global TNCs leading to industry-estimated losses that saw revenues halve in a decade. Large corporations reacted strongly to the increased accessibility of music downloads, as lawsuits in the early 2000s against peer to peer download sites such as Napster and Kazaa exemplified. The firms moved rapidly into digital download technologies in an attempt to regain control of the market. A new player, Apple, through its iTunes product, dominated the market for downloaded music in the early 2010s and delivered this through hardware that was tailor-made for such purposes. The industry headed, to use the terminology of Chapter 4, to a phase of more flexible and footloose accumulation where formerly dominant delivery technologies such as the CD began to decline in importance (Osbourne, 2013). Furthermore the music production companies saw a decline in their hold over artists as technology had evolved that could produce quality home recording that could be distributed across the internet. New communications infrastructure has already created space for alternative artists and labels. For example, there are tens of thousands of independent bands currently listed at the website GarageBand.com. Larger acts such as Radiohead and David Bowie also experimented with online release and greater control of their product. However, as long as access to this technology is stratified, it may serve only to perpetuate the geographical asymmetries in power that have always characterized the global music industry.

One of the most visible geographical outcomes of the online revolution in music has been the closure of high-street music stores all across the world and the shift of sales to large supermarkets such as Walmart and Tesco as well as online through dedicated sites. There has, at the same time, been a slight 'retro' revival involving purchase of music on vinyl in the UK, New Zealand and Australia in part in reaction to this (see Plate 6.10). A further impact has been the rising importance of performing live, and the 2000s saw the reformation of many major Western acts such as the Police and the Eagles in part due to this. Some, then, have talked of the democratizing implications of the online revolution in music. However, the other side to this has been that, in the face of increased pressures, concentration has risen in the sector (as evidenced above). Furthermore, many of the live music venues and ticket sales are controlled oligopsonistically by companies such as Ticketek. This, combined with highly stage-managed production of globally consumable pop through so-called 'reality' platforms as *American Idol* and *Britain's Got Talent*, has put the large players firmly back in control of an industry that has always been subject to highly imbalanced power relations.

Conclusion – towards progressive cultural spaces?

A global cultural shift has taken place, and this is reflected in the academy by the evolution of perspectives that give greater precedence to cultural interpretations and explanations. The 'cultural turn' seemed to suggest that culture underpins change in all spheres. A more convincing argument is that the determining powers of culture, politics and economics are intertwined inseparably.

Culture gives people a sense of community and belonging – and is therefore one of the principal means through which identity is constructed and sustained. Until recently, cultures have been thought of as relatively stable and spatially bounded, although historical evidence does not necessarily support this. Contemporary globalization has led to the growing recognition of ties between distant and disparate places, symbols and ideas. The emphasis in cultural geography has thus shifted from consideration of bounded spaces to flows of commodities, people, ideas, images and beliefs. This is an unsettling time for cultural identities as globalization has proceeded at a rate unmatched in any other sphere, and the recent rise of social media such as Twitter, Facebook and YouTube seems to be further perpetuating this velocity. Centrally, does this imply cultural homogenization and the deterritorialization of identity, or does it increase the space for a new assertion of the local sometimes as resistance to the former?

Homogenization arguments are based on dated views of cultural interaction. Although the Westernization and Americanization of global culture are powerful processes that really do exist, the arguments that support these theses are usually anecdotal and often focus on the most readily visible aspects of cultural change. Homogenization is wittingly and unwittingly resisted by individuals and the cultures and nation-states they inhabit on a daily basis. Thus the thesis of cultural imperialism is greatly exaggerated and Eurocentric. Rather, contemporary globalization has led to the hybridization of culture to a greater degree than ever before. Notwithstanding this, the continued rise of global corporations and the accelerated reduction in barriers to trade and investment, married with global political governance that favours this outcome and communications media that facilitates it (see Chapters 4 and 7 in this regard), could in theory act as ever-more powerful vehicles for future homogenization and will have to be resisted actively. In the context of the reaction to homogenization, the reassertion of national culture is a trend that should not be underestimated and one which, arguably, is being facilitated through new globalized social media. In this regard, globalized social media is in fact neutral in terms of its impact on culture – and it depends crucially on how it is utilized and regulated. This reminds us that it is people and not technology that drives globalization. What makes this process all the more complicated is that 'national' identity and culture in general is increasingly 'located' in dispersed networks. The increasing incidence of diaspora is challenging the very concepts around which globalization has been constructed. Diaspora may plant the seeds of a *progressive* globalization of culture – so sorely required in today's society – which celebrates and builds from difference and diversity.

Further reading

Anderson (2010): This is an excellent overview of cultural geography that provides in-roads to many of the issues discussed here.

Crang (2008): This is the second edition of an excellent introduction to cultural geography that is lively and accessible, yet challenging.

Creswell (2013): This chapter introduces the concept and debates concerning place and cultural landscapes.

Gibson and Waitt (2009): This is a succinct yet broad-ranging entry on the nature and evolution of cultural geography that provides signposts to further work in this area.

Johnston and Sidaway (2014): The chapter on cultural geography traces the rise of the 'cultural turn' in geography and is the best sourced entrance point into the broader literature.

McEwan and Daya (2013): This chapter provides a succinct overview of definitions of culture, cultural interaction and the impacts of global change.

Miller and Shahriari (2013): This is one of the most popular and thorough texts on world music. See also Connell and Gibson (2003).

Nederveen Pieterse (2004): This work is a classic outlining the thesis of cultural imperialism in a compelling way. It represents relatively advanced reading however.

Osbourne (2013): This provides a fascinating history of the vinyl record and how this relates to structural and cultural changes in the music industry.

Ritzer (2008): This is the fifth edition of this important book on the globalization of culture and so-called McDonaldization.

Swyngedouw (2006): This is the most complete discussion of glocalization available – though it is a challenging read.

Part III

Globalization and three crises

Neoliberalism, instability and globalization

- Circuits of capital
- Neoliberalism and globalization
- Free trade
- The Global Financial Crisis – a critical geographical look
- Neoliberalism and the globalization of agriculture: A case study
- Reforming the global system
- Towards fair trade
- Conclusion – a crisis in regulation

In this chapter, we consider the first of three global challenges – the inherent instability of neoliberal globalization and the unjust impacts of this model. At the time of writing, the global economy remained in the throes of one of the most serious financial crises ever experienced (see David Harvey's view on this crisis below). The crisis was having serious social, economic and political consequences globally. We seek to trace the roots of this crisis making the two central points that such instability is not unprecedented and that ultimately it reflects the pursuit of the flawed model of globalization that has been followed during the second phase of the second wave. By critically reviewing neoliberalism, free trade, and the institutions that promote it, we see that the crisis is something that should have been expected and something that will occur again if it is not reformed. We explore the globalization of agriculture in order to illustrate some of the negative impacts of neoliberalism. We discuss some ideas with respect to progressive reform, such as fairtrade, and conclude that solving instability and injustice requires consideration of the other two crises – in the environment (Chapter 9) and development (Chapter 8) – simultaneously.

> This was undoubtedly the mother of all crises. Yet it must also be seen as the culmination of a pattern of financial crises that had become both more frequent and deeper over the years . . . there have been hundreds of financial crises around the world since 1973, compared to very few between 1945 and 1973. . . . There is therefore nothing unprecedented except its size and scope. There is therefore some inherent connectivity at work here that requires careful reconstruction.
>
> (Harvey, 2011, p. 10)

Circuits of capital

The concept of circuits of capital is a Marxist-influenced idea which helps us understand globalization. Under this schema the capitalist process is seen as an infinite circuit (See Figure 7.1). The capitalist imperative is to ensure that the money at the end of each circuit is greater than the money at the beginning (i.e. that surplus capital is produced). There are three types of capital: Money (M), Commodities (C) (these refer here to raw materials and labour) and the means of production (P). Commodities acquire increased value (C') through the production process and are exchanged for money (M'), this creates profit and this value can be used to purchase another round of inputs. The central idea with respect to globalization is that each of these circuits has become expanded spatially in the following ways:

● *Commodity capital circuit* – This was the first to become globalized through trade.

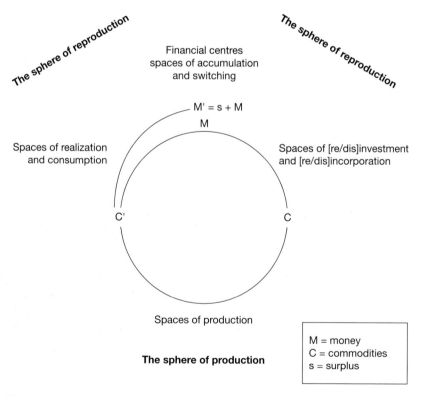

Figure 7.1 Circuits of capital

Source: Adapted from Cloke *et al.* (2005)

- *Money capital circuit* – This globalized subsequently through the rise of investment capital (see Box 7.1 and Plate 7.1).
- *Productive capital circuit* – This globalized most recently with the rise of TNCs and global networks and chains.

Other circuits could be added to this model. For example, it is feasible to argue that cultural capital is now being globalized through the technological revolution. From a geographical point of view, the global capitalist system and the circuits that constitute it have explicit spatiality (see Box 7.1). This is to say that the creation of value or surplus takes place in specific economic spaces. These may be broken down into spaces of production, investment and consumption, all of which are highly complex and have particular geographies. As noted in Chapter 2, Harvey (1989) provides the most compelling explanation for globalization related to the circuit of capital. Inherent to capitalism is the desire to make the circuit operate at the highest possible velocity, maximizing the accumulation of capital and surplus in any given period. This requires the development of space-shrinking and time-saving technologies which reduce the turnover time of capital. It is the development of such technologies, based on the capitalist imperative of maximizing profit that, according to Harvey, has led to time–space compression and globalization as we know it.

Plate 7.1 *Nescafé in Africa*
Nescafé – owned by the Swiss TNC Nestlé – is a major brand of instant coffee that is found worldwide, even in countries such as Kenya which is a large producer of its own coffee beans, illustrating the central role of circuits of capital in conditioning local geographies.

Box 7.1

Hyperglobalization in the financial sector – myth or reality?

The global financial sector is often held to be the epitome of globalization, and there can be little doubt that since the 1970s, relative to other sectors, transnational flows of finance have blossomed in unprecedented ways. The financialization of the economy has been profound as the total proportion of economic activity accounted for by activities such as banking, insurance, currency exchange and investment services has risen sharply as regulations have evolved and instruments have been developed to facilitate the increasingly rapid flow of money across the global economy. In the United States, at the present time, financial services income accounts for approximately 8 per cent of total income. Approximately half of all non-farm corporate income comes from financial services. Financial service companies such as Berkshire Hathaway, AXA and ING are among the largest companies in the world with annual revenues exceeding US$120 billion per annum. The US stock market is worth over two-fifths of the total GDP of the same country.

As Pollard (2012) points out, a number of theorists talk of the role of money as a vehicle for the homogenization of space given its ability to transform different commodities and experiences into an abstract entity – a monetary value. A high level of integration between the different parts of the financial system has developed, embodied in apparently disembedded transnational banks that transfer money with extraordinary fluidity. These flows have woven local and national markets into what Martin (1994) refers to as a 'new supranational system'. In this new system, transactions – for example, in currency markets – in one part of the world can have almost instant material ramifications in another, making the regulation of financial flows by state institutions increasingly difficult and economic change in any one place less predictable.

The globalization of finance has come about through the facilitating role of neoliberal reform, the development of communications technology, and the evolution of tradable financial products that have made increasingly rapid movements possible. In particular, the creation of financial derivatives bundle together products and sell them on in order to allow hedging of financial liability. The prevalence of this type of product, which in total are worth thousands of billions of US dollars, partly explains the vulnerability that led to the Global Financial Crisis of 2007 (Ritzer, 2011). Underlying the development of these regulations and instruments is the capitalist imperative to reduce the turnover time of capital through the shrinking of space (Harvey, 2011). However, is the financial sector as 'globalized' as many presume? Pollard (2012) debunks this myth through a number of observations. First, the global financial system is actually concentrated in

Box 7.1
continued

a handful of Western economies, and the bulk of flows move between specialized districts in world cities, such as Wall Street, New York and the City in London. Second, there are in fact distinct geographical circuits of money – or a 'patchwork of different financial spaces' (Leyshon, 1996, p. 62) – which are only loosely integrated. Finally, networks of money are actually constituted of 'materiality' and 'practices'. In this context, Pollard argues, 'although the global financial system may sometimes appear to be a rather intangible "thing", it is really an assemblage of consumers, workers, computers, telephones, office buildings, bits of paper, financial reports, and so forth' (2012, p. 392). In this sense, it occurs in specific places and is far from omnipresent. Overall, then, the global financial system is best conceived of as 'an intersecting web of networks – some fast, some slow, some long, some short . . . we are not talking about a machine operating "out there", but a socially, economically, culturally and politically constructed network of relations' (Pollard, 2012, p. 393). This was illustrated to devastating effect during the financial crisis of 2007 onwards where particular localities suffered unemployment and recession given the particular intricacies of the networks through which the signals had been transmitted.

Neoliberalism and globalization

Neoliberalism is the term used to refer to the economic paradigm and cultural ideology that is now virtually hegemonic across the world (see Chapter 4). Neoliberalism recommends the elimination of government intervention in the economy, arguing that 'governments fail' – causing inefficiency, crowding out private investment, and ultimately lowering competitiveness on the global stage. Some neoliberal arguments have linked state developmentalism and corruption. According to this argument, 'rolling back the state' will foster efficient and sustainable economic growth. A deregulated, privatized, economy is argued to be the best way to maximize welfare at global level, and global free trade based on the exploitation of comparative advantage is given precedence (see Chapter 4). The most important point from a globalization perspective is that neoliberalism opens the doors to flows in investment and trade, facilitating the penetration of TNCs, which allows the greater articulation of global networks into localities. Neoliberalism is argued by some to be an emerging culture, which forms part of the wider mode of regulation of capitalism (Peck and Tickell, 2002). It is ideologically grounded in Western values of modernization, 'civilization', individualism, materialism, accumulation and rationality. In this sense, it forms part of what Cowen and Shenton (1996) term 'doctrines of development' which are rooted in the Enlightenment period. The application of neoliberal policies has important cultural impacts in terms of the rhythms and spatiality of everyday life, leading to the increased flow of Western cultural symbols especially.

Neoliberalism emanates from the University of Chicago and is associated with Milton Friedman's monetarist economics. It was first applied in Chile, after the military government which seized power in 1973 contracted Chicago economists to develop a model that would totally reverse the socialist path taken hitherto (Barton and Murray, 2002; Murray, 2009a). The 'Chicago boys', as they became known, effectively used the country as a laboratory for their theories, aggressively privatizing and opening up the economy, and reducing the size of the state to a bare minimum. Because neoliberalism requires austerity, which hits the poor the hardest, neoliberalism could only be applied in Chile, and later Latin America in general, under military rule. Neoliberalism is not just part of the socio-economic history of the poorer world; from the early 1980s it was also applied across the OECD as the post-Second World War 'welfare consensus' was eroded following the oil crises of the 1970s. Neoliberal principles were applied in the UK under Thatcher from 1979, and this was followed in the USA under Reagan from 1980. New Zealand adopted such reform with particular purity from 1984 onwards, and, despite rising income inequality, marginalization and deprivation since that time (Kelsey, 1995), is seen as something of a model in neoliberal and hyperglobalist circles (Harvey, 2005).

The rapid economic development of East Asia over the past 30 years is sometimes claimed as a victory for neoliberalism and outward orientation, although many have argued that what sets East Asia apart is its special combination of neoliberal and structuralist/developmentalist policies (Coe *et al.*, 2003). In human geography, there is a general consensus that neoliberal policy is in fact tantamount to jungle law (McKenna and Murray, 2002; Peck and Tickell, 1994), which increases inequality both within and between countries. Thinkers are divided on the extent to which 'pure' neoliberalism has diffused across the world; there is no such thing as a pure free-market economy and all are 'mixed' to a certain extent. However, there can be no doubt that there has been a paradigmatic shift towards the market end of the political spectrum and that this has facilitated heightened globalization and increasingly intense restructuring impacts on the ground.

Free trade

It was during the second half of the twentieth century that the global free-trade agenda reached its zenith. During the latter half of the nineteenth century, trade was relatively spatially limited, involving a handful of colonies and their imperial superiors. By the second half of the twentieth century, however, virtually all economies were enmeshed in the global system. This was heightened after the end of the Cold War as many former socialist economies were launched into the orbit of the global capitalist economy. The growth in world trade between 1950 and 2000 outstripped the growth in world production significantly (Gywnne *et al.*, 2003). The Bretton Woods institutions were critical in the pursuit of free-trade goals in the post-war era (see Chapter 3), GATT being the major initiative in this area. This was set up in 1947 with the goal of reducing tariffs and removing other barriers to free trade. It had four major principles (Gwynne *et al.*, 2003):

1 *Non-discrimination* – no one country should be given preference over another.
2 *Reciprocity* – tariff reductions should be reciprocal between countries.
3 *Transparency* – trade measures should be made clear.
4 *Fairness* – practices such as anti-dumping and unfair protectionism were regulated against.

GATT increased its influence steadily throughout the latter half of the twentieth century with membership rising from 30 in 1950 to 120 in 1995, with seven trade negotiation rounds having taken place. Often the sticking point in these negotiations was agriculture. Richer countries maintained, and continue to maintain, high direct and indirect subsidies for their agricultural sectors, citing political, social and cultural reasons. In 1995, GATT was replaced by the World Trade Organization (WTO) which was more powerful (see subsequent section). Despite the fraught debate over the socio-economic impacts of these institutions, they have been relatively successful in achieving their stated objectives. By the end of the twentieth century, tariffs were at an all-time low and barriers to trade in services had been reduced.

The pursuit of free trade is the cornerstone of neoliberalism, the now-dominant economic paradigm. This approach, rooted in the neoclassical economics of the eighteenth century, argues that markets allocate resources in the most efficient way and that state intervention distorts allocative processes and, ultimately, reduces total global welfare. Neoliberalism also implies political and cultural shifts, arguing for smaller government, unfettered capital flows and TNC penetration, and the benefits of individualistic welfare maximization. Adopted first in dictatorial Chile in the mid-1970s and then in the form of Reaganomics in the USA (1980) and Thatcherism in the UK (1979), it has spread rapidly subsequently. There is a growing literature within geography which analyses and describes contemporary spaces of neoliberalism and the impacts of this from economic, cultural and political perspectives (see Brenner and Theodore, 2003). In this sense, then, the discourse of neoliberalism has become synonymous with globalization as currently practised, and it is against neoliberalism that the anti-globalization protests are aimed.

The World Trade Organization

Together with the IMF and the WB, the major international institution driving the neoliberal agenda is the WTO. The WTO was formed in 1995 after the completion of the Uruguay round of GATT talks. The WTO covers trade in manufactured goods, raw materials, agricultural services and intellectual property rights, and it has over 150 member countries. The most important recent addition was China, whose accession in December 2001 was considered a major moment in the history of global capitalism given the government's continued adherence to state communism. The organization is ostensibly democratic and each country receives one vote in major matters – although it is well known that the larger economies do have a greater say in practice. In addition, as new countries apply to join the WTO they find that they have to adhere to stricter conditions than those imposed on the earlier members.

Essentially, the WTO monitors whether member countries are following free-trade rules. If they do not, the organization has the power to impose fines and sanctions. The WTO has the right to make legally binding agreements, which sets it apart from the GATT regime. As noted above, agricultural discussions have been the most fraught of all. In 2004, the WTO announced significant reductions in barriers to agriculture in core economies, which if they are implemented will come into force by 2012. We are yet to see if the new WTO will move to see countries in the periphery and semi-periphery gain non-discriminatory access to agricultural markets in the West, however, and post the Global Financial Crisis there has been a stalling in WTO reform.

The World Trade Organization is extremely powerful and pro-neoliberal. It has become the source of major controversy and there is increasing criticism, in academia and civil society more generally, that it is too deterministic and too skewed towards the interests of TNCs. As such, it has been the focus of many protests by anti-globalization groups and has even been criticized by insiders, such as Joseph Stiglitz (a former World Bank chief economist), who claim that it has become undemocratic, bureaucratic and is undermining the sovereign rights of national governments (Stiglitz, 2002, 2010). A number of areas of policy have caused special concern; under WTO rules there have been a number of cases where environmental protections and welfare subsidies within any given state have been ruled as obstructive of free trade and have been removed (see e.g. Box 7.2).

Arguments for and against free trade

The argument for free trade is based on ideas developed by two leading neoclassical economists, David Ricardo and Adam Smith. The former developed international trade theory based around the concept of comparative advantage. Although the nature of the world economy has shifted dramatically, the arguments of the proponents of free trade, such as the WTO, are remarkably similar to early neoclassical ideas. It is argued that where the opportunity cost of producing goods varies between two parties (countries) trade will result in welfare gains for both. A country has comparative advantage in producing a good or service relative to others if the relative cost of producing that good (i.e. its opportunity cost in terms of other goods forgone) is lower than in the other country. Note that this refers to the relative cost of producing the item and not the absolute cost; therefore, even if one country can produce all goods at a cost which is absolutely lower than another, there will still be gains from trade. If all countries specialize in producing the commodities in which they have a comparative advantage, global living standards will be raised. Essentially then, in theory, free trade allows a maximization of world production, making it possible for every household to consume more goods than it could without free trade. Proponents argue that the long-run welfare maximizing potential of free trade offsets short- and medium-term restructuring costs.

Free trade has certainly raised consumption possibilities in some cases, but it is clearly not the case that it has raised global living standards for all (see also Chapter 7). This is the case for a range of reasons:

1 Many poorer countries have often been unable to persuade richer ones to open their markets to the extent that they open theirs.

Box 7.2

The WTO in action: banana wars

Bananas caused one of the largest trade disputes in history in 1997 and tested the WTO in action. In terms of trade, that of bananas is the largest fruit sector and the fifth largest agricultural sector. Three agribusiness TNCs dominate the industry – Chiquita, Dole and Del Monte. They are vertically integrated, owning farms, transport and distribution networks. The commodity chains linking the plantations to the consumers are top-heavy in terms of the distribution of profits along the chain; approximately 90 per cent of the final price stays in the destination countries. Wages in plantations are very low – in Guatemala, for example, workers receive around US$0.60 an hour.

There are two major production zones in the Americas: the Caribbean (e.g. St Vincent, Dominican Republic) and Central America (e.g. Guatemala, Honduras). Many of these countries, especially the Caribbean ones, are extremely reliant on these exports. Higher production costs in the Caribbean – due to smaller production sizes and less intensive input – mean that they are not able to compete in the global free market.

Aware of these competitive pressures, the EU granted the Caribbean special access to their market under the Lomé Convention of 1975. This convention was established ostensibly to assist former colonies and represented an extension of 'imperial preference'.

Under the new WTO rules, US firm Chiquita, the major banana TNC in Latin America, appealed for this development assistance to be removed. The firm asked the USA to represent it in the WTO tribunal and President Clinton himself headed up the protest. The EU was ruled against, and the USA threatened $500 million dollars-worth of sanctions if preferential access was not withdrawn. The WTO ruled that $191 million-worth of sanctions could be applied.

The Lomé Convention has been replaced by the Cotonou Agreement – which substantially reduces EU assistance to former colonies. As a result of this case, a number of Caribbean national banana sectors and hundreds of thousands of individual livelihoods have been undermined.

2 Free trade can lock natural resource-exporting nations into a primary product export trap from which it is difficult to evade the long-run decline in the terms of trade (relative cost of exports to imports).

3 There are high restructuring costs of moving to a free-trade regime which often result in structural, and often localized, unemployment and other social impacts.

4 Many of the theoretical assumptions underlying the argument that free trade maximizes global welfare are not met in the real world. Factors of production are not mobile and markets do not operate perfectly. Monopoly capitalism practised by corporations in already wealthy nations can erode any benefits of free trade.

5 Free-trade theory does not take into account environmental factors and the long-run implications of any given nation specializing in resource-intensive activities (as is the case in much of the periphery). Furthermore, the concept of maximizing resource consumption may not be compatible with the limits of global ecosystems (see Chapter 8).

6 Specialization brings its own dangers and makes economies, especially small ones, highly vulnerable to changes in global conditions. Linking world economies has the potential to spread growth, but it also has the potential to spread downturns (such as during the Great Depression, the Asian Crisis of 1997, as well as the GFC of 2007).

7 The current policy dogma of free trade does not take into account the historical evolution of global capitalism. Richer industrialized countries built up their economies over centuries, often behind protectionist walls. Poorer economies are not afforded the same luxury and thus the wealth gap is structural/historic in nature.

Proponents of free trade argue that the problem with the current global economic complex is that there is not *enough* free trade and that the long-run benefits of reform will take time to trickle through. Sceptics argue that the model is flawed and that free trade in itself brings socio-economic and environmental costs which outweigh the benefits. There are of course many shades of opinion in between these two perspectives. In reality, global trade is anything but free, and the system is highly skewed to benefit core economies which maintain high levels of protectionism while encouraging, and sometimes coercing, poorer nations to open up their economies.

The 'Global' Financial Crisis – a critical geographical look

In 2007, debatably the most serious crisis to hit the global economy since the Great Depression of the 1930s began. This was not the first so-called 'crisis' since the 1930s; other notable episodes include, for example, the 'debt crisis' of the early 1980s and the so-called Asian Crisis of the late 1990s. In the case of the latter, commentators began to talk of the contagion effect of global downturns and consequent linked recessions given the unprecedented increased interconnectedness of the global economy. This was shown to great effect, emanating from the capital sector, but spilling over into all sectors from 2007 onwards. Stiglitz emphasizes the importance of the crisis:

> In the great recession that began in 2008 millions of people in America and all over the world lost their homes and jobs. Many more suffered the anxiety and fear of doing so and almost anyone who put money away for retirement or a child's education saw that figure dwindle to a fraction of that value. A crisis that began in America soon turned global, as tens of millions lost their jobs worldwide – 20 million in China alone – and tens of millions fell into poverty. This is not the way it was supposed to be modern economics with its faith in free markets and globalization had promised prosperity for all. . . . The great recession has shattered these illusions.
>
> (Stiglitz, 2010, p. 3)

The economic social and political fallout of the crisis is being felt acutely at the time of writing in a range of social, political and economic ways across the world. Although commentators are divided as to the structural roots of the crisis and subsequent recession as we shall see, the immediate cause of the crisis was the speculative housing price bubble that developed in the USA in the early to mid-2000s. In the USA there had been political pressure for many decades from government administrations upon lenders to increase capital flows to low- and middle-income families to make house ownership possible. Low interest rates at the start of the 2000s fuelled this and the tax cuts of the Bush years further stimulated credit accumulation. Given poorly regulated credit markets, mortgage lenders in the USA – such as Fanny Mac and Fanny Mae – lent to customers that had little or no hope of servicing mortgage payments. This predatory lending came around in part due to deregulation of the financial system through the neoliberal 1980s and 1990s. This led to inflation in house prices combined with very high levels of personal debt and further rounds of lending based on property collateral that was not backed by 'real' assets and was thus unsustainable. One of the consequences of this was that house prices escalated to levels that were over four-and-a-half times the median income. As Harvey argues: 'Fictitious financial capital took control and nobody wanted to stop it because everybody who mattered seemed to be making lots of money' (2011, p. 17).

Much of the so-called 'toxic' debt had been repackaged and sold as financial products and recycled within the circuits of financial capital.

When the extent of the fragility became clear, there was a loss in confidence in the banks and lending between them stopped. This led to what became known as the 'credit crunch' where 'liquidity' was exhausted. Banks in advanced capitalist economies simply stopped lending to each other, leading to the failure of a number of large institutions such as the Lehmann Brothers bank. As these banks collapsed, given the general squeeze on credit together with the fall in house prices, many people experienced negative equity whereby what they owed was greater than their assets. Hundreds of thousands of people had homes repossessed by financially strapped banks (see Figure 7.2).

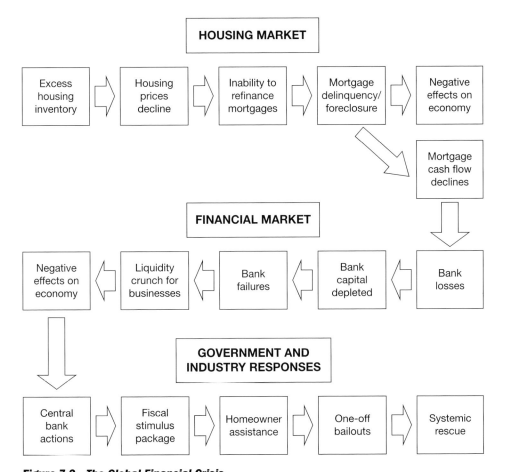

Figure 7.2 *The Global Financial Crisis*
The diagram shows the ripple effects of the crisis rather than its underlying causes.

Source: adapted from Farcaster's diagram on Wikipedia – http://en.wikipedia.org/wiki/File:Subprime_Crisis_Diagram_-_X1.png accessed 15 September 2013

The credit crunch was rapidly transmitted through various mechanisms including stock markets and trade flows across the world economy, and similar overstretching and associated bubbles had been built up in other countries. A number of countries, principally those in the global North, entered periods of recession – this was accentuated in the case of those whose GDP was heavily reliant on capital flows. Ultimately the multiplier effect of this was to damage overall economic growth at the macroeconomic level, particularly though not exclusively across the Western world. What the housing and credit crisis revealed was the intricate web of interlinkages that constitutes the contemporary financial system. Overall, world GDP per capita growth rates declined by 3.3 per cent in 2009 (Table 7.1), but rates of decline were greatest in high-income countries (4.2 per cent decline in 2009), especially in the European Union (4.6 per cent) and North America (4.0 per cent). There was some recovery in 2010 and 2011 but economic growth rates there remain sluggish (Table 7.1). Significantly, much of the rest of the world was less affected (see also Map 7.1). Low- and middle-income economies kept growing during the crisis and this growth was most marked in South Asia and China: though there were aggregate falls in 2009 in Sub-Saharan Africa and Latin America. Due to the impacts of the crisis, the OECD lagged behind the rest of the word in terms of economic expansion. A number of countries in Europe were severely impacted. Among the worst hit in terms of recession and consequent unemployment were Italy, Spain, Portugal, Iceland, Ireland and Greece. In 2012 the EU as whole shrank again (by 0.5 per cent) partly as a consequence of the sovereign debt crisis that was precipitated by the

Table 7.1 *The impact of the Global Financial Crisis (growth rate of per capita GDP – annual % at constant $US)*

	2007	2008	2009	2010	2011
East Asia and Pacific	5.4	2.2	−0.2	6.2	3.2
European Union	2.8	−0.1	−4.6	1.8	1.4
Latin America and Caribbean	4.1	2.6	−3.3	4.4	2.7
Middle East and North Africa	2.8	2.7	−0.4	2.6	3.6
North America	1.0	−1.2	−4.0	1.6	1.1
South Asia	7.6	2.4	6.3	8.1	4.7
Sub-Saharan Africa	4.1	2.3	−0.7	2.2	1.8
World	2.8	0.2	−3.3	2.8	1.6

Source: World Bank (http://data.worldbank.org/ accessed 3 September 2013)

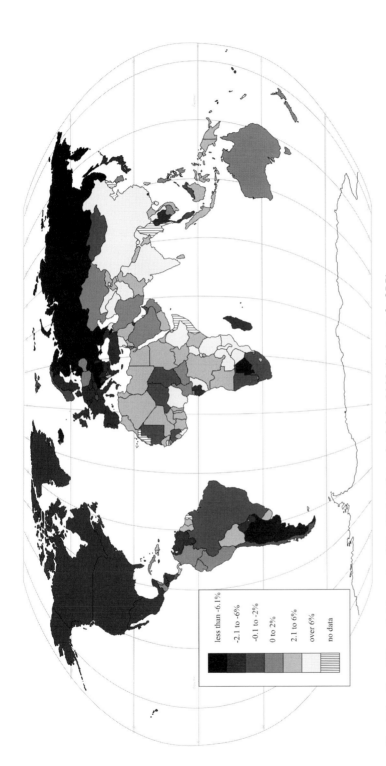

Map 7.1 *The differential impacts of the Global Financial Crisis, 2008 (% change in GDP)*

less than -6.1%
-2.1 to -6%
-0.1 to -2%
0 to 2%
2.1 to 6%
over 6%
no data

Source: World Bank (http://data.worldbank.org/ accessed 3 September 2013)

earlier financial downturn and cost of the stimulus packages that were launched in order try and salvage something from the economic wreck. Whilst the financial impacts of the GFC should not be underestimated and continue to resonate, the continued social and political problems brought about through direct and indirect impacts across the world will continue for many decades.

The geographic effect of this downturn was uneven as is shown in Map 7.1. In general, those economies that were most heavily reliant upon financial flows were impacted the most profoundly. In Iceland, for example, the impacts were devastating, leading to a major banking crisis in 2008, the collapse of the government in 2009 and the enactment of profound austerity measures and economic restructuring. In countries where the housing bubble had been especially pronounced, the effect was also very serious – Ireland, for example, entered a period of recession that reversed a decade of relatively spectacular economic growth which had earned it the name the 'Celtic Tiger'. There were similar knock-on political impacts across the world and in many countries governments changed in part due to the fallout of the crash – as was the case in places as diverse as the United Kingdom, Chile and New Zealand. In the Global South, the impacts were felt differently as we shall see in Chapter 8 – reductions in some aid flows, increased trade protection in the North, reduced remittances due to more harshly imposed migratory rules between poorer counties and richer economies together with the direct financial implications of the downturn in terms of capital flows took their toll; but the relative slowdown was not as great (see Map 7.2). In this sense, the GFC has led to a closing in the economic gap at the global scale in a way that is likely to have profound effects in the future. Notwithstanding, many of the negative impacts in the Global South are likely to be intergenerational and have a large lag time – a reduction in education expenditure has long-term impacts that are difficult to predict. Thus while the impacts were more acutely felt in the advanced capitalist world, it is likely that those impacts will be less long-lasting as is often the case during economic downturns, as happened in the Great Depression and was noted by the structuralist thinkers there (Kay, 1989). (See also Bello, 2007; Aalbers, 2009.)

China continued to grow during the general downturn, and the demand for commodities to fuel that growth increased; a number of countries where commodities played an important role in overall economic activity were protected from the worst impacts of the downturn. Some economies in Latin America, for example, that export mineral and agricultural products to East Asia, were spared the worst impacts – as happened in the case of Brazil. Australia was protected from the worst impacts of the crisis given a boom in energy and mineral prices due to Chinese demand. Some have argued that commodity price increases – especially in the case of oil – were partly to blame for the crisis in the first place, given the rise of speculative investment that saw these prices rise in the early 2000s (McMichael, 2009). What is certain is that those countries that have seen increase in commodity exports have developed a dangerous reliance on a relatively narrow economic base, and this is having untold environmental impacts. In this sense the concept of the resource curse (Auty, 1993; Barton and Murray, 2008) remains relevant (see Chapter 8).

The response to the crisis varied from country to country. In some, large bailouts of the banking sector were enacted that sought to ensure that the flow of credit would continue. Other countries nationalized some or all of their banking sector. The 'bailout' in the USA enacted under George W. Bush was the largest single governmental financial intervention in history. The stimulus package of President Barack Obama signed in 2009 – the American Recovery and Reinvestment Act – has been estimated to cost $US831 billion. There were similar bailouts and stimulus packages in other economies including the UK, France and Australia. Other governments chose to reduce government expenditure and did not undertake large wholesale bailouts such, as in New Zealand for example. Critics of the bailouts argue that financial institutions were rewarded and protected from the impacts of their own failure. Furthermore, the rapid return to the high bonus culture that characterized the late 1990s and 2000s financial sector was argued to perpetuate the same incentives that led to the crisis in the first place. In this regard Harvey is scathing:

> The US political class has so far caved in to financial pragmatism and not touched the roots of the problem. . . . Will the powers that currently hold sway seek merely to clean up the problem at popular expense and then give the banks back the class interests that got us into the mess? This is almost certainly where we are headed unless a surge of political opposition dictates otherwise.
>
> (2011, pp.11–12)

One of the consequences of large bailouts was to increase the sovereign debt of countries undertaking this route. In 2008, the USA spent over $US5.2 trillion which represented over 40 per cent of GDP for that year. The USA now has foreign debt of $US16.7 trillion and much of this is owed to China – an outcome that is leading to a profound restructuring of the power relations in the global economy. In a number of countries, the huge debt burden has become unsustainable. The sovereign debt of Greece was partly financed by the European Union on the condition of harsh austerity measures, an outcome that led to enormous tension within the EU with regard to the flow of capital funds from the larger economies such as Germany and the United Kingdom. The austerity cuts led to large-scale protests within Greece and led to the changing of the political order, as noted previously. Similar problems were being felt in Portugal, Italy and Spain at the time of writing. One consequence of this European crisis was the rise of scepticism in some countries with regard to EU membership. The UK's membership was being openly questioned in 2013 and anti-EU parties such as the UK Independence Party made impressive gains in local elections.

In many countries, undertakings were made to re-regulate the financial sector so that the process that led to the downturn would be prevented in the future and the signals detected early. In the USA these measures were seen by some as woefully inadequate (Stiglitz, 2010) and represented a return to the status quo.

> One might have thought that with the crisis of 2008, the debate over market fundamentalism – the notion that unfettered markets by themselves can ensure economic prosperity and growth – would be over. One might have thought that no one ever again – or at least until memories of this crisis have receded into the distant past – would argue that markets are self-correcting and that we can rely on the self-interested behaviour of market participants to ensure that everything works well.
>
> (Stiglitz, 2010, p. xv)

In general, however, there has been a return to interest in Keynesian ideas in terms of macroeconomic management that sees an active role for the government in stimulating aggregate demand. This contrasts with the neoliberal approach that has dominated since the early 1980s – where monetarist policy, principally enacted through control of the money supply through interest rates, dominated. However, the manner in which the debt burden has been passed from the banks to the state and ultimately to the people is a point that is of moral outrage according to Harvey and others that think like him.

> The term national bail out is inaccurate. Tax payers are simply bailing out the banks, the capitalist class, forgiving them their transgressions, their debts and only theirs. The money goes to the banks but so far in the US not to the homeowners who have been foreclosed upon. And the banks

are using the money, not to lend to anybody but to reduce their leveraging and to buy other banks. They are busy consolidating their power.

(Harvey, 2011, p. 31)

One of the most profound outcomes from a social viewpoint has been a widespread movement that has grown partly in response to the kinds of solutions that have been pursed in the wake of the GFC. The Occupy movement, which grew out of an earlier anti-globalization movement, has taken as part of its cause the response to the GFC. The initial occupation of Wall Street in 2011 (see Chapter 5) and the reclaiming of that space on behalf of 'the 99%', as the slogan read, was a direct response to the perceived injustice of solutions that seems to sustain the status quo and reward a sector for the activity that had created the problem in the first place (see Chapter 5 for more detail on the roots and nature of the Occupy movement and how this links to the anti-globalization movement more generally). There were widespread protests across the world in the wake of the Occupy movement and these are also related to the anti-austerity movements in places as diverse as Greece and Brazil. Some have even credited the spark for the Arab Spring as a reaction to the recessive impacts of the GFC.

From a critical point of view, there are some interesting points that can be made in terms of the terminology used to describe and analyse the 'crisis'. The source and major impacts of the downturn were felt in advanced capitalist nations, especially those that had a high reliance on financial flows for their GDP. Some commentators have referred to it as the 'moral' crisis (Stiglitz, 2010) or the American Crisis (Sidaway, 2008). Harvey (2011) makes the case that the institutions that caused the crisis made out that the consequences would be truly global and lead to 'the end of the world as we know it' p. 5, in order to receive more assistance in terms of state bailouts using public money and money borrowed on overseas markets in countries that had been affected. This reminds us that terminology of the 'global' is in fact political and socially constructed. It may also be the case that the hyperbole caused by the institutions themselves in their search for bailouts led to a self-fulfilling prophecy and economic forecasts of the crisis precipitated, or at least made worse, that very thing.

> [T]he treasury secretary, who is a former resident of Goldman Sachs, emerged with a three-page document demanding a $700 billion bailout of the banking system while threatening Armageddon in the markets. It seemed like Wall Street had launched a financial coup against the government and the people of the United States. A few weeks later with

caveats here and there and a lot of rhetoric, Congress and then George
Bush, caved in and the money was sent flooding off without any controls
whatsoever, to all those institutions deemed too 'big to fail'.

(Harvey, 2011, p. 5)

So, what caused the crisis? This is a question that vexed even the Queen
of England:

When Her Majesty Queen Elizabeth II asked the economists at the
London School of Economics in November 2008 how come they had not
seen the current crisis coming (a question which was surely on
everyone's lips but which only a feudal monarch could so simply pose
and expect some answer), the economists had no ready response.
Assembled together under the aegis of the British Academy,
they could only confess in a collective letter to Her Majesty, after six
months of study, rumination and deep consultation with key policy
makers, that they had somehow lost sight of what they called
'systemic risks', that they, like everyone else, had been lost in a
'politics of denial'.

(Harvey, 2011, p. vii)

In terms of the short-term impact, research undertaken in the USA
principally have sought to explain the initial financial crisis associated
with the housing bubble and credit markets. A number of reports have
argued that poor regulation, overreliance on complex financial
instruments and incorrect pricing of risk are to blame. For example, the
Financial Crisis Inquiry Commission of the United States concluded:

the crisis was avoidable and was caused by: widespread failures in
financial regulation, including the Federal Reserve's failure to stem the
tide of toxic mortgages; dramatic breakdowns in corporate governance
including too many financial firms acting recklessly and taking on too
much risk; an explosive mix of excessive borrowing and risk by
households and Wall Street that put the financial system on a collision
course with crisis; key policy makers ill prepared for the crisis, lacking a
full understanding of the financial system they oversaw; and systemic
breaches in accountability and ethics at all levels.

(Financial Crisis Inquiry Commission, 2011)

The lack of consensus concerning the short-term impacts pales into
insignificance when compared to the debate concerning the deeper
underlying causes of the downturn. Whilst the short-term factors have
been identified, the longer-term causes are matters of intense debate. For
Stiglitz (2010), the issue is one of 'unreal economies', that is to say an
overwhelming reliance of the world economy on capital flows that have

no direct root in the real economy and are thus fragile and subject to inherent instability. The solution, then, is to move away from reliance on speculation in capital and currency markets and to return to core economic pursuits in production and trade. For McMichael (2009), the problem is one of the coming together of a number of crises including a food crisis and an oil crisis all of which are symptomatic of a deeper misplaced economic strategy for global progress, and this a point that we echo in this book continually (see also Taylor, 2009). There are those such as Krugman (2010) that argue that fundamental to the crisis is the process of inequality – that those on lower incomes wanting the vast consumer wealth they see others consume precipitated the credit boom, and predatory lending led to the speculative bubble. Inequality, sustainability and financial instability are linked in this sense.

Joseph Stiglitz, a former World Bank insider who, at the time of writing, has been critical of the neoliberal orthodoxy and bank policy for over a decade and has become very influential in that regard, argues that this is nothing short of a watershed moment in human history. The final chapter of his book *Freefall,* on the GFC, is titled 'Towards a new society' no less; and in it he argues for a wholesale reconfiguration of the values and norms that drive society and culture and the interaction of humans and the environment. All of this is attainable through a financial system that works for people on the basis of decisions made with the community in mind; that recognizes the consistent faults in the markets; that rebuilds trust; that values the right things; and that places a moral agenda at the core of all these pursuits. This is a long way from the technocentric responses to the GFC that others have favoured and talks of a much deeper problem and set of possibilities that are associated with the nature of globalization in a much more general sense. In this context Stiglitz argues:

> The rules of the game have changed. The Washington consensus policies and the underlying ideology of market fundamentalism are dead. In the past there might have been a debate over whether there was a level playing field between developed and less developed countries; now there can be no debate. The poor countries simply can't back up their enterprises in the way that the rich do, and this alters the risks that they can undertake. They have seen the risks of globalization badly managed. But the hoped for reforms in how globalization is managed still seem on the distant horizon.
>
> (2010, p. 296)

Notwithstanding the size of the challenge, Stiglitz is relatively optimistic on the possibilities that lie ahead and argues:

> We now have the opportunity to create a new financial system that does what human beings need a financial system to do; to create a new economic system that will create meaningful jobs, decent work for all those that want it, one in which the divide between the haves and have-nots is narrowing rather than widening; and most importantly of all to create a new society in which each individual is able to fulfil his potential, in which we have created citizens who live up to shared ideals and values, in which we have created a community that treats our planet with the respect that in the long run it will surely demand. These are the opportunities. The real danger now is that we will not seize them.
>
> (2010, p. 297)

Harvey is less optimistic and even more profound in his analysis of the underlying causes. He argues that the ultimate cause of the crisis is the need to reduce the turnover time of capital in order maximize the velocity of circuits of capital to maintain profits. In order to do this, argues Harvey in *The Enigma of Capital* (2011), there has been a shift over previous decades away from real production into financial speculation and capital flows that have become dangerously precarious and thus subject to bubbles. This also acted as a way of solving the absorption of capital problem whereby increased profits needed to be recycled through investment in the system. Thus in order to sustain the growth of the economy in the face of nearing the limits of real production, credit has been extended in predatory ways to people who are unable to afford it, stimulated by a marketing industry and cultural change that is increasingly materialistic. This has led to damage to the environment and to the eventual collapse of the system. With increasingly scarce real foundation to the economy, the instability is likely to become increasingly problematic:

> Does this crisis signal . . . the end of free market neoliberalism. . .? The answer depends on what is meant by that word. My view is that refers to a class project that collapsed in the crisis of the 1970s. Masked by a lot of rhetoric about individual freedom, liberty, personal responsibility, and the virtues of privatisation, the free market and free trade, it legitimised Draconian policies designed to restore and consolidate capitalist class power. The project has been successful judging by the incredible centralisation of wealth and power in all those countries that took the neoliberal road. And there is no evidence that it is dead.
>
> (Harvey, 2011, p. 10)

Harvey is clear on the enormity of the political challenge that lies ahead:

> The logic of endless capital accumulation and endless growth is always
> with us. It internalises hidden imperatives, to which we either willingly or
> mindlessly submit, no matter our ethical inclinations. The problem of
> endless compound growth through endless capital accumulation will have
> to be confronted and overcome. That is the political necessity of our
> times.
>
> (Harvey, 2011, p. 27)

We believe that the current crisis represents just one further chapter in an
inherently unstable and unsustainable global economic system. Without
meaningful regulation, it is unlikely that such crises will be averted in the
future. Such regulation includes shifting assets towards the real economy
and regulating the activities of banks so that safeguards are put in place
that prevent speculative bubbles that wreak such havoc. Furthermore, the
systems of bonuses and payment incentives for the banking system also
require overview. There was an opportunity to move beyond the
neoliberal financial system following the crisis – but this was not taken
and as such it is only a matter of time before the next one (below we
consider some other solutions to the unstable financial system). In a
broader and deeper sense we see the GFC as part of an unfolding three-
part crisis, none of which are solvable in isolation – there is an
environmental crisis and an inequality crisis, both of which in part have
been precipitated by the very practices that led to the financial crisis.
Terming them 'crises' in many ways moves attention away from the fact
that we are in a deeper and a perpetual malaise that is associated with the
pursuit of globalization as currently practised.

Neoliberalism and the globalization of agriculture: A case study

Neoliberalism has had a profound effect on the nature of global
agriculture, in ways that undermine both equity and sustainability. By
way of a case study of the impacts of neoliberalism, we outline the
evolution of the global agricultural sector in what follows. Agriculture is
transforming as rapidly as any other sector in the global economy. Over
the past century, the definition and nature of farming and food production
has moved from the traditional model of family-based production for
local and national markets to a very complex activity which is global in
its extent (Whatmore, 2009, 2002). This has forged what some have

called a 'new political economy of agriculture' both driving and driven by processes of globalization where systems, chains and networks link the four corners of the planet in unprecedented ways (LeHeron, 1993). The new global agriculture, however, is – as is the case with all sectors touched by globalization – not homogeneous or on a predetermined path to a particular end-state. If anything, the processes that have been unleashed by globalization have made global agriculture a more uneven and contested terrain than ever. For, while in a number of cases we are seeing the evolution of industrialized and truly globalized agricultural production complexes (for example, in the case of fresh fruit), paradoxically we are seeing the evolution of new niche and localized forms (sometimes referred to as post-productive agriculture). At the same time, subsistence agriculture remains of critical importance to the livelihoods of many in poorer countries (Murray, 2012). This section looks at one of the most understudied of the world's sectors in terms of the nature and impacts of globalization.

Plate 7.2 *The Lautoka food market, Fiji*
For many farmers in the developing world, local markets are critical. In this market in Fiji a wide variety of foodstuffs produced by local farmers and fishers (alongside a small amount of imported food) provides good fresh and affordable food for urban consumers.

Whatmore (2002, pp. 57–58) conceptualizes the contemporary agricultural system as an *agro-food complex*. Drawing on an OECD definition, this is defined as the set of activities and relationships that interact to determine what, how much, by what method and for whom food is produced and distributed. This complex has been transformed by two important, and overlapping, trends: the growth of 'agribusiness' and its globalization.

The growth of agribusiness

In an attempt to increase profits the agricultural sector has industrialized and commercialized, with the effect of subsuming non-capitalist agriculture. This expansion has taken place through processes of horizontal and, especially, vertical integration (see section above on TNC growth). The latter has involved the linking up of the different nodes of the agro-food complex – from research laboratory, to the field, in transport and to the supermarket (see Figure 7.3). This process has allowed companies to internalize risk and reduce transaction costs, thereby increasing profits. In general, integration has led to oligopsonistic competition as TNCs take over significant proportions of the industry to control where production, marketing and distribution function. There are numerous examples of large-scale agribusiness TNCs that play an overwhelming role in their respective agricultural sectors, including Monsanto (technological innovation and GM), Nestlé (food production), Fonterra (dairy production) and Bulmers (cider production). Although, in theory, the efficiency gains of industrialized agriculture can be transferred to consumers as lower prices, generally the resulting non-competitive structures result in higher mark-up levels. Some also argue that food choice is eroded also.

Globalization of agribusiness

The agribusiness model has diffused globally over the past 50 years and is associated with the broader expansion of capitalism and modernization following the Second World War. Over the past two decades, facilitated by the liberalization of the world's economies, agribusiness TNCs have sought direct and indirect links into sectors outside their home countries.

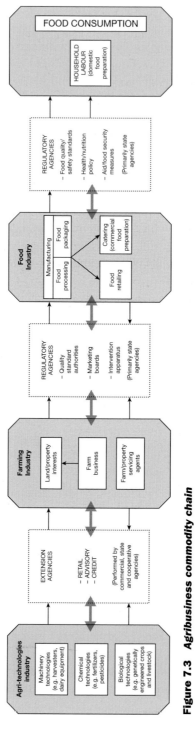

Figure 7.3 Agribusiness commodity chain

Source: Adapted from Whatmore (2009a, 2009b)

Plate 7.3 *A large dairy factory in New Zealand*
This factory owned by Fonterra – a company which is responsible for 30 per cent of the world's dairy exports – processes milk from dairy farms and produces butter, milk powder and cream products largely for export.

This may be conceptualized as an agricultural NIDL, with the aim of creating a mass food production system globally. Given relatively cheap land and labour costs combined with lenient environmental regulation, this has led to the diffusion of agribusiness to the Third World. This has been especially pronounced in fruit and horticultural complexes, for example, since these are labour intensive at the production and packing stage and because TNCs seek out localities where items can be produced counter-seasonally (i.e. harvested and supplied to affluent Northern Hemisphere markets in their winter). This has led to the production of what may be termed 'Fordist' fruit (McKenna and Murray, 2002; Murray, 2009). In general, neoliberal agriculture is drawing rural localities and communities in the Third World into the ambit of globalized circuits of capital as never before.

How is the globalized agro-food complex structured? Basically, different nodes of any given network are located in different nation-states, linked through horizontal and vertical integration and regulated by agents in between those nodes. Vertical integration may involve *ownership* of the different nodes of production by the same agribusiness TNC (direct vertical), or it may consist of the linking of different functions through *contract relations* (indirect vertical). Regulating agents – which may be state, quasi-state or privately owned – interact with commodity flows and circuits and intervene to govern the complex (Whatmore, 2009b). For example, in New Zealand, the Ministry of Agriculture, Forestry and Fisheries (a state body) and Agri-quality (a commercial company) play a role in maintaining food safety and quality for retailers. Until recently, the global marketing of New Zealand fruit and dairy products was undertaken by state boards that have now been privatized (see McKenna and Murray, 2002; Shaw-Taylor, 2009). Agricultural regulatory bodies may also be involved in: research and development of agro-technology; the extension of technology, knowledge and finance; health monitoring; and food security/aid issues. Given the move to open up economies, it has been argued that the nation-states are less effective regulating agents than they once were. Thus, the rise of global agricultural networks raises the need for globalized regulation of supply. This is made more urgent by the rising concern in the West over food safety. The worldwide spread of diseases such as CJD alerts us to the vulnerabilities that are created by transnational food networks.

The globalization of agriculture also has a cultural globalization effect as populations are increasingly exposed to agro-produce from distant localities. This effect works in many directions. In the West, the new networks supply items that were hitherto unavailable (for example, fruit in the wintertime in the UK was relatively unheard of until the globalization of fruit began in the mid-1970s) (see Murray, 1998; Maye and Ilbury, 2012). On the other hand, non-Western countries are increasingly exposed to Western diets. McDonald's and Coca-Cola are the most prominent examples of TNCs supplying relatively homogeneous Western products around the globe.

In order to make sense of the history of the global agro-food complex the idea of *food regimes* has been developed. This theorizes the transition of agriculture in the context of the evolution of global capitalism in general. Under this schema, which is influenced by the French regulation school of thought, the history of the agro-food complex is divided into three phases, separated by restructuring crises (see LeHeron, 1993; Whatmore, 2002; Murray, 2012):

- *First food regime* (1780s–1945) – This is based on the colonial division of labour whereby colonial territories supply imperial and former/neo-imperial powers with unprocessed and semi-processed items. The major trade during this period was in grains and meat, a classic example being the beef production system of Argentina in the 1800s. This phase corresponds to the industrial phase of the first wave of globalization identified in Chapter 3. The Great Depression and the First World War led to the end of this regulatory regime.
- *Second food regime* (1945–1980) – This is sometimes referred to as the 'productivist' phase in agriculture, and is characterized by the rise of intensive agriculture. Grain-fed livestock systems and the production of fats/durable foods played a central role, the UK dairy system being an example. The regime saw the rise of globalized capitalist agriculture in the periphery as well as large-scale subsidy systems in the core (especially the EU and USA). This phase corresponds broadly to the modernization phase of the second wave of globalization and created a NIDL in agriculture. The oil crises and the inability of nation-states to maintain subsidies, together with oversupply and safety concerns, are leading to the eclipsing of this regulatory regime.
- *Third food regime* (1980–) – This is referred to as 'post-productivist' agriculture, which is less intense and more diversified than the preceding regime (see section below). It involves the search for quality and safety by consumers. The central importance of biotechnology, as well as the rise of flexible and small-scale niche agriculture in the core, characterizes this phase. It corresponds with the neoliberal phase of the latest wave of globalization and more generally mirrors the rise of flexible accumulation. The rise of organic farming worldwide is one of the principal outcomes of this regime. The rise of counter-seasonal fruit and vegetable exports is also sometimes associated with this phase (see Box 7.3).

Overall, despite its analytical utility, global agriculture is more fragmented and uneven than this conceptual model can possibly capture.

Although food regimes are useful concepts for organizing thought about the globalizing agro-economy they are macro scale and require qualification at the local scale. There are many remaining controversies. For example, although some have seen the rise of the global fruit and vegetable system as symptomatic of the third food regime, much of the complex is intensive in its nature and standardized in terms of output. In reality, as is the case with many economic sectors, a mixture of Fordism and flexibility more accurately reflects the nature of the agro-food complex. Notwithstanding this, we look at some evidence for the rise of 'post-productivist' agriculture in the subsequent section.

Box 7.3

Fruitful globalization?

The global fruit sector is one of the most truly globalized of all the agricultural subsectors. Over the past 100 years, the flows that constitute the complex have intensified and widened in geographical scope, creating a configuration that sees different nodes of supply networks spread across many continents (see Murray, 1998, 1999, 2009). Trade in fresh fruit has grown considerably in volume and value terms over the past four decades. In the years 1961 to 1963 the average total global export value of the ten most important fruits stood at US$1.56 million; by 1990 to 2001 this figure had risen to US$19.5 million (nominal figures). In 2012 the figure stood at over US$40 billion (Food and Agriculture Organization of the United Nations, 2013). In 1969 the total volume of fruit exports was approximately 19 million metric tonnes per annum; over 600 million metric tonnes per annum are now traded. The role of fruit in the agricultural economy has also grown. In 1980, fruit and vegetable exports represented 11 per cent of total global agricultural exports; this figure had risen to 16.5 per cent by 2001 and over 20 per cent at the present time. It is this evolution which allows citizens of the Northern Hemisphere

to eat Chilean grapes in December, New Zealand apples in November and South African passion fruit in January. The fruit system illustrates many of the points made above concerning restructuring, spatiality and equity in global agriculture. Using a regimes framework the evolution of the sector may be traced as follows.

The first fruit regime

The origins of globalization in the fruit complex are found in the period of the first food regime, when the dominant form of fruit trade involved the export of a relatively small range of then 'exotic' products from colonies to imperial cores. Trade in bananas, a product which could not generally be grown in the core, and was relatively non-perishable, was most important. The fruit was produced in parts of Southeast Asia, Latin America, the Caribbean and some parts of Africa, shipped to Europe and subsequently to the USA. This system was almost fully extractive and based around plantation agriculture, leading to local enclaves where backward and forward linkages into the wider economy were limited.

Box 7.3
continued

Second fruit regime

It was during the intensive phase that the global fruit system, as we know it today, developed. An NIDL in fruit was forged,

involving the 'going abroad' of fruit TNCs. The spatiality of this is illustrated in Figure 7.4, which conceptualizes the global fruit chain. Fruit TNCs, such as Del Monte, Standard Trading, Dole and Chiquita, played

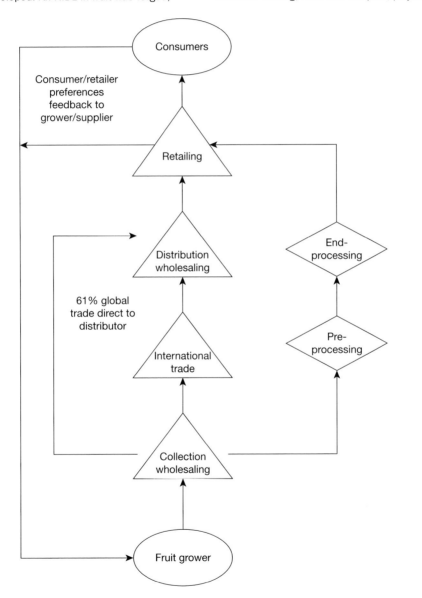

Figure 7.4   *The global fruit chain*

Source: McKenna and Murray (2002, p. 506)

Box 7.3
continued

Plate 7.4 *Global fruit – Chilean* temporeras *pack apples for export to the EU*
The evolution of the Chilean fruit export economy has created demand for female labour which, in many cases, is employed outside the home for the first time.

a central role in this globalization as they actively searched out cheap pools of labour and natural resources. Geopolitics also played a role in the diffusion of the complex as the USA supported its fruit TNCs in Central American countries such as Honduras and Nicaragua through economic and, during the Cold War, military intervention (see section on geopolitics in Chapter 5). Many Third World countries, especially in Latin America, reoriented their agricultural production systems and became geared towards the satisfying of 'luxury' fruit demands in the core, often at the expense of domestic production of food. This threat to food security is one of the most damaging legacies of the rise of 'non-traditional agricultural exports'.

The factors driving globalization during the second fruit regime were complex and variable. They may be divided, somewhat artificially, into socialcultural, economic and technological factors with many of the impulses coming from the demand end of the chain.

1 Social factors
   ● *Evolution of middle class*. Higher disposable income and changes in lifestyles – especially travel – bring members of this class into contact with new fruits which are now affordable. Rising dietary concerns stimulate increased per capita fruit consumption. Furthermore, immigrants who arrive to take the place of the former working classes in Western

Box 7.3
continued

economies bring new dietary habits with them.

- *Ageing populations*. Older populations accumulate greater knowledge, have more leisure time to inform themselves, use more resources on travel and thus spend a greater proportion of their income on fruit.

2 Technological factors

- *Development of cool chains*. These are integrated systems of supply from the orchard to the supermarket involving shipment in refrigerated vessels. They were developed essentially in the 1960s, and facilitated the transport of perishable items over large distances.
- *Transfer of technology*. Fruit growing is a highly technical operation, with each locality needing special appraisal and adapted technology. The source of knowledge and hardware has been agribusiness TNCs, as Third World governments are often constrained in supplying such factors.

3 Economic factors

- *Rising disposable incomes*. Growth of the world economy leads to a more than proportional increase in demand for fruit given the 'income elasticity of demand' for this relatively luxury product. This may change, however, as fruit becomes a standard product and luxury status fades.
- *Increased mobility of capital*. Going abroad to establish foreign fruit supply began in the 1960s. In the neoliberal 1980s, foreign investment of this type increased dramatically.

Third fruit regime

According to regulation theorists we are now in the throes of a crisis in the global fruit system, involving oversupply in core countries, combined with safety issues and environmental concerns in general. It has been suggested that the second regime will be replaced by a third which bears much in common with post-Fordism in general (Whatmore, 2002; Maye and Ilbery, 2012). The possible nature of this regime in the context of fruit is contested. LeHeron (1993) has termed the period to which we are headed the 'integrated stage'. This will be characterized by a greater importance of biotechnology, employed in a flexible way to satisfy 'niche' demands. This will increase the possibility of research-intensive production in Western economies themselves and thus may reduce the need to go abroad. It may also be characterized by a renewed search and construction of 'exotics' – meaning that large-scale exploitation of vast amounts of stock fruits, such as grapes from Chile or apples from South Africa, may become less marketable. These trends are visible in some fruit-supplying localities – although the extent to which the basic Fordist model is becoming undermined is uncertain. In reality the future global fruit complex is likely to be characterized by multiple strategies by suppliers, with some following the standardized price/quantity route (the 'commodity' fruit), while others are more likely to aim for the quality end of the market, investing more in product differentiation and innovation (the premium or 'luxury' fruit). The Southern Hemisphere's two largest fruit exporters, New Zealand and Chile, have been differentiating their production on this basis for over a decade (see Figure 7.5 for a model of this).

Box 7.3
continucd

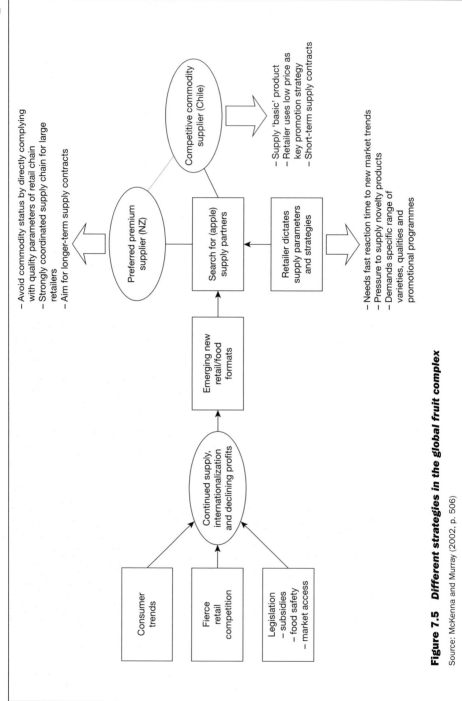

Figure 7.5 *Different strategies in the global fruit complex*

Source: McKenna and Murray (2002, p. 506)

Post-productivist agriculture – new food networks?

Although an increasing proportion of the world's agricultural production is undertaken by TNCs, localized non-corporate agriculture still remains very important (Woods, 2007). Millions of farmers across the world, the majority in terms of absolute numbers, are not linked explicitly into globalized networks and supply local markets only. There are also many millions of peasant farmers who are bypassed by commercialism entirely, and subsistence agriculture remains central in many societies in Africa, Asia, Latin America, and especially the Pacific Islands (Wilson and Rigg, 2003). This is not to say that global processes do not impact upon such groups of farmers. There is mounting evidence that such agriculture is being squeezed out by agribusiness as smaller farmers lose their means of production, including land, and are increasingly employed as labourers on commercial farms or migrate to large cities (Bryceson *et al.*, 2000; Murray, 2002a). Simultaneously, we are seeing the return to the practice of smaller-scale agriculture in some places that is in direct

Plate 7.5 *Post-productivist agriculture – lifestyle development and vineyards in New Zealand*
This hoarding advertises a property development in the wine-producing district Hawke's Bay, New Zealand. The development aims to turn a former (productivist) pastoral farm into an area that combines residences, small-scale vineyards and pastoral grazing.

resistance to globalization. In the core, it is possible to discern a partial shift to new alternative forms of flexible agriculture which some see as evidence for the third food regime. This has been labelled 'post-productivist' agriculture (see McCarthy, 2007; Murray, 2012) and is characterized by a move away from intensive, concentrated and specialized agriculture towards diversified, extensive and niche-oriented production. The nature of this agriculture is outlined in Table 7.2 (based on Maye and Ilbery, 2012) and is sometimes referred to as 'multifunctional'.

The transition to post-productive agriculture is explained by a range of factors linked to the perceived costs of the intensive globalized model of agriculture and the wider regulatory crisis (Wilson, 2005). First, government intervention is being rolled back, meaning that large-scale productivist sectors can no longer be supported. Although agriculture is the last sector to receive the WTO-inspired deregulation, there are have been signs that the largest intervention systems of all, such as the EU's Common Agricultural Policy, are being dismantled. Second, increased concern for the environment has fuelled the promotion of less resource-intensive agriculture based on principles of permanence and sustainability. For example, the number of permaculture communities currently operating in New Zealand and Australia has increased of late as some people choose to drop out of the industrial agricultural system for reasons of both lifestyle choice and food integrity (Argent, 2008). Finally, consumer demand has been transformed; and consumers in the West, and elites in developing countries, are increasingly motivated by quality over

Table 7.2 *Post-productivist agriculture*

Major characteristics

- A reduction in food output and greater emphasis on food quality
- Progressive withdrawal of state subsidies for agriculture
- The production of food within an increasingly competitive and liberalized international market
- The growing international regulation of agriculture
- The creation of a more sustainable agricultural system

Farm restructuring

- Diversification of farm resources into new agricultural products
- Diversification of farm resources into new non-agricultural products (e.g. tourism, retailing)
- Off-farm employment
- Hobby and semi-retired farming

price. This has led to a worldwide boom in organic farming, for example, and safety and equity concerns have led to the development of sectors where consumers are able to trace the exact chain that delivers their food. Post-productive agriculture is generally thought to be applicable in richer economies only, which have markets where consumers are sufficiently affluent to trade price for other product traits (Woods, 2007). Ironically, in countries that have been largely bypassed by the globalization of agriculture, the implementation of such production may be easier to establish, as it broadly mirrors practices that have existed for centuries. For example, virtually all fruit production in the Pacific Islands region is 'organic'. However, the problem for peripheral societies is gaining access to networks of finance and marketing which will allow them to trade on their new comparative advantage in the context of post-productivist demand shifts in the affluent North.

Rather like flexible technology concepts, it may be argued that the notion of post-productivism is exaggerated. The bulk of the world's agricultural production is, and is likely to remain, intensively produced and standardized. In New Zealand agriculture – one of the world's most 'cutting edge' agronomies – the majority of production is still 'productivist' (see Plate 7.3). In reality we are likely to see a mixture of the two in any given situation.

In theory however, post-productivist agriculture has the potential to partly reverse the agricultural NIDL. Whether this is seen as a positive or negative outcome depends on one's opinion of the impact of agribusiness TNC investment. A further complication is that post-productivist agriculture is arising at a time when genetically modified varieties are spreading across the planet in a 'second green revolution'. This has the potential to revive the agricultural NIDL, and some Third World countries, such as India, are rapidly adopting such technology in order to capture export markets. What this points to is that global agriculture remains highly uneven and, despite penetration of national and local agricultures by 'global' capital, the patterns on the ground have increasingly diverse multifunctional geographies in the developed world and increasingly monocultural geographies in the periphery (see Murray, 2009a, 2012).

Reforming the global system

Does free trade have the potential for universal benefits as promised? Or is it likely to go on concentrating wealth? Free-trade proponents argue that globalization will increase the participation of poorer nations in the world economy. However, the share of world trade accounted for by poorer nations remains very low (see above section on contemporary trade patterns). There is no doubt that the gap between rich and poor has increased (as is elaborated in Chapter 7). The open question is: Is this a result of World Bank and WTO strategies, or is it because they have not been pursued with enough vigour?

There is growing consensus that the inequality of the world economy is not inevitable and that the system can be reformed in order to distribute the benefits in a more equitable and, possibly, efficient manner (Amin, 2004; Stiglitz, 2010) The World Bank and WTO argue counter to this that it is isolation from global capital circuits that causes poverty and that peripheral countries that have become enmeshed have improved their economic performance. The crisis should certainly lead to a reappraisal of this perspective. The question is not whether globalization should be allowed to occur; it is not possible to roll back the networks, ideologies and trends that have unfurled. However, '[t]he debate is instead over the forms that globalization should take. It is in other words over the priorities, rules and regulations governing globalization. It is also over which bodies should have authority over the process' (Gwynne *et al.*, 2003, p. 226). Although a considerable amount of literature has been drawn up with respect to such reform (see Amin, 2004; Bello, 2002; Stiglitz, 2002, 2004, 2010, for example), alternative 'bottom-up' regulation remains poorly articulated. This debate will be a defining feature of global politics over the coming decades. To those who doubt the possibility of building an alternative to the current hegemony Amin retorts, 'just look at the spell cast by neoliberalism' (2004, p. 231).

In a groundbreaking book, *Making Globalization Work* (2006), Joseph Stiglitz offers a new 'contract' that will deliver the change that is required. He claims that to make globalization work requires an international economic regime in which the needs of developing and developed countries are better balanced – a new global social contract that sees an increased role for the council of social and economic development of the United Nations. This contract includes the following:

1 a fairer trade regime that actually assists poorer countries (see later section);
2 a new approach to Intellectual property (IP) that allows access to lifesaving medicines and technologies for poorer countries;
3 an agreement to pay the poorer countries for environmental services, such as carbon sequestration;

4 acknowledgement that climate change is real and requires a global solution;
5 to pay fairly for natural resources and not leave behind environmental damage;
6 to achieve the goal of 0.7 per cent of GDP in aid to be spent on the poorest of the poor;
7 to increase the debt forgiveness package agreed at Gleneagles in 2005;
8 reform of the global financial system to make it more stable including a global reserve system;
9 establishment of a new institution to manage the cross-border activities of TNCs and resist monopolization and enforce liability;
10 moving beyond lip service in promoting democracy by not selling arms and working for genuine representation.

Towards fair trade

The Dictionary of Human Geography (p. 238) defines fair trade as:

> the demand that producers from poorer countries should not be denied the legitimate maximum rewards from their sales by the action of richer countries or other powerful agents.

A concept that eventually led to fair trade was first conceived of in the 1950s and applied to handicrafts; two decades later fair trade movements existed in a large range of commodity markets including coffee, cocoa and sugar. The rise of fair trade forms one part of a wider shift towards ethical consumption that takes into account the social and environmental impacts of global production systems, particularly where these are controlled by oligopsonistic interests in asymmetric commodity networks that see large agribusinesses interacting with small-scale producers. Fair trade refers, then, to a movement for justice in trading for producers from around the global as well as a range of products associated with that. Oxfam was instrumental in establishing the concept when it introduced the 'helping by selling' promotion in the 1950s. 'Trade not aid' became a popular slogan in the 1960s, and indeed it was adopted as the central phrase by the first Geneva conference of the UN Commission on Trade and Development (UNCTAD) in 1964. New Zealand was one of the pioneers in fair trade and a small not-for-profit society named Trade Aid was formed in 1973, after importing and selling carpets made by Tibetan refugees in Northern India and seeing the potential to link more directly producers in the developing world with consumers in the West.

The goals of fair trade are ultimately to reduce inequalities in the global trading system. Within this context there are a number of more specific points; better prices, more stable markets, greater flows of information such as prices, the promotion of ethical consumption, as well as the promotion of ethical production. Many fair trade campaigns have argued for a better price for producers. Others however have focused on the global institutional regime and how this is biased against producers' interests, especially in the case of small-scale operations in poorer countries. Issues related to this have included, for example, export dumping by groupings such as the EU due to subsidization as well as the pressure to adopt neoliberal reforms in economies where industry can only compete on low wages and poor environmental standards. Attention has also been drawn to the condradication of institutions based in the global north suggesting the latter reforms while at the same time continuing to protect agriclture.

Fair trade networks have been notable not just for the premiums that they pay to producers, but they have also seen the investment of the premiums in local and social development projects and capacity building including education, health, food security and sovereignty. According to Raynolds *et al.* (2007), research shows that the long-term benefits of these endeavours actually exceed the short-term benefits of higher prices.

In the application of fair trade policy and projects, NGOs have played an overwhelming role; and the rise of NGO activity and involvement in development should be seen as part of the unfolding of this story. A number of institutions have evolved from the 1980s onwards to monitor and certify fair trade. Fairtrade Labelling Organisations International (FLO) was formed in 1997 as an amalgamation of institutions and certification agencies and FLO owns and controls the fair trade certification mark that is recognized globally. In 1998, several agencies also banded together to form FINE, an informal association of fair trade organizations involved in advocacy. FLO certification guidelines are based on a set of core requirements for producers and buyers. Fair trade producers must be organized into democratic groups of small growers or estate workers who uphold high social standards and more effective environmental standards.

Fair trade buyers must guarantee payment of FLO minimum prices, pay a social premium, offer producer credit and long-term contracts, and trade as directly as possible with producers. While fair trade certification resembles other labelling initiatives, such as regulating ecological conditions in food or timber or labour conditions in garments or textiles, fair trade is the only such initiative that goes beyond establishing production standards to address trade conditions (Raynolds, 2009, p. 9).

The growth in fair trade consumption has been rapid over the last 20 years. In the 2000s it was rising at around 30 per cent per annum, and in 2010 sales of fairtrade products amounted to €4.3 billion. The vast majority of total trade is accounted for the by the FLO. The highest per capita consumption is in the EU and the UK is a major market (Figure 7.6). In 2012 the UK market for fair trade products was estimated at £1.53 billion, with (in order) sugar products, cocoa, bananas, coffee and tea the most popular items. USA and Canada are also very important markets for fair trade products, though it is important to note that fair trade markets are small outside of Europe, North America and Australasia (Figure 7.6). It is uncertain how the GFC has impacted the uptake of fair trade products however. Marketing campaigns for fair trade products are widespread, can be very creative and often younger

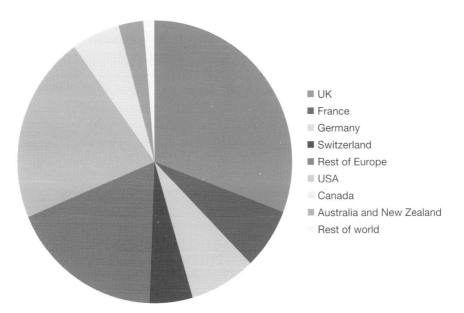

Figure 7.6 *Fair trade markets, 2010 (% of global fair trade sales)*

Source: fairtrade.net *Supplement to Annual Review 2010–11* (www.fairtrade.net/fileadmin/user_upload/content/2009/about_us/FLO_Annual-Financials-Sales_2010.pdf accessed 20 September 2013)

groups are targeted. Over the very recent past there has been a shift in universities across the West, many of whose campuses have adopted fair trade policies. Leading popular entertainers such as Coldplay and Neil Finn have also played a role in promoting the fair trade approach to development.

The major sector in terms of fair trade is coffee. This was the first sector to be developed and accounts for over 80 per cent of all fair trade in terms of value. In 2011 some 98,000 metric tonnes of fair trade coffee were sold worldwide. Bananas, sugar and cocoa have also emerged as important fair trade commodities and new certified products such as wine (see Box 2.2), flowers and recently timber are being traded increasingly. In terms of the source of production, Latin America is by far the most important supplier. Coffee fair trade was begun in Mexico and there are hundreds of institutions involved across the continent. Latin American fair trade coffee supplies over 80 per cent of the market but in Africa and Asia many coffee producers are now being certified.

Fair trade has been criticized on a number of fronts. Establishing what exactly is meant by 'fair' is difficult. Some have argued that by raising prices the fair trade system is often inefficient as it causes producers to stay in sectors that are uncompetitive and discourages their diversification into other activities – meaning that in the long run it is unfair. Some authors have criticized the power that the certifying agencies have over supply and others have noted how private companies in some cases have come to monopolize supply and extract surplus benefits. In some regions it is argued that the 'one size fits all' rule with regard to what fair trade means does not apply, and thus there has been a growing trend to circumvent the certification agencies and establish routes that bypass them. This has been happening widely in the Pacific where some have argued that the very specific and small scale nature of agriculture means that fair trade networks as practised by the FLO are not appropriate – producers are dispersed and attaining economies of scale is very difficult. Given the premium prices associated with fair trade products, others have commented that it has become a luxury only attainable by the wealthy middle class.

Other criticisms include the notion that the networks are focused on export and not production and as such undermine food security and sovereignty – goals which they are supposed to support. The way in which growers participate in the decisions that are made with respect to the distribution of the surplus has been controversial and, despite the rhetoric of participation, some research has shown that top-down structures often exist in this regard.

In support of fair trade, Raynolds (2009) argues that in the most successful examples it does represent a significant shift in the nature of commodity systems and chains that sees a significant reconfiguring of power, which has also taken place in the organic sector (see Chapter 9 for examples of ethical consumption in the environmental arena). Fair trade has some very high profile adherents (Stiglitz and Chappel, 2006). In his book '*Making Globalization Work*', Stiglitz (2006) dedicates a whole chapter to the strategy of fair trade as a means of correcting economic power imbalances and restoring justice to the global economy that favours developing economies. Demonstrating strong support, the *Encyclopaedia of Human Geography* (2009b) sees fair trade as part of the future of alternative globalization and the new cartography of justice. It argues (p. 311):

> The true significance of fair trade lies not in its albeit impressive markets, but in the extent to which this movement de-stabilizes conventional market principles. Conventional capitalist markets are guided by prices that understate the full ecological and social costs of production, and thus encourage the degradation of environmental and human resources, particularly in the global South. The fair trade movement makes visible the ecological and social relations embedded within a commodity and asks that consumers shoulder the true costs of production.

Conclusion – a crisis in regulation

The Global Financial Crisis that the world economy faced beginning in 2007 was not unprecedented, indeed there have been crises that wreaked havoc across the world economy since the beginnings of capitalism. It is clear that these have increased in magnitude and, debatably, severity since the shift to the neoliberal era. The global system has been redesigned by the institutions that support the free market to facilitate the flow of capital. Its deregulation allows the more rapid movement of capital and thus achieves the desire for time–space compression. There are, as such, unprecedented levels of economic interconnectedness and a heavy reliance on financial flows. Deregulation has sped up circuits of capital and neoliberal thought has sought to justify this in a range of economic political and even cultural arguments. The fragility that this creates was laid bare by the financial crisis – without a guiding hand from the state that sets rules that prevent a velocity of accumulation that is too rapid and without rules that ensure that the fruits of that accumulation are distributed in a just way, it is highly likely such crises will be repeated.

In reality the global crisis was not 'global' as such – it touched the very heart of the capitalist world, especially those economies that have become overreliant on capital flows. The opportunity to redesign the system so that such vulnerabilities might be protected against and so that the economy delivers what it should – welfare for human beings – has been missed at the time of writing. There have been shifts in the model at the margins, but what is required is a fundamental redesign that places people not capital at the centre of the economic problem. The problem is not globalization *per se*, but the way that globalization has been practised. As the case study of agriculture shows, the outcome is increased inequality. There are potential alternative and practical solutions as explored here, but there are powerful forces preventing the uptake of such progressive ideas. At present this shows little sign of changing and therefore we should expect more and ever deeper crises in the global economy over the coming decades. We make the case in the remaining two chapters and the final chapter of this book that solving this instability requires a simultaneous concern with the environmental and development challenges that are inherently interlinked with the economic one.

Further reading

Harvey (2011): This is one of the most compelling discussions of the roots, consequences and causes of the financial crash. It also considers what can be done to avert such an event in the future.

Harvey (2005): This brief history of neoliberalism outlines the economic, social, cultural and political roots and implications of this model and traces its irresistible rise up to the mid-2000s.

LeHeron (1993): Although 20 years old, this book is still the leading contribution to the globalization of agriculture literature.

McMichael (2012): In this, the fifth edition of the highly influential *Development and Social Change*, the author outlines the nature of the food crisis in the context of development patterns more broadly including resource and financial markets.

Maye and Ilbery (2012): This is a comprehensive chapter on issues of rural change and rural geography.

Murray (2012): This chapter reviews rural change in a global context, making the point that understanding the impacts of globalization in the Global North requires understanding of what is occurring in the South. It reviews regimes, post-productivism and multifunctionality.

Pollard (2012): In this chapter the geography of money and finance is discussed, including the fallacy that the financial sector has no embodied geography.

Raynolds *et al.* (2007): This has become the classic book on fair trade and its relationship with globalization.

Stiglitz (2010): Stiglitz has become one of the most prominent thinkers on the progressive reform of the global system. In this book he looks at the causes and impacts of the GFC and is relatively optimistic on the possibilities for reform.

Stiglitz (2006): This is a must-read for those who would like to know what can be done to alter the current model of globalization to make it more inclusive, sustainable and egalitarian. It is the natural sequel to *Globalization and its Discontents* (2004) by the same author.

8 ▶ Development, inequality and globalization

As people wake up in the world this morning they do so in very different conditions. Some are healthy, well clothed, employed and secure. Three-quarters of the world's population are not so lucky. The majority find themselves in unhealthy, poor conditions with little economic security. Many of the world's poor live in Third World rural areas, but it is the ranks of the urban poor that are swelling most rapidly. Clearly, living standards vary widely between countries; it is less obvious that they vary widely within countries (see Plate 8.1). Latin American countries, for example, have among the highest income inequality in the world. There are also pockets of deprivation in 'rich' countries. The principal argument of this chapter is that accelerated globalization has exacerbated unevenness throughout the world. Development geography studies patterns of inequality at various geographic scales – globally, within and between nation-states and localities and, more recently, within and between networks. A relatively young sub-discipline, it has a number of overlapping aims:

- to describe and map inequalities, especially as they relate to poorer regions;
- to explain and interpret the factors giving rise to such inequalities;
- to uncover historical trends of inequality across space;
- to study competing discourses of development and their spatial manifestations at various levels;

Plate 8.1 *Inequality on Tongatapu Island, Kingdom of Tonga*
These starkly contrasting royal and commoner homes are located within a
kilometre of each other. Feudalism persists in this highly hierarchical Pacific
Island society.

- to propose and assess frameworks for the reduction of geographic
  inequalities.

Development geography is an integrative sub-discipline in that it deals
with multiple spheres of human activity – economic, cultural, political
and environmental – in the context of understanding, interpreting and
analysing differentiated patterns of well-being. It has changed

considerably since its transformation from 'tropical geography' in the 1960s through the work of pioneers such as Keith Buchanan (1963, 1964) and Harold Brookfield (1975) (Power and Sidaway, 2004). Broadly speaking, the subject has evolved a greater concern for the political and cultural as opposed to the purely economic; it has shifted from positivism to a broader range of epistemologies and methods, including radical approaches and discourse analysis; and the focus on the nation-state has moved in part to networks. Notwithstanding this transformation, elements of the old and the new permeate the way development geography is practised across the world. Excellent, though differing, overviews are offered in Potter *et al.* (2008), Power (2003) and Willis (2009, 2011). Increasingly, development geography is moving from standard views which divide the world in various binary ways (rich/poor, North/South, First World/Second World/Third World, developed/developing, etc.) to a more varied understanding of the complexities of poverty and inequality within as well as between countries. Some of the old divides are still valid in many ways, and the global aid industry is still predicated on such understandings (for example, between rich donor countries and poor recipient countries) so it still important to understand and analyse global patterns of inequalities. This chapter argues that we need to maintain a focus on such global patterns whilst simultaneously examining more micro-level and societal inequalities.

Development and globalization – an uneven world

The proposition that globalization leads to uneven development has focused the attention of human geographers. To investigate this claim, we need to carefully establish how globalization and development might be related and to assemble empirical evidence hung within coherent theoretical frameworks. There have been few attempts to link theses of globalization and theories of development explicitly, and Table 8.1 offers some first thoughts. Broadly speaking, three views with respect to the developmental implications of globalization may be envisaged – *neoliberal, neostructuralist* and *dependency/post-development*, all of which build upon established theoretical traditions in development studies. The former, which draws on *hyperglobalist* views, sees globalization as a positive force for development. The second, which bears much in common with the *transformationalist* thesis, argues that the impacts of globalization on development depend on the way it is regulated. The final perspective, which links with *sceptical* and radical *hyperglobalist* views, posits that globalization perpetuates underdevelopment, arguing that discourses of both 'globalization' and 'development' represent extensions of imperial strategies of dominance and are thus much the same thing. As is clear from Table 8.1, these perspectives turn largely on the definition of development – something to which we turn subsequently.

As discussed in Chapter 4, the rhetoric of *neoliberal globalization* is that world income differentials are conflating and that prosperity will eventually be equalized. Evidence elsewhere in this book and beyond suggests that this is far from the case, and this is particularly so in terms of patterns of well-being and 'development'. Despite great absolute advances in poorer regions since the Second World War, the relative distribution of welfare, measured both between nation-states and individuals, is more uneven than ever before (Potter *et al.*, 2008; Willis, 2009). These 'gaps' have a long heritage, rooted in particular in the period of colonial globalization. Arguably, however, it is during the postcolonial wave, and especially the neoliberal phase, that welfare inequalities have crystallized as never before. The opportunities for those plugged into new networks of prosperity are not inconsiderable, but the vast majority of the planet's population are not linked in this way. This has the potential to fuel global conflict. Indeed, following 9/11 a number of commentators argued that the root cause of so-called religious fundamentalism was in fact inequality (Chomsky, 2004, 2012).

As this book has constantly argued, the nation-state is not necessarily the most appropriate unit for comparative purposes. Although national boundaries still play an important function in terms of regulating and containing politics, culture and economy, wider spaces, flows and networks have become increasingly important. These networks of power and privilege increasingly transcend traditional boundaries. In general, such networks are concentrated most obviously in richer countries. Conversely, 'black holes' of marginalization and deprivation – that is, gaps in the net – are more common in the poorer world. Notwithstanding this, elements of both exist in both 'worlds'; there are pockets of privilege in the Third World and pockets of poverty in the West. As such, it could be argued that 'worlds' exist within 'worlds'. However, in the face of the processes of globalization it is increasingly meaningless to talk of separate 'worlds' in a territorial sense (although this was never *really* feasible, as *comprador* dependency theory elaborated). Networks of inclusion/exclusion are part of the same global system and the spatial implications of this are complex. It is these networks that geographers need to get inside in order to understand the relationships between

Table 8.1 Theses of globalization and development theories – a schema

Development theory	Perspective on globalization	Definition of development	Explanation for lack of development	Outcome of development	Main strategy and policy
Neoliberal	Pro-globalization hyperglobalist	Market-based economic growth, modernization	State intervention corruption isolation	Convergence in incomes. Liberal democracy	Liberalization, deregulation, marketization
Structuralist/ neostructuralist	Alter-globalization transformationalist	Holistic income growth that is sustainable	Nature of insertion into global system	Depends on how practised and regulated	Selective intervention for equity and sustainability
Dependency/post development	Anti-globalization sceptical	Discourse to perpetuate capitalism	Exploitation by imperial and neo-imperial elites	Perpetuation of underdevelopment and marginality	Withdraw from capitalism, alternative lifestyles

inequality, development and globalization. In the remainder of this chapter we explore definitions, meanings and histories of development and assess these issues in the context of two Third World regional case studies that consider the impacts of globalized networks.

Discourses and measurements of development

Not surprisingly, perspectives on development are wide-ranging and often conflicting. Box 8.1 offers four contesting options flowing from the highly orthodox to the very radical. Given the many meanings and discourses of development, no universal measure exists. When 'development' was first conceptualized following the Second World War, it was assumed that social progress would trickle down through economic gains. The most common measurements were thus economic: GNP per capita; real GNP per capita; economic growth; the level of industrialization and/or urbanization; and export structure. Clearly there are problems with these measures; they are not always comparable due to data inadequacies and, more importantly, they do not take social factors and inequality into account. By the end of the 1970s, there was wide agreement that such yardsticks were not broad enough to capture what development *should* mean. For example, Edgar Owens said:

> Development has been treated by the economists as if it were nothing more than an exercise in applied economics, unrelated to political ideas, forms of government, and the role of people in society. It is high time we combine political and economic theory to consider not just ways in which society can become more productive but the quality of the societies which are supposed to become more productive – the development of people rather than the development of things.
>
> (in Todaro, 1997, p. 15)

Given these criticisms, there was an attempt to incorporate broader factors, and a number of *indices of development* were produced. Often, it was found that little correlation existed between GNP per capita and 'social' development. Although there are many more complex measures, the most broadly used is the UN's *Human Development Index* which combines the following three factors weighted equally to produce a score out of one (see also Map 8.1):

1 *Longevity* (life expectancy at birth);
2 *Knowledge* (expected years of schooling [1/3] and years of schooling [2/3]);

Box 8.1

Four contesting perspectives on development

The process by which a traditional society employing primitive techniques, and therefore capable of sustaining only a modest level of per capita income is transformed into a modern, high-technology, high-income economy.

(Williamson and Milner, 1991)

The process of improving the quality of all humans' lives. Three important aspects of development are (1) raising people's living levels, through relevant economic growth processes; (2) creating conditions conducive to the growth of people's self-esteem through the promotion of institutions which promote human dignity and respect; and (3) increasing people's freedom by enlarging their choice variables.

(Todaro, 1997)

Development has evaporated. The metaphor opened up a field of knowledge and for a while gave scientists something to believe in. After some decades, it is clear that this field of knowledge is a mined, unexplorable land. Neither in nature nor in society does there exist an evolution that imposes transformations towards 'ever more perfect forms' as a law. Reality is open to surprise. Modern man has failed in his effort to be god.

(Esteva, 1992)

It is not just the poor but also the rich, and their economy as well, that have to be called into question. At any rate, the quest for fairness in a finite world means in the first place changing the rich, not the poor. Poverty alleviation, in other words, cannot be separated from wealth alleviation.

(Sachs, 2012)

3 *Standard of living* (real GNP per capita converted to purchasing power parity).

Disillusionment with trickle-down approaches to development spurred new ways of thinking about and practising development over recent decades. Drawing inspiration from feminist writers on development, issues of inclusion and exclusion from development based on factors such as gender, ethnicity and age have been highlighted, and these have called for participatory approaches to development. Similarly, rights-based approaches have seen development primarily as a process which should build and protect people's basic rights to survival and freedom, rather than assume that increased material wealth will secure these rights. Furthermore, and inspired by the work of the economist Amartya Sen, thinking about development has shifted from solely a deficit model (what people lack) to an assets-based capablities approach which identifies what resources people have, what they seek to achieve and how their livelihoods might be improved in different ways.

In academia, as well as these moves to explore alternative ways of conceiving and practicing development, usually with a stronger sense of social justice, inclusion and equity, the debate over what development is and should be has broadened substantially over the past 20 years. In particular this has followed the 'cultural turn' in geography and the social sciences (see Chapter 6) and criticized the very concept of 'development'. Grand narratives and theories such as those espoused in the modernization and neoliberal periods have been roundly criticized (see discussion below). Influenced by postmodernism, some local activities, institutions and solutions have been identified as alternatives *to* development. Some have come to question the concept of development itself, arguing that it forms part of the neo-imperial discourse of the West (Power, 2003, 2012). The anti-development school, for example, sees development as part of the effort by the economically dominant West to expand its interests through the diffusion of market-based policy. In this sense globalization, if defined as an agenda of this nature, and development are much the same thing. Postcolonial thinkers argue the case further, claiming that development is merely an extension of discourses of development instrumental during the formal period of imperialism and, as such, represents a further wave of globalization. Today's colonialism is less overt and not always 'economic' in nature. It is increasingly cultural – thus the calls by postcolonial thinkers for the 'decolonization of the mind'. It is important to distinguish between development as an ideal construct and development *as currently practised*. In this sense, measuring development is increasingly meaningless to a new wave of development geographers.

Despite the richness of the academic debate (see Willis, 2011; Potter, 2013; Brohman, 1996; Crush, 1995; Potter *et al.*, 2008; Power, 2003; Preston, 1996), development in practice is largely still a top-down, modernist project. The bodies that are instrumental in the diffusion of orthodox development – principally the Bretton Woods institutions (IMF, WB and WTO) – are also the main bodies that espouse the virtues of economic globalization. The currently dominant discourse is that development may be achieved only by opening up borders to the processes of globalization – and this involves neoliberal reform designed towards that end. Fundamentally, the goals of orthodox development are much the same as they were after the Second World War. It is true that a number of ideas from the reformist schools, such as sustainable development, participation and empowerment, have been coopted by mainstream development trustees including transnational financial institutions and aid agencies, although policies often pay lip service only to such concepts. As we will see, there were some major changes to aid strategies during the early 2000s involving a subtle shift from 'pure' neoliberalism to what can be termed 'neostructuralism'. This involved a greater concern for poverty alleviation and it called for a greater role for the state in providing welfare services, infrastructure and general law and order. Yet it retained a central focus on market-led economic growth and globalization.

The hegemonic discourse at present then is still 'globalize or perish', measuring development in terms of economic prosperity.

Patterns of global inequality

This section presents some empirical evidence to back up the claims of considerable and growing spatial inequalities of well-being at the global level. In the absence of comparable smaller-scale data, the nation-state and regions are focused upon. This information gives us a broad idea of the contours of inequality, but it must be viewed as a broad-brush outline only. Whether such inequality is a direct result of the processes of globalization *per se* is a proposition with which geographers and allied social scientists are currently engaged. This has been a consensus in human geography for over 15 years; Potter *et al.*, in their book *Geographies of Development* (1999, p. 103), argued that:

> We can conclude that uneven and unequal development are still characteristic of the global capitalist system. Globalization is not all-encompassing and there is much that remains uneven about global relationships and global processes. All of these aspects of dynamic change are strongly skewed towards the developed North.

The use of statistics to describe patterns of development is problematic; they can hide difference, lead to labelling, may be unreliable and, crucially, miss very important *perceptions* of well-being. However, selective use can be instructive. Approximate absolute figures speak of vast global differences. At the start of the new millennium 1.3 billion people lived on less than US$1 a day; over 800 million were malnourished; 840 million illiterate; and 1.2 billion lacked access to safe water (UNDP, 2001). The latest available figures at the time of writing

(2013) on global inequality, measured in terms of income, life expectancy and education, are shown in Table 8.2.

Table 8.2 reveals a number of important spatial and empirical patterns. Incomes in the OECD nations are around six times that of developing countries as a whole. On average the least developed countries have GDP per capita incomes approximately six times smaller than the world average. There is significant variation between 'developing' regions, with Latin America, the Arab States and East Asia being relatively wealthy although if China were removed from this measure, the latter region would appear much wealthier. Far and away the most deprived region is Sub-Saharan Africa. Here the historically high levels of poverty have been exacerbated by the rise of AIDS which has further lowered life expectancy and detrimentally affected livelihoods. The United Nations divides the world into a three-part 'human development' classification, which illustrates the spatial inequalities that exist (see Map 8.1). Based purely on income, the World Bank's classification system reveals even deeper differentiation (see Map 8.2). Maps 8.1 and 8.2 reveal a clear spatial concentration of deprivation and affluence

Table 8.2 *Global inequality in material terms – income, life expectancy and education*

	GDP per capita 2011 (current US$)	Life expectancy at birth 2011 total (years)	Adult literacy rate 2010 (% of people aged 15+)
Arab World	7,018	70.6	73.6
East Asia and Pacific	8,493	73.4	94.2
Europe and Central Asia	24,806	76.2	98.6
Latin America and Caribbean	9,693	74.4	91.4
North America	48,463	78.9	99.0
Sub-Saharan Africa	1,440	54.7	62.6
South Asia	1,410	65.7	61.6
Least developed countries	776	59.1	59.7
OECD countries	36,944	79.6	97.8
World Bank Classification			
Low income	566	59.4	62.9
Middle income	4,359	69.3	83.3
High income	37,703	78.8	98.1
World	**10,102**	**69.9**	**84.1**

Source: World Bank (http://data.worldbank.org/ accessed 15 August 2013)

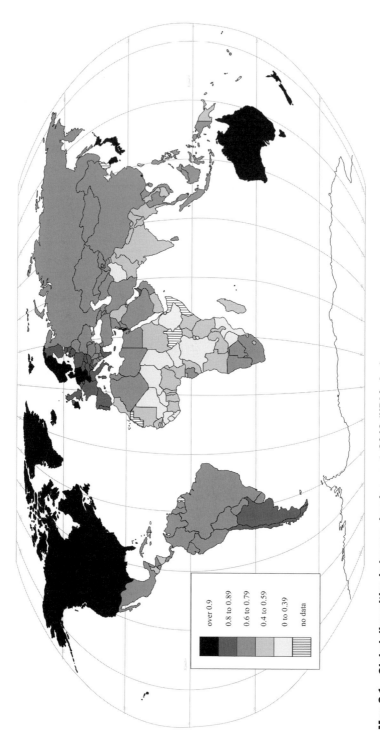

Map 8.1 _Global disparities in human development, 2012 (HDI index)_

over 0.9

0.8 to 0.89

0.6 to 0.79

0.4 to 0.59

0 to 0.39

no data

Source: UNDP (http://hdr.undp.org/en/statistics/ accessed 20 August 2013)

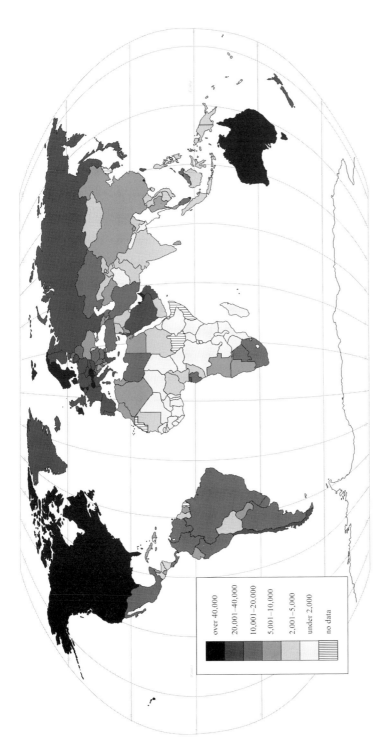

Map 8.2 Global disparities in income, 2011, GDP per capita ($US 2005 PPP)

over 40,000
20,001–40,000
10,001–20,000
5,001–10,000
2,001–5,000
under 2,000
no data

Source: World Bank (http://data.worldbank.org/ accessed 20 August 2013)

measured at the scale of the state. North America, Western Europe, parts of East Asia and Australasia are relatively prosperous, while Latin America, the Middle East, Eastern Europe, South and Southeast Asia, Africa and the Pacific are relatively deprived. The majority of the world's people (close to four-fifths) live in these latter regions.

Global poverty, in an absolute sense, is concentrated in South Asia, where half of the world's poor live; but as a proportion of the population, there are more people in poverty in Africa than anywhere else. Poverty has declined dramatically in East Asia (falling by 125 million between 1987 and 1998), particularly in China. However, over the past decade, poverty has been rising in South Asia, Eastern Europe during the transition from communism, and especially Sub-Saharan Africa where over 140 million people fell into poverty between 1965 and 2000 (UNCTAD, 2002). There is a huge gap between rich and poor countries across a range of health measures. The number of children who die before reaching one year of age is around seven times higher than in high income economies (UNDP, 2004). In many Third World regions gender inequalities remain stubbornly persistent, most tangibly in terms of access to education and political systems. The *technological gap* is especially troubling, though it may be changing rapidly. Internet access is 'paralleling class systems of stratification, and threatens to splinter the globe into haves and have nots based on access to information/communications technologies' (Potter *et al.*, 2008, p. 103). Such a marked digital divide is revealed in Map 8.3 which shows that there are marked disparities across the globe in access to the internet, as measured by broadband connection. Many countries have no fixed broadband services at all whilst most countries in Western Europe have a third or more of their populations connected.

Interestingly, this observation may have become obsolete, at least in terms of telephone access. Although a digital divide is strongly in evidence in terms of broadband access (mirroring fixed-line telephone access) (Map 8.3), the meteoric growth of cell phone usage worldwide has redrawn the map of global connectivity (Map 8.4): for many people now, an affordable mobile phone has allowed them to leapfrog technologies such as fixed-line telephones and computers. Mobile phone usage rates indeed are higher in countries such as Botswana and Vietnam than they are in USA.

But these static measures, however troubling, do not necessarily imply a worsening in the distribution of well-being over time. Figure 8.1 illustrates that, in terms of real GDP per capita, some Third World regions

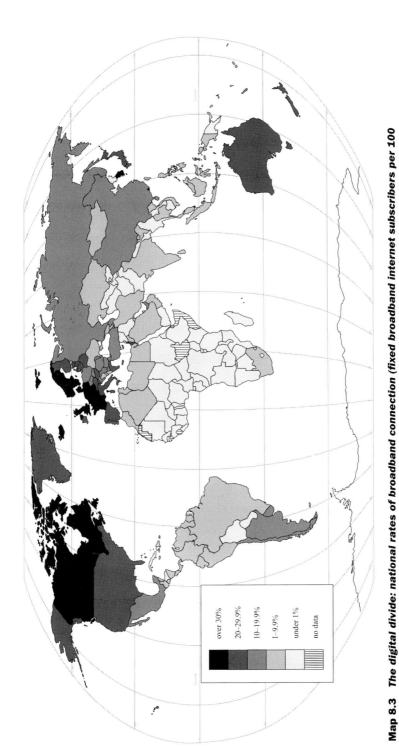

Map 8.3 The digital divide: national rates of broadband connection (fixed broadband internet subscribers per 100 people)

over 30%
20–29.9%
10–19.9%
1–9.9%
under 1%
no data

Source: World Bank (http://data.worldbank.org/ accessed 20 August 2013)

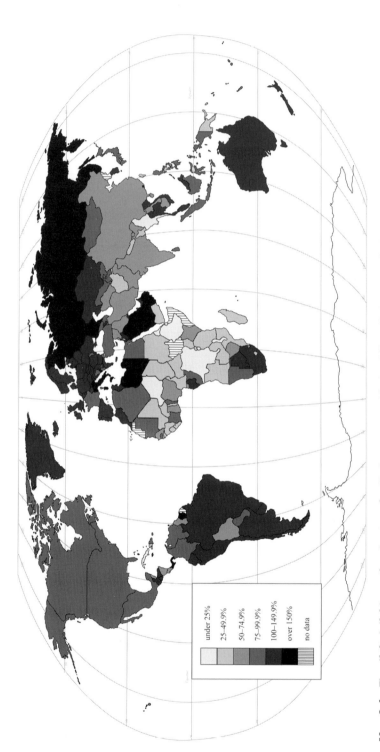

Map 8.4 *The digital divide: national rates of cell phone ownership (mobile cellular subscriptions per 100 people)*

under 25%
25–49.9%
50–74.9%
75–99.9%
100–149.9%
over 150%
no data

Source: World Bank (http://data.worldbank.org/ accessed 20 August 2013)

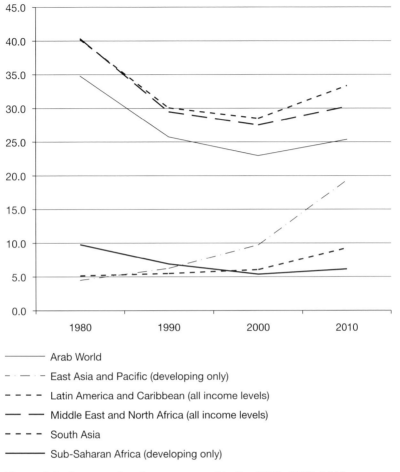

- Arab World
- · — · — · — East Asia and Pacific (developing only)
- – – – – Latin America and Caribbean (all income levels)
- —— —— Middle East and North Africa (all income levels)
- - - - - South Asia
- Sub-Saharan Africa (developing only)

Figure 8.1 *Income of regions compared to the OECD, 1980–2010 (percentage of OECD average GNI per capita)*

Source: World Bank (http://data.worldbank.org/ accessed 12 July 2013)

have converged with the OECD economies. This is particularly the case in East Asia (including China) where in 1980 GDP per capita stood at 13.3 per cent of the OECD level; by 2010 this figure had risen to 21.5 per cent. Over the same two decades there was also a marginal income convergence between the OECD and Southeast Asia/Pacific, South Asia and East Asia (including China). Latin American incomes fell dramatically relative the OECD between 1980 and 1990 but since then have recovered. South Asia also fell but recovered slightly though the picture for Sub-Saharan Africa is much more bleak: despite a slight recovery between 2000 and 2010, incomes there fell from 8.2 per cent of

OECD incomes in 1980 to a meagre 3.8 per cent in 2010. When we compare the poorest countries with the richest countries, however, the evidence for global income divergence becomes clearer. Figure 8.2 shows income per capita (at constant prices) for developed and developing countries (using World Bank classifications). Over the period 1960–2010, the real per capita incomes of developed countries more than doubled (to over $US34,000 per head). Least developed country economies also grew, though only at 42 per cent over the period and remained at a very low average of $US534. Thus, not only did the two sets of countries diverge in relative terms but the absolute gap between them grew massively. The only slight bright spot in the picture is the growth of

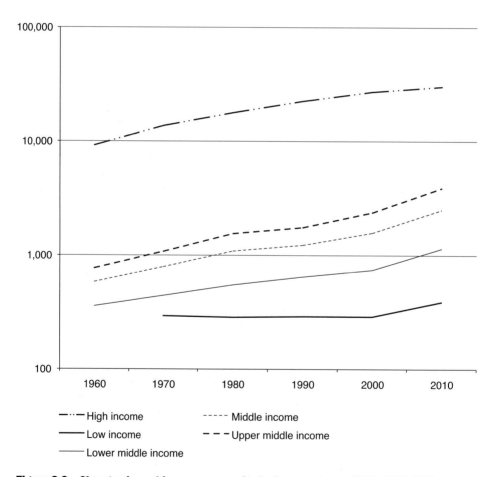

Figure 8.2 Changes in real income per capita by income group, 1960–2010 (GDP per capita (constant 2005 US$), log scale)

Source: World Bank (http://data.worldbank.org/ accessed 12 July 2013)

middle-income developing economies: their incomes grew sixfold over the period but still remained at less than 10 per cent of the rich countries (and the gap in real dollar terms still widened). Non oil-exporting poor countries performed particularly badly during this period. Potter *et al.* (2008) estimate that between 1820 and 2000 the income ratio between the richest 20 per cent and poorest 20 per cent of nation-states has risen from 3:1 to approximately 70:1. In the past decade and since the Global Financial Crisis (see Chapter 7), there has been some change in this pattern, especially with the emergence of China and other East Asian economies and the faltering of North American and European economies, but, in the long term, the extremes of inequality are likely to remain.

When we consider trends in poverty in the poorest countries of the world, the crisis of inequality becomes apparent. The proportion of people living on less than a dollar a day (the standard measure of poverty) increased from 48 per cent to just over 50 per cent between the periods 1965 to 1969 and 1995 to 1999. In absolute terms this represents a doubling in the numbers in poverty from 123 million to 279 million people. The situation is particularly poor in African least developed countries where rates have risen from 55.8 per cent to 64.9 per cent over the same period (see Figure 8.3). In the other 22 developing countries, however, the ratio of those in poverty has fallen consistently and stood at less than 8 per cent at the turn of the century. This corresponded to an absolute fall in poverty from around 760 million to 290 million.

There have therefore been some gains in certain development measures over the past 30 years. Potter *et al.* (2004, p. 32) argue:

> One of the reactions to anti-developmentalist thinking is the argument that, in overall terms, impressive gains have been made to conditions in developing nations over the past thirty years . . . a child born today can on average expect to live eight years longer . . . the level of adult literacy has increased. . . . Average incomes in developing countries have almost doubled in real terms between 1975 and 1998.

The spectacular growth of the Chinese economy in the past 20 years has also lifted millions of people out of poverty. However, the main point in the context of the discussion here is that absolute deprivation levels have risen to crisis point and that the gap between the upper and lower end has increased markedly. The Millennium Development Goals (MDGs) were launched in 2000 with ambitious targets set for achievement by 2015. As this deadline nears, data point to some impressive gains but also some critical areas of failure. Aggregate improvement has been achieved but

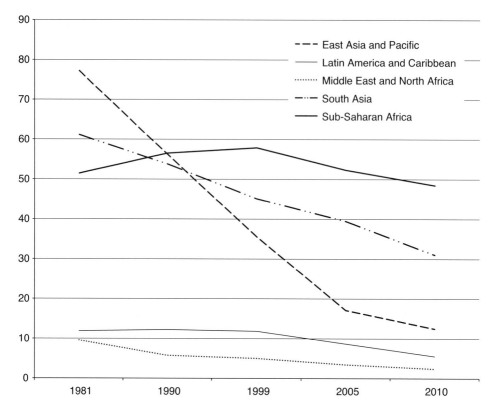

Figure 8.3 *Relative levels of poverty, 1980–2010 (Poverty headcount ratio at $1.25 a day (PPP) (% of population))*

Source: World Bank (http://data.worldbank.org/ accessed 2 September 2013)

absolute poverty persists and inequalities have widened. In 2013 a review of the MDGs began by noting considerable progress:

> Together, [the MDGs] have contributed to remarkable achievements; half a billion fewer people in extreme poverty; about three million children's lives saved each year. Four out of five children now get vaccinated for a range of diseases. Maternal mortality gets the focused attention it deserves. Deaths from malaria have fallen by one-quarter. Contracting HIV is no longer an automatic death sentence. In 2011, 590 million children in developing countries – a record number – attended primary school.
>
> (United Nations, 2013, p. 1)

However the review report drew a picture of new patterns of inequality. On the positive side, it noted that low- and middle-income countries are growing at a faster rate than high-income ones (probably a consequence of the GFC), thus lessening inequality amongst states. Yet it noted:

> At the same time we are struck by the level of inequality in the world, both among and within countries. Of all the goods and services consumed in the world each year, the 1.2 billion people living in extreme poverty only account for one per cent, while the richest 1 billion people consume 72 per cent.

Despite notable achievements in East Asia, extreme poverty persists in Sub-Saharan Africa and persistent poverty is in evidence in most countries even when there has been overall economic growth. Poverty remains a stubborn feature worldwide. On the other hand, the gains of economic growth have not been shared equally so that the richest segments of most societies have continued to accumulate wealth – notwithstanding some setbacks with the GFC.

History of global development – from modernization to poverty alleviation

Before going on to look at particular regional cases, we need to understand how the current development orthodoxy was arrived at and how this relates to discourses and concepts of globalization. The contention here is that development as conceived by mainstream thinkers in the post-Second World War era and the concept of globalization are wrapped up in the same desire to expand and speed up capitalism. In this sense the concept and goal of development is a defining feature of the second wave of globalization – although specific strategies have shifted across time. The agents instrumental in setting up the development agenda were the World Bank and IMF created after the Second World War in their effort to restructure the global economy and reform the capitalist mode of regulation (see Chapter 5). We focus first on the modernization phase of this endeavour, and neoliberalism is considered subsequently. Note that, in this context, we are exploring the development concepts that have explicitly informed orthodox policy.

Modernization in the post-war development era

Modernization theory evolved from two real world events: the Great Depression (1930s) and the end of the Second World War. The latter provided resolve for liberal-capitalist powers, while the former supplied an example of how to lift nation-states out of recession through state intervention (based on the US New Deal). These factors converged in the form of the Marshall Plan for the reconstruction of post-war Europe. In general, the approach was a 'grand' paradigm (collection of linked theories) which argued that development equalled 'modernization' and that the 'level' of any given nation-state could be measured by comparison to 'developed' Western countries. Poorer societies had to throw off 'tradition-bound' ways to progress. In order to advance the vital processes of economic growth, urbanization and industrialization, 'modern' attitudes such as competition and individualism were seen as essential. Rostow's Stages of Growth model is the epitome of such a theory, involving five steps through which countries should progress in order to reach the end-goal of 'high mass consumption' (Rostow, 1960). It was in the context of the elaboration of the modernization argument that President Truman first used the terminology 'development' and 'underdevelopment' to refer to the challenges facing what became known as the 'Third World'.

What did the modernization paradigm imply in terms of policy for the Third World? Economically, the crux of Rostow's 'take-off' stage is stimulating investment. To do this, it was argued, urbanization and industrialization were required as these processes would raise savings through increased incomes and profits, feeding back into investment. These social changes would yield the necessary conditions for this, since backward and forward linkages, and thus multiplier effects, were higher in urban areas and industry than in rural zones and agriculture respectively. Industrialization, in particular, could be achieved through state intervention from national governments supported by external credit and aid (based on the Marshall Plan model). In this sense, modernization promoted a state-centred 'developmentalist' approach, representing a form of state-assisted capitalism. It drew on Keynesian notions of the social responsibility of the state to regulate yet encourage capitalism. Foreign capital played a central role in investment and flows increased markedly after the Second World War, especially to East Asia and Latin America, underpinning subsequent industrialization in the former at least. There was an important ideological component to modernization, forming part of the West's attempt to prevent the spread of communism. In this sense, it was the economic backbone of the West's Cold War geopolitics. It also suited a post-war world of decolonization: political independence and nationalism were matched by economic nationalism (state-assisted economic growth). In this sense, modernization theories were close to the ideas of the Latin American structuralists, led by Prebisch, who advocated import substitution industrialization, albeit with more self-sufficiency and a more interventionist state.

By the 1970s modernization approaches were largely discredited. In academia, the major criticism was that the paradigm represented an ahistoric and non-geographical interpretation of the world that was Eurocentric. The approach was overly economic and universalizing – failing to take into account cultural diversity and contesting perspectives on development. In the policy arena the approach was refuted because it did not work – economies did not grow as expected when pursuing the policies that were espoused. There was a transition some time in the late 1970s from the state developmentalist/aid modernization model to the neoliberal model. It is important to stress, however, that the ultimate objective of these two approaches – in terms of what development *should* be – are the same. It is the route taken to stimulate economic growth and modernity that varies.

The debt crisis – a transition point

From the end of the Second World War until the early 1970s the global economy as a whole boomed. This was interrupted in the early 1970s by the *oil crisis* which saw a large hike in the price per barrel of oil charged by the OPEC nations. Inflated oil prices left the global economy awash with 'petrodollars' (windfall profits of the OPEC group) and vast amounts were channelled through the western banks to the Third World in the form of loans for modernization. When the second oil crisis hit in 1979 to 1980, it had a devastating effect on the whole global financial system. Cost-push inflation rose sharply in the OECD and, to fight it, the monetarist solution of raising interest rates was pursued. This had an impact on Third World countries' debt to both transnational institutions and richer countries, exemplifying the increased interconnectedness of the global economy. In many cases, given extensive borrowing in the 1970s, it was not possible for countries to pay back the interest on their loans (never mind the capital). In some Latin American countries, for example, debt rose to levels higher than total GDP per annum. Led by

Mexico in 1982, there was widespread default on debt to the World Bank especially, resulting in what became known as the Debt Crisis (see Box 8.2). In response, the World Bank and IMF developed and applied 'structural adjustment programmes' (SAPs) based on neoliberal principles. Ostensibly, these were designed to foster economic self-sufficiency and to make the most of globalization. In reality, they were designed to stabilize the global financial system, which underpinned the economic prosperity of the West (Murray, 2009b).

Box 8.2

Drop the debt and make poverty history?

The issue of debt in the Global South together with famine has exercised a good deal of global attention in terms of public perspectives on development. Current attempts to alleviate debt as embodied in the Make Poverty History campaign can be traced back to the events of the original debt crisis of the 1980s, which itself has roots in the widespread lending that took place following the first oil hike in 1973. Virtually all developing countries that were not oil-exporting nations, which is the vast majority, were hit hard by the oil crises and the ensuing interest rate hikes and general recession of the early 1980s. In an absolute sense, Latin America shouldered the highest burden; by 1982 countries in the region had accumulated debt of over US$315 billion (Gwynne, 1985). In Sub-Saharan Africa by the same year, total debt equalled US$200 billion. However, absolute figures are not the most revealing; it is the ratio of GDP and debt that is the most relevant. In 1981, in Latin America as a whole, debt stood at around 30 per cent of GDP, and this figure

was substantially higher in Sub-Saharan Africa. In individual countries, such as Bolivia, debt rose to levels that outstripped GDP. The ratio of debt to export earnings is also important as the latter provide the foreign exchange to service the former. In the early 1980s in Latin America, debt represented approximately 220 per cent of export earnings, whereas in East Asia it was closer to 50 per cent. The lower ratio in East Asia, given far higher per capita export earnings there, formed part of the explanation for the region's subsequent economic success. In the Third World in general a 'lost decade' of development during the 1980s resulted from the debt crisis, which, it could be argued, is continuing into the present (see Murray, 2009b; Power, 2012). The 'lost decade' was characterized by the following processes:

● Direct foreign investment declined markedly and an outflow of profits to the OECD took place.

Box 8.1
continued

- Continued high interest rates in the 1980s meant that countries were paying out more to cover interest on debts than they were actually repaying in terms of original capital.
- Non-oil commodity prices continued to decline, further lowering poor countries' terms of trade.
- Aid and lending declined as the recession cut into the global financial system.
- Protectionism grew in the OECD – despite the fact that this organization recommended that the Third World open its doors and reduce regulation.
- Austere structural adjustment policies were introduced, raising poverty and unemployment.

This combination of factors had major economic, social and political ramifications:

- *Economic crisis* – Many Third World countries were plunged into negative growth and declining exports. This was particularly the case in countries extremely reliant on primary product exports (apart from oil), i.e. most of Africa, Latin America, South and Southeast Asia. Latin American incomes fell by 10 per cent in the 1980s for example (Gwynne and Kay, 2004)
- *Social crisis* – Income inequality increased as government spending was cut and other social programmes were pulled back through the application of SAPs. Poverty rose dramatically, with women and children being the hardest hit (Corbridge, 2002).
- *Political crisis* – The ensuing socio-economic chaos provided the rationale for military interventions in many countries. Latin America, for example, was entirely governed by dictatorships by the mid-1980s (Silva, 2004).

The debt crisis is only over in the sense that debts have been rescheduled. In absolute terms, the foreign debt of developing countries has continued to rise. From US$658 billion in 1980, it had soared to US$2,560 billion by 1999 (World Bank, 2002). It currently stands at over US$4,150 billion, and costs developing countries approximately US$1.5 billion a day merely to service. Between 1980 and 1999 Sub-Saharan Africa's debt tripled to around US$220 billion, and throughout the decade Africa returned US$81.6 billion to its creditors. By 2011, despite some writing off the debt of some of the most heavily indebted nations, it had risen to nearly US$300 billion. Latin America's debt had grown to around US$820 billion by the start of the twenty-first century, and currently stands at over US$1,120 billion. Individual country figures point to the overwhelming challenges which many poorer countries still face. If we compare total external to Gross National Income in a range of countries in Sub-Saharan Africa and Latin America, it stood at a proportion of over 100 per cent (UNDP, 1999; World Bank, 2012). The Mexican crisis of 1994, the Asian crisis of 1997 and the Argentinean crisis and the more recent GFC have all had the effect of increasing debt. The World Bank considers that there are 39 heavily indebted countries, 33 of which are in Sub-Saharan Africa. In all of these, debt is equal to 200 per cent of export earnings per annum and in many cases over 100 per cent of total GNI.

Box 8.1
continued

The total cost of debt relief for these worst impacted is thought to be approximately US$70 billion – a small figure compared to the recent stimulus packages undertaken in the EU and the USA. More than any other single factor, it is debt burden that prevents meaningful progress in poorer countries.

Calls for the cancellation of Third World debt have grown from across the NGO sector, and some European governments have been actively involved in debt-relief initiatives. Global consciousness has steadily grown at the popular level over recent decades. Live Aid – a charity relief concert designed to raise money and awareness of African famine took place in 1985, based in spirit on the George Harrison concert for Bangladesh in 1971. As the decades passed, debt relief became the central concern of the popular and NGO movement associated with alleviating poverty in the Global South. Oxfam, a transnational NGO, has fought for two decades on this front, for example. In the 1990s, a significant global movement pulling together anti-debt groups and powerful figures from politics and entertainment was launched. Jubilee 2000 aimed to 'cancel the unpayable debts of the poorest countries by the year 2000 under a fair and transparent process' (Jubilee 2000, 2002, p. 2). A global petition signed by 24 million people was assembled to support this objective, giving birth, in 2001, to the Jubilee Movement International. Partly in response to this general groundswell, the World Bank and IMF established an enhanced 'heavily indebted poor country' (HIPC) initiative, referred to above, and the issue of debt was placed on the agenda at G7 summits. This has seen debt relief

extended to over 35 countries to date, with decisions pending in other cases. In the case of Bolivia, for example, total external debt has been reduced by 50 per cent. In the 2000s the UK government, led by the efforts of Chancellor Gordon Brown, lobbied the G8 and international financial institutions for 100 per cent debt relief for 37 of the worst hit cases, as part of Labour's Marshall Plan for Africa promoted under the slogan of 'Make Poverty History' which became the catch-cry of the Jubilee 2000 programme. This slogan became widespread and focused much popular attention on the issue in the West together with a large charity concert, Live 8 (named after the G8), which was responsible for raising considerable awareness. In February 2005, G8 ministers committed to these objectives and confirmed immediate debt relief for the countries affected by the 2004 Boxing Day Tsunami in the Indian Ocean. They remained divided on how such initiatives should be funded, however. The USA, in particular, was sceptical of Gordon Brown's proposal for an international financial facility that would channel US$10 billion per annum into debt relief over the subsequent decade, and rejected plans for increased overseas aid. Over the last decade, aid as a proportion of GNI in OECD countries has risen slightly – although the nature of that aid has shifted to economic development and infrastructural projects together with private sector initiatives playing an increased role (Murray and Overton, 2011a).

In 2006 the multilateral debt reduction strategy (MDRI), involving the major multilateral donors was signed which aimed

Box 8.1
continued

to fund the HIPC scheme. By 2011 under the HIPC scheme 32 of the 40 countries had had their debts cancelled and further countries were being considered based on debt and economic performance criteria. Further countries were also being considered for inclusion. Whilst these are laudable gains, criticisms have been widespread. Some countries in Latin America did not qualify as their per capita income was too high. Economists such as Jeffrey Sachs have also criticized the austerity requirements of the policies that were required as a condition. By focusing on the poorest of the poor, debts in low-income and middle-income countries have been given relatively less attention and as such the amounts involved represent but a small proportion of total debt. Low-income and middle-income countries will remain locked in the vicious cycle of debt for the foreseeable future, as they are obliged to pursue neoliberal solutions for their relative deprivation whilst even in those countries that have received relief debts are continuing to rise as a proportion of GNI.

A significant development in the 2010s was the entry of China as a creditor, especially in Africa – and it was noted that not all of the credit flows were publicly known, implying that total debt might well be underestimated. At the present, the focus has shifted away from the issue of developing world debt, and international attention has turned to the sovereign debts of wealthier countries in the wake of the GFC. The UK's external debt of over US$9,000 billon is over double that of all developing world debt, and the US debt of approximately US$14,500 billion has moved attention away from Third World debt. In the case of richer nations, the money they are owed offsets this debt at least in part (Guardian, 2012). Furthermore the debt relief and stimulus packages undertaken by the West make developing world debt relief look very small in comparison, and critics have been quick to note just how readily corporations and governments were bailed out compared to the difficulty encountered in reducing developing world debt. Frustrated by progress on the international front, forward-thinking countries such as Norway had developed significant programmes of cancellation of all developing world debt owed to it by 2013. Global justice movements such as Make Poverty History continue to campaign for more meaningful reform of the global system to ease the burden of debt at the time of writing, but remained a long way from achieving this objective.

From structural adjustment programmes to the poverty agenda

As noted previously, neoliberalism dominates development policy and has eclipsed a range of alternative development theories from academia and beyond (see Box 8.3). The major development instrument of neoliberalism was seen in structural adjustment programmes (SAPs). These were put in place by developing country governments under pressure from the World Bank and IMF. These latter institutions were able to enforce such changes because many governments were facing severe debt crises and needed to reschedule their loans or possibly face default and bankruptcy. The IMF and World Bank were able to bail out highly indebted countries in the 1980s and 1990s but only on condition they imposed harsh and deep-seated economic reforms in packages known as SAPs. SAPs diffused widely across the Third World as a consequence. By 1990, for example, nearly every country in Latin America had been subject to such a programme.

SAPs have been criticized for a one-size-fits-all approach although they do vary over time and space to a certain extent (see Stiglitz, 2002, 2010). They generally include the following sets of measures:

- *Downsize* – reduce public expenditure (e.g. education, health and public infrastructure) and taxation to raise incentives for private investment and enterprise, and eliminate budget deficits.
- *Privatize* – sell off state-owned enterprises and privatize government functions where possible to release an efficiency-inducing profit motive and reduce government expenditure and debt.
- *Deregulate* – reduce the intervention of the state in the economy, such as price controls or sector subsidies, to allow the market to find its natural equilibrium. Reduce the red tape required to establish private enterprise (see Cartoon 8.1).
- *Globalize* – reduce tariffs and any other protectionist measures in order to open borders to inward TNC investment; to facilitate the transfer of technology, to stimulate competition with world producers; and to attain economies of scale from larger market size in order to enhance efficiency and exports. Devaluation of the domestic currency in order to stimulate the latter is often a further characteristic.

The impacts of SAPs were hotly contested. In general, it was recognized by proponents and detractors alike that the austerity required of structural adjustment would bring regressive impacts in the short term. All countries adopting SAPs experienced increases in poverty, income inequality, unemployment and informal sector activity (see Plates 8.2 and 8.3). Empirical studies of SAPs revealed that it was women, children and the poor who bore the heaviest burden, particularly as state support mechanisms were withdrawn (Radcliffe, 2004). In the case of women in Latin America, for example, double or triple days became increasingly common in the reform period of the 1980s as women took on both formal and informal employment in order to make ends meet. This was

Cartoon 8.1 *Deregulation*

Source: Kirk Anderson

Plate 8.2 *Cape Flats, South Africa*

combined with the fact that female-headed households were becoming
increasingly common, due in part to the poverty unleashed by SAPs
(Chant, 2004). Neoliberalism in general privileges capital over all else
and creates networks that are difficult for the poor and already
marginalized to access.

Plate 8.3 *Informal economy – roadside stalls in Kenya*

As a result of structural adjustment, some countries were able to develop some competitive industries, engaged in resource extraction and export or based on low-wage industrialization. Others, such as small island states, struggled to find any viable economic alternatives and migration and remittances, along with greater aid dependency, became the basis for their economies. In all cases, following structural adjustment, there was a shift from inward-looking self-sufficiency to much greater engagement with the global economy.

This phase of neoliberalism, built upon SAPs, began to change in the mid-1990s. There was some recognition of the negative social and political impacts of severe reforms and also a recognition that poverty had increased. There was also concern that heavy cuts in state expenditure had often threatened the very viability of many states, raising the spectre of absolute collapse (so-called 'failed states'), capture by criminal elements (for use as tax havens or money laundering), or general breakdown of law and order. Aid and development agencies were stung by criticism from respected observers such as Joseph Stiglitz and began programmes to rebuild state capacity. This was often in the form of 'good governance' programmes. These were ostensibly about protecting human rights, law and order and democracy and also attacking corruption and lack of transparency but they also had strong elements of facilitating and protecting private property, foreign investment and free trade.

There was also a shift towards the new poverty agenda. Concerned with rising levels of poverty and aware that top-down and enforced programmes of reform could not be sustained, the IMF and World Bank moved to end the harsh SAPs and replace them with a new tool, Poverty Reduction Strategy Papers (PRSPs), in the late 1990s. These asked states that received aid to develop their own strategies to reduce poverty, thus 'owning' their programmes of reform. It seemed to signal a shift away from neoliberalism. Yet, in practice PRSPs still had to conform to international guidelines and gain the approval of international agencies before aid would follow. Conditionalities and the neoliberal intent of aid remained firmly in place.

Alongside the introduction of PRSPs, the international community put greater focus on poverty alleviation, and in 2000 the United Nations promoted a new vision for development: the MDGs. These encompassed eight major goals and focused attention on reducing poverty and hunger, promoting primary education and maternal and child health, and striving for gender equality, especially in education. The MDGs formed the cornerstone of this new poverty agenda and poverty became the central objective for most development and aid agencies in the first decade of the 2000s. The promotion of the MDGs was spurred also by popular awareness and public campaigns such as Live 8 concerts in 2005 and the Make Poverty History campaign (see discussion in Box 8.2 above) which lobbied politicians in wealthier countries to support increased aid budgets that focused on poverty alleviation.

This new development agenda was broadened through a reconsideration of the role of the state. It came to form what might be termed a neostructuralist approach to development, an approach which reached its zenith in the decade prior to about 2008. This was built upon the foundations of the MDGs and the 2005 Paris Declaration – an agreement by major aid donors to improve aid effectiveness by encouraging recipient government ownership of development. It also built upon the good governance approach of the earlier decade and adopted the populist language of participation, human rights and democracy. The neostructuralist approach helped focus aid efforts on alleviating poverty (as identified by the MDGs), it promoted a state-centred approach to providing health and education services, and it was accompanied by a substantial increase in aid funding by major donors. Aid donors moved to fund state-run programmes in health and education through mechanisms such as Sector Wide Approaches (SWAps) or General Budget Support (GBS) and these put considerable emphasis on recipient governments and helped build their capacity to deliver services. Yet, although this approach marked a change from the former early neoliberal approach of 'rolling back the state' to one that built state capacity in certain areas (effectively 'rolling out the state'), it retained the neoliberal objectives of promoting globalization. States might have seemed to be more concerned with tackling the worst aspects of poverty and inequality but they retained faith in economic growth and trade liberalization as the primary means to achieve social as well as economic goals.

Responses from the 'periphery'

In spite of the hegemonic position attained by modernization and, later, neoliberalism, there is a rich tradition of alternative development theory. Some of this has emanated from the periphery itself, especially Latin America but also Africa and Asia. Structuralist and dependency theories, two important Latin American contributions, are implicitly tied up with globalization (see Box 8.3). A unifying characteristic of these linked approaches is the anti-modernization idea that all development is 'contingent'. That is to say, development turns in part on the historical, social, cultural and economic specificity of place and nation-state. Central in both accounts is how globalization has unfolded according to these specificities. Although these ideas are considered outdated in Western policy circles there has been a revival of academic interest recently, as thinkers search for realistic ways of interpreting the developmental failure of neoliberalism. In the places that they were evolved, such perspectives are becoming increasingly relevant. The current economic model in Chile can be termed 'neostructuralist', for example (see Gwynne and Kay, 2004). Indeed the neostructuralist approach outlined above, that became the centrepiece of the global poverty agenda, owed much to its Latin American origins and its implementation by countries such as Brazil.

Despite innovative theorizations from the periphery such as those outlined in Box 8.3 – and to these could be added many more examples such as marginalization and anti-development – they have been largely superseded in orthodox circles. Structuralism and dependency influenced policy for a few short decades between the Second World War and the

Box 8.3

From the margins? Globalization and dependency/structuralism

Two theoretical innovations from Latin America conceptualize the relationship between the global, the national and the local in much more explicit ways than modernization or neoliberalism and have informed policy there, and beyond, at points in recent history.

Dependency

Dependency analysis is a collection of theories that emphasize the 'dependence' of the Third World on international capitalism. Based on Marxist analysis – which emphasizes class struggle – exploitation is theorized at different geographical levels. This approach became popular during the 1970s, as Cold War struggles continued and the post-war boom faded (Kay, 1989). At the global scale Frank (1969) argued that connections between 'metropolitan' and 'satellite' states (trade flows, aid flows, technology transfer, capital flows, cultural exchange) were channels through which rich countries extracted 'surplus' (profit) from the colonized and formerly colonized periphery. This was referred to as 'unequal exchange', as occurred in Marxist theory between capitalists and the working classes. As such, dependency is actually an early example of network theory. TNCs were argued to be the major agents in this process and the neocolonial relations that resulted would persist indefinitely, thus perpetuating the 'development of underdevelopment'. Similar relationships could evolve within nation-states also. The suggested action to break this cycle was, in radical forms of dependency analysis at least, political revolution.

Structuralism

Dependency theory is a more radical take on an earlier structuralist paradigm. This collection of associated theories traces the impact of historical evolution on the structure of the global economy and the resultant patterns of well-being. Raul Prebisch of the Economic Commission for Latin America (ECLA) was the leading structuralist (Prebisch, 1950, 1964). His major contribution was the 'secular decline' hypothesis, which related to the problems of specialization in natural resource exports created by the historical insertion of peripheral economies into the colonial system as suppliers of resources. It was argued that primary products face income-inelastic demand, while manufactured products face income-elastic demand. Therefore, as global income increases, the demand for, and therefore the price of, primary products will decline relative to that of manufactured products. In countries that specialize in primary product exports this has serious implications. Over time they will have to export more primary products to pay for the same amount of manufactured

Box 8.3
continued

products. This vicious circle is referred to as a decline in the terms of trade. In the long run, secular decline will tend to increase the gap between the 'core' and 'periphery', as the centre captures the benefits of boom, and growth in general. On the basis of this, Prebisch argued for rapid state-sponsored industrialization in order to escape the 'primary product trap'. Import Substitution Industrialization (ISI) was the policy prescription which would involve the replacement of imported items, moving through progressive stages. In many ways, structuralism represented an early elaboration of ideas concerning resource peripheries and state developmentalism (Hayter et al., 2003; Murray, 2002a). In the late 1990s and early 2000s, following the shift away from pure neoliberalism in

Latin America which in some cases had been adopted under conditions of dictatorship, the new centre left began to posit the idea of 'neostructuralism'. This adaptation of the structuralist model was intended to give a more central place to environmental concerns and equality whilst embracing globalization. A core slogan for the approach was 'equity with growth'. Critics argued that this approach was little more than neoliberalism with window dressing. Others were less critical and pointed to the great strides taken in some Latin American countries under neostructural administrations – such as Chile, Brazil and Argentina – where poverty reduction was considerable (see Gwynne and Kay, 2004; Murray and Overton, 2011a)

debt crisis; economic policy in Latin America and East Asia built on structuralist ISI models, for instance, and a number of revolutions in Latin America and Africa in the 1970s were informed by dependency analysis. Generally, however, doctrines of development have diffused from the core to the periphery. There has been widespread resistance to, and calls for reform of, the dominant globalist model (see Power, 2003; Routledge, 2002). Collective and individual resistance to Western development in Latin America has a long and violent history, for example (Kay, 2001). It could be argued that ongoing conflicts across the Middle East form part of this collective resistance to the penetration of Westernization. There has been significant work undertaken to attempt to explode the myth that Third World populations are passive victims of 'development' – practising everyday resistance in the form of livelihood strategies and life-cycle shifts (Bebbington, 2004). There have been a number of institutionalized attempts to counter the transnational institutions of the OECD and Bretton Woods. Significant examples include OPEC, the non-aligned movement and the G7 (see McGrew, 2008). It is possible to interpret increased regionalization efforts in poorer regions as a form of resistance (see Chapter 5), with examples including MERCOSUR, SADC, COMESA and ACS. In general, however, the dominant ideology of development has been articulated so powerfully and imposed so forcefully that it is difficult to see how it can be reversed. This proposition is illustrated through two regional case studies in the following section.

Regional case studies of globalization and development

There have been relatively few human geographies which seek to explicitly trace out the impacts of globalization on poorer regions. It is suggested here that an ideal framework for the exploration of these issues is provided by new regional geography (see Bradshaw, 1990). This approach, which builds on the criticisms of 'traditional' regional geography and takes on some of the points made by both radical and cultural geographers, provides a multi-scaled analysis of the outcomes of globalization. New regional geography searches for links, rather than breaks, within and between regions. Recently, geographers in this field have elaborated the concept of 'resource peripheries' in the global economy, partly building on structuralist ideas (Hayter *et al.*, 2003).

The two case studies offered below reflect the research interests of the authors and focus on economic globalization in particular. In both cases, regional trends and outcomes relating to globalization – both historical and contemporary – are identified, and neoliberalism is placed in a long-run context. These transformations are then illustrated in more detail with reference to examples of change at national and local levels. These studies illustrate that the unfolding of globalization is path-dependent; that is to say, it evolves in distinctive ways across space according to historical, environmental and geographical contingencies. Those who are interested in the impacts of globalization on development in East and Southeast Asia should see Rigg (2003), Lin Sien (2003) or McGregor (2009). Geographies of globalization and development in Africa have been understudied; an entrance point is Aryeetey-Attoh (2003), Cline-Cole and Robson (2003) or Binns *et al.* (2011).

The Pacific Islands – off the global map?

The Pacific region is the largest region on Earth in terms of surface area (see Map 8.5). In terms of land mass, however, it is one of the smallest.

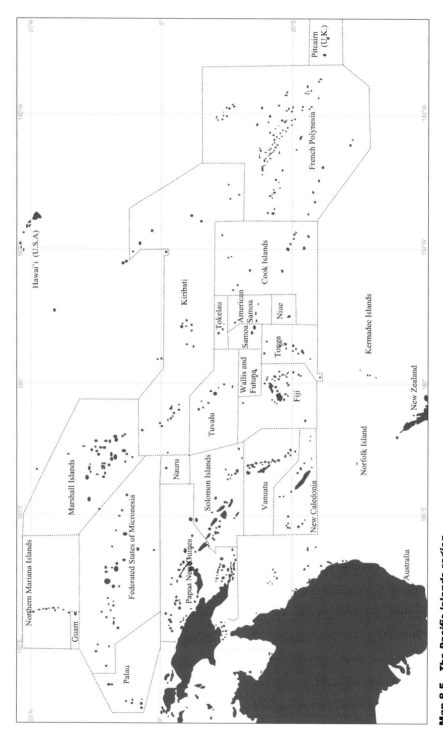

Map 8.5 The Pacific Islands region

Some of its nation-states and territories are among the most diminutive on Earth in demographic and economic terms. Despite this, it is a complex region in terms of history, economy, culture and politics, which is divided into three subregions that exhibit some, albeit limited, internal homogeneity: Polynesia, Melanesia and Micronesia. The processes which have unfolded in the region tell us much about the relationship between globalization as an agenda and a set of processes, and development as both a discourse and a process. It is a region little known or understood by the rest of the world (it is often invisible in world atlases, for example), including those countries nearest to it in a territorial sense (Murray, 2009c; Murray and Storey, 2003; Overton and Scheyvens, 1999). Constructions of an exotic and friendly paradise, perpetuated through the tourist sector, led to a pervasive 'orientalism' of the Pacific (Nicole, 2000) (see also Plate 8.1).

At a global level, the Pacific Island region is marginalized in economic and political terms. Material standards are especially low in parts of Melanesia and Micronesia. They are highest in French colonies, especially French Polynesia, although income distribution in these territories is highly regressive. Since the start of decolonization in the 1960s, Pacific Island Countries (PICs) have found it difficult to prosper in the competitive, globalizing world – many of the problems of its special geography plus its subordinate position in the global economic hierarchy weigh against it. Modernization is causing massive social, economic and environmental tensions as new systems of resource management, and governance systems, interact with traditional ones. Absolute poverty is high in some countries (e.g. Fiji 25 per cent), although it is not as high in relative terms as it is in Latin America or Pacific Asia. The region is sometimes argued to be characterized by 'subsistence affluence', a concept which suggests that abundant land and sea resources confer a level of prosperity far greater than economic indicators might suggest. To a large extent this is an invention of the Western imagination, which idealizes daily life in the region. Economic marginalization and environmental degradation make livelihoods as precarious as they are anywhere else in the Third World. Notwithstanding this, it is true to say that traditional systems of land ownership have led to a relatively equal income distribution, in the independent states at least. Some human development indicators for the independent countries of the region are set out in Table 8.3, illustrating the relative impoverishment of conflict-torn Melanesia. (See also Rappaport, 2013.)

Plate 8.4 *Contrasting images of the Pacific Islands*
The top photo shows a tourist resort on the island of Moorea in Tahiti – what many
tourists seek on a Pacific Island vacation - whilst the lower picture is of a squatter
settlement in Suva, Fiji, the reality of life for large numbers of Pacific Island people.

Table 8.3 *Human development indicators for Oceania*

	Population at last census (000s)	Annual population growth rate (%)	GNI per capita ($US) 2011	Life expectancy (years) 2010	Urban population at last census (%)	HDI (2010)
American Samoa	55.5	0.6	nd	72.5	50	nd
Cook Islands	15.0	0.5	13,478	76.4	74	0.822*
Federated States of Micronesia	102.8	0.0	2,992	70.0	22	0.569
Fiji	837.3	0.5	4,252	65.4	51	0.702
French Polynesia	268.3	0.2	26,290	74.1	51	nd
Guam	159.4	2.6	nd	73.6	94	nd
Kiribati	103.1	2.1	2,155	62.2	54	0.629
Marshall Islands	53.2	0.8	4,370	70.0	74	0.563*
Nauru	10.0	1.7	6,746	60.4	100	0.663*
New Caledonia	245.6	1.3	38,690	77.4	67	nd
Niue	1.6	−2.4	nd	69.4	na	0.774*
Northern Mariana Islands	53.8	1.1	nd	75.3	90	nd
Palau	17.4	0.5	9,790	69.0	77	0.791
Papua New Guinea	7,059.7	2.2	1,638	54.2	13	0.466
Samoa	187.8	0.0	3,440	74.2	20	0.702
Solomon Islands	515.8	2.5	1,337	70.2	20	0.530
Tokelau Islands	1.2	−0.9	nd	69.1	na	nd
Tonga	103.3	0.1	4,426	70.6	23	0.710
Tuvalu	10.6	1.6	5,537	69.6	47	0.583*
Vanuatu	234.0	2.5	2,994	71.1	24	0.626
Wallis and Futuna	13.4	−0.4	nd	74.3	nd	nd

Sources: Secretariat of the Pacific Community (www.spc.int/sdd/ accessed 20 September 2013) for population, population growth rates, life expectancy, urban population; United Nations (http://data.un.org/ accessed 20 September 2013) for GNI; UNDP (http://hdr.undp.org/en/data/profiles/ accessed 20 September 2013) for HDI

Notes: * HDI for 1998 (using Secretariat of the Pacific Community data – see Murray, 2009b)
nd – no data available
na – not applicable (no defined urban areas)

The region's unique geography has created five important challenges in terms of the participation of its constituent nation-states in the global economy.

- *Territorial* – isolation in terms of absolute distance;
- *Environmental* – vulnerability to extreme weather and global environmental change;
- *Political* – low political bargaining power due to diminutive nation-state size;
- *Economic* – limited economies of scale and low diversity due to small economic size;
- *Factor endowment* – limited land and land-based natural resources in some cases, especially on atoll islands.

Orthodox analysis has often blamed these 'inherent' problems for the economic deprivation of the region. Along with these constraints to development, some also add 'traditionalism' and 'communalism' – although this approach has been widely criticized. Although the above five factors are of influence, the underlying explanation for the region's relative economic deprivation is far more complex. Since colonial penetration, the region has served as a resource periphery and has been understood through an analysis of the way the Island region has fitted into the wider world over time.

Waves of globalization in Oceania

The history of globalization in the Pacific Islands has been divided into two waves (Firth, 2000; Murray, 2001), punctuated by a period referred to as the MIRAB era (Bertram and Watters, 1985). The following periodization is being suggested here:

1 the first wave – colonial globalization (1830s–1960s);
2 MIRAB era – 'indigenous' globalization (1960s–1980s);
3 the second wave – postcolonial neoliberal globalization (mid-1980s–).

Colonial globalization

The first wave began with the arrival of the 'buccaneers of global capitalism' (Firth, 2000, p. 180) who were part of the evolving global

Box 8.4

Our Sea of Islands: alternative development perspectives from the Pacific

Analyses of Pacific Island countries have often stressed their handicaps: their vulnerability to hazards and climate change, their remoteness, the smallness of their lands, populations and resource base, and their very islandness, fragmenting economic activity over scattered islands in a vast ocean. Such analyses invariably see the region and its people as passive, as victims of circumstance, and in need of external assistance. The MIRAB model (see below Box 8.5) is one such model and was strongly criticized by some Pacific Island academics.

Tired of the very negative tenor of these approaches, the Tongan academic Epeli Hau'ofa countered with an alternative vision of the region. Called 'a sea of islands' it argued that Pacific Island people see the ocean not as empty and a constraint to development but as a vast resource and the home of a multitude of islands and cultures. Full of vitality, Hau'ofa argued, the ocean is criss-crossed by many lines of interaction. Migration, trade and kin networks span the Pacific Islands and Pacific island people have proved themselves over thousands of years to be highly adept at oceanic voyaging, at surviving and thriving in what appear to others to be very limited and fragile environments, and in making the ocean a welcoming home, not a debilitating obstacle.

Not only is this perspective grounded in the history of the region and its peoples – the way Pacific Island people explored and colonized a vast area of the globe well before European maritime technology was capable of such achievements – it also sees the present as an extension of this worldliness and mobility. Today, Pacific Island people are found in many parts of the globe: there are sizable expatriate Pacific communities in mainland USA, in Australia and New Zealand, and in France. Pacific Island peacekeepers and nurses work overseas in many locations and Pacific Island athletes are highly visible worldwide in sports such as rugby, gridiron and boxing.

Yet such movement is not just a matter of moving away from the region. Pacific island migration is usually associated with a high degree of contact and interaction with 'home': people move easily to and from their locations over the sea, kin come and stay with relatives whilst they find work, and remittances of money are sent back to support relatives at home. It has been suggested that such networks of Pacific Island families operate as 'transnational corporations of kin', spanning the globe, organising people and resources and moving them around to secure good livelihood outcomes for the wider kin network. In some ways, we might see this phenomenon as a form of alternative globalization, one that is integrated into the global market economy but one run by communities and families rather than

Box 8.4
continued

commercial companies, and one that has strong social and cultural objectives, not just profit maximization.

If we accept Hau'ofa's view of Pacific Island vitality, adaptability and enterprise, we begin to see a very different view of development in the region. Globalization in this framework is a process that involves foremost the movement of people. They live and work within separate states and have to conform to local laws about citizenship and work entitlement yet, like the movement of capital investment, people seek to move outside and beyond state boundaries, exploiting opportunities wherever they might occur. Unlike most standard analyses of globalization, however, this Pacific Island approach sees cultural, communal, and family networks as key resources for, not constraints to, development and well-being. Furthermore, Hau'ofa and others might argue that Pacific Island globalization has, in effect, been occurring for many centuries and is an enduring feature of life in the region (see Hau'ofa, 1993).

division of labour made possible through improved maritime technology. Thus initial contact was made through whalers, missionaries and traders who were largely British and French, beginning the process of the diffusion of Western sociocultural values, such as monetization, Christianity and Eurocentric constructs of progress and civilization. Political rivalries meant that the South Pacific became an important strategic location (Rapaport, 1999). There were also economic benefits to be extracted. Based on calls from settlers and traders for the establishment of law and order in their new peripheral economies, formal colonization began in the mid-1800s in the French colonies first, followed by British and German territories 30 to 50 years later. The German territories were lost following the First World War, ran as mandates and then handed over to the UK, Australia or New Zealand (see Table 8.4). The first wave of globalization drew the PICs into the mercantilist circuit of production, placed firmly at the periphery of the global economy. As Firth (2000, p. 182) argues:

> Globalisation when combined with colonial rule, meant incorporation into the global economy on terms that suited the interests of the colonial powers. . . . The place of the tropical world in the first globalisation was to be subordinate to the temperate and developed world.

The classic colonial era in the region persisted until the 1960s, although a number of territories remain colonies into the present. Independence came late to the PICs relative to the rest of the Third World. This late decolonization can be explained in terms of two factors – economic

Table 8.4 *Colonization and independence in the Pacific Islands*

Colonial power	Incorporated into colonial power*	Dependent colony/ territory	Self-governing in free association	Independent
UK and France				Vanuatu (1980)
United Kingdom		Pitcairn (c1838)		Fiji (c1874, i1970) Tonga (i1970)* Tuvalu (c1916, i1978)** Solomon Islands (c1920, i1978)** Kiribati (c1916, i1979)**
France		French Polynesia (c1904) New Caledonia (c1853) Wallis and Futuna (c1842)		
New Zealand		Tokelau (c1916)	Cook Islands (c1900, sg1965) Niue (c1901, sg1974)	Samoa (c1900, i1962)
Australia		Norfolk Island (c1788)		Nauru (c1920, i1968) Papua New Guinea (c1884, i1975)
USA	Hawai'i (1898, in 1959)	American Samoa (c1899) Guam (c1898)	Marshall Is (c1885, sg1986) FSM (c1885, sg1986) Palau (c1885, sg1994)	
Chile	Rapa Nui (Easter Island) (in 1888)			

Notes: *
**

non-viability and the geopolitical advantage to US-allied powers of maintaining Pacific territories in the Cold War context. In those countries that became independent in the 1960s and 1970s, the influence of Keynesian and structuralist policies was strong and many turned to inward-oriented development strategies, including ISI (Chandra, 1992). These were often funded by aid flows from former colonial powers in attempts to modernize the Pacific economies. This, combined with increased flows of remittances from expatriate islanders who migrated in large numbers to the Pacific Rim in the 1960s and 1970s, led to the development of networked MIRAB economies (see Box 8.5). The development strategies of newly independent Pacific Island states

---

Box 8.5

MIRAB – indigenous globalization?

MIRAB is an acronym used to describe countries where migration, aid, remittances and bureaucracy play a central role. The hypothesis, developed by Bertram and Watters (1985) to refer to postcolonial Pacific Island economies, seeks to explain why Pacific Islanders have higher average standards of living than per capita GDP measures predict. Following independence, a number of linked processes were observed. Migration, from the islands to former colonial and other powers on the Pacific Rim (predominantly New Zealand, Australia and the USA) established 'transnational corporations of kin', through which individuals, acting in a utilitarian manner, channel remittances both financially and in kind. Aid formed a significant proportion of national income, driven originally by the strategic imperatives of the Cold War and also by postcolonial 'obligations', sometimes constitutionally enshrined, of former powers. The combined magnitude of aid and remittances allowed

consumption at levels well above GDP. Having inherited relatively large colonial bureaucracies, operating in small labour markets, the state sector became the major employer.

The model is applicable to varying degrees in the different PICs, and has been applied less successfully to other island regions. It remains relatively salient in Polynesia where remittances and aid play significant roles. In Niue, for example, aid accounts for over 50 per cent of GDP, while in Samoa remittances can account for up to 40 per cent of GDP (Bertram, 1999). In 1985, Bertram and Watters argued that MIRAB constituted a form of dependent development that was both rational and sustainable. This has been called into question, however, as third and fourth generation migrants lose physical and psychological ties with their 'homelands' (although there is, as yet, no recorded decline in remittances); aid from traditional

Box 8.5
continued

sources declined as the Cold War geopolitical comparative advantage of the region faded in the 1990s (although aid from new donors such as China began to replace this); and public sectors were radically downsized on the advice of regional and global development institutions. The MIRAB model, however, was revived somewhat in the early 2000s. Following earlier neoliberal cutbacks of aid and enforced and substantial reductions in state sector employment, Western donors returned and aid levels increased markedly. Such aid went to the old MIRAB states that had retained strong ties with their former colonial masters but also to larger states, such as Solomon Islands and Papua New Guinea, where indices of poverty revealed pressing need for resources. Furthermore, this new aid inflow was guided by the neostructural turn and it used state-centred funding mechanisms such as SWAps and GBS to deliver aid, thus boosting state involvement and employment. Another

aspect of this new aid approach was a move to encourage migration and remittances by allowing Pacific Island people to enter Australia and New Zealand to engage in limited-term seasonal employment, often in agricultural industries before returning home with their savings.

Within the region itself, MIRAB remains a highly controversial topic that still frames much development and globalization debate at academic and policy levels (see Bertram, 2006). Interestingly, although there was opposition to the MIRAB model by Epeli Hau'ofa and others – mainly because it seemed to argue for a continuation of dependence on external patronage and assistance, the two approaches have much in common. They both stress the importance of globalization for the region, though forms that involve a high degree of local agency particularly through mobility, employment and remittances.

involved a high degree of state involvement, both internally and from metropolitan patron states, usually former colonial masters. Former colonial powers, the United Kingdom and France, offered trade preferences so that the products of the Pacific could enter protected European markets at subsidized rates. The key example here was the sugar industry which thrived following 1970 as it was able to receive prices considerably above world market rates and the industry expanded to become the region's most successful export industry. Later Fiji and others were able to develop garment industries because they were able to use their cheap labour to manufacture goods which could enter the Australian, New Zealand and American markets with low or zero tariffs and compete successfully against much larger and lower-cost Asian manufacturers.

For states smaller than Fiji, where economies of scale meant that even subsidized industries struggled to survive, the only alternatives seemed to be tourism or aid. Tourism proved very successful in Tahiti, New

Caledonia, Fiji, Samoa and the Cook Islands but it was patronage through the MIRAB system that was critical for others. According to Bertram and Watters, such economies represented a form of 'dependent development' that offered long-term sustainability. This was rapidly eclipsed however.

Postcolonial globalization

Aid and finance donors did not share the optimism of Bertram and Watters with regard to the sustainability of MIRAB economies. Furthermore, in the 1990s with moves to liberalize world trade, the old subsidies and protection for Pacific agricultural and garment exports began to come to an end. Across the Pacific Island region there was a notable trend towards neoliberal economic restructuring. This change mirrored patterns elsewhere in the Third World – although it came relatively late to the Pacific Island region. Following the end of the Cold War, economic policies of donor and creditor countries shifted from the traditional aid/modernization approach towards that which was intended to stimulate autonomous economic development through free trade. The end of the Cold War also meant that donors could impose conditions on aid without fear of being replaced by Soviet aid. The major donors to the region (France, Australia, New Zealand and the Asian Development Bank in particular) cut back aid considerably but also withdrew direct funding for state agencies, instead moving to fund individual projects.

Governments in the region had to cut staff, their services were reduced or privatized and local economies were hit hard. Furthermore, there were pressures on PICs to reduce tariffs, deregulate the economies, open them up to foreign investment and seek membership of the WTO (itself with a range of strict compliance measures for economic liberalization).

Plate 8.5 *The changing role of women in the Pacific Islands*
Female seasonal workers in an export tuna cannery, Levuka, Fiji

The first country to adopt this strategy was Fiji, following the coup of 1987. The inward-oriented strategy of the state, which represented the most advanced attempt at industrialization in the region, was reversed and reforms intended to stimulate export-oriented growth enacted (Murray, 2000). Policies included a number of devaluations, the establishment of export-processing zones, widespread privatizations, and a reduction in the government sector and social expenditure (Chandra, 1992). The Cook Islands also received a severe shock when the major donors, especially New Zealand, enforced a dramatic downsizing of the public sector, with some two-thirds of government employees being reported as losing their jobs in 1996. Other reductions occurred in Tonga, French Polynesia and Samoa. The MIRAB model thus lost or reduced its aid and bureaucracy elements though many people resorted still to migration and remittances as local economies suffered. The PICs had little choice but to adopt such restructuring, and in this context Firth (2000, p. 186) laments:

> Just as the place of the Pacific Islands in the first globalization was to be subordinate to the temperate, developed world, so their place in the second globalization is also to be subordinate, this time to a set of international institutions that have set the rules of the global economy. The new globalization, *now combined with independent sovereign rule*, means incorporation into the global economy on terms that suit the interests of the financial markets, the aid donors, and those relatively few Pacific Islanders who are in a position to benefit from the new situation.

What are the impacts of the diffusion of neoliberal globalization for the region? A number of points can be made:

- Studies have shown that neoliberalism has led to the development of inequitable and environmentally and socially unsustainable export systems, especially in the agricultural sector (Murray, 2000; Storey and Murray, 2009c).
- In urban areas, unemployment and marginality have risen as state support mechanisms and traditional networks are eroded (Connell, 2003a).
- In general, the economic policy of the region is determined by outside powers with a vested interest in exploiting the region, and island governments themselves have very little role in decision-making (Henderson, 2003).
- Poverty has risen sharply in urban and rural areas affecting children and women disproportionately (Sriskandarajah, 2003).

- Privatization has led to a number of scandals in the public sector, and buyouts by government.
- The development of enclave economies, such as those which characterize tourism sectors, have proliferated.
- Pressures to reform land structures from communal systems to private property regimes have grown, despite the role these have played in maintaining relative welfare equality (Overton, 2000; Murray, 2009c).
- Linked to this latter point, and in general, some have argued that Pacific Island cultures, which are millennia-old, are increasingly unsustainable as Western constructs become increasingly pervasive (see Plate 8.6).
- Finally, some have argued that neoliberalism and the inequalities it has produced have led to overt political conflict (Murray and Storey, 2003; Overton, 2003).

Neoliberal reform in the PICS and the postcolonial wave of globalization underwent important changes in the past decade. First, the very severe reductions in state activities enforced by aid cuts soon led to the realization that the states themselves were no longer viable. As well as the coups in Fiji in 1987, 2000 and 2006, there was an armed separatist movement on the island of Bougainville in Papua New Guinea and more

Plate 8.6 *Coconuts and Coca-Cola – market vendors in Sigatoka, Fiji*

widespread political instability in that country, and the Solomon Islands experienced a virtual collapse in 2003. Violence in Solomon Islands and the inability of authorities to cope led to the request for help from neighbours and the Regional Assistance Mission to the Solomon Islands (RAMSI) followed with troops and police officers from the region restoring order. This programme, largely funded and run by Australia, marked a new chapter in aid and development in the region for it involved the virtual reconstruction of the state – RAMSI not only ended the violence but also aimed to build the institutions that had collapsed: police, customs, the judiciary, etc. Donors had become concerned that Solomon Islands, along with Papua New Guinea and Fiji constituted an 'arc of instability' in the region and, in the climate of post-9/11 fears of terrorism, they tried to bolster local states. This along with greatly increased aid for education, health and infrastructure seemed to mark the end of neoliberal restructuring. There was more focus on the MDGs and, for the smaller states, a return to the MIRAB model in practice as the state sectors were rebuilt.

However, the underlying objectives of neoliberalism remained and this postcolonial wave of globalization continued. States were supported and reconstructed but they were so only in ways that donors wanted: they did not return to a regulated and protected public sector but instead instituted policies that encouraged further globalization. Trade liberalization measures continued, foreign investment was encouraged and protected, and there were moves to convert communal land tenure systems towards more individualized private property. As we saw elsewhere, this neostructural development approach adopted a friendlier face, concerned with poverty and public services, but it still saw the market and engagement with the global economy as the central strategy for change.

Latin America – perpetual resource periphery?

Despite improvement in some development indicators in the 1990s, Latin America remains one of the poorest regions in the world. The average GDP per capita at purchasing power parity stood at approximately US$8,500 per annum in 2010, and very high levels of inequality, relative poverty and debt plague the continent (see Plate 8.7). Globalization has affected the region profoundly (Gwynne and Kay, 2004; Barton and Murray, 2009). Within the global economy, Latin America's role

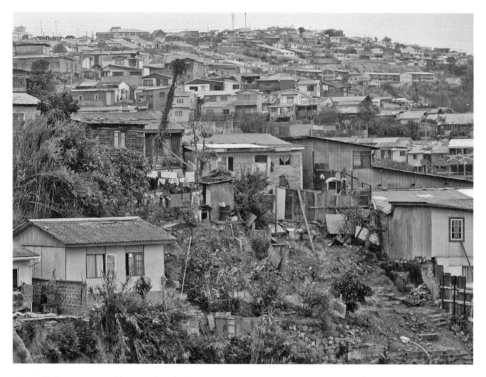

Plate 8.7 *Poverty and inequality – low cost housing in Valparaíso, Chile*

has long been that of a 'resource periphery' and this largely remains the case. It is a fascinating place to study development models and globalization as policy reversals have been common and have often moved between extremes. Nowhere is the search for new and exotic resources to fuel European development both now and in the past better exemplified than in Latin America; to early explorers, the continent was the 'land of gold'. Latin America has proceeded through various periods with respect to its insertion into the global capitalist economy, and these are explored briefly in the discussion that follows (Murray, 2004).

1 Colonialism and mercantilism (1500–1820s)
2 Independence and neoclassical theory (1820s–1930)
3 The Great Depression and structuralism (1930s)
4 Revolutions and dependency (1960s)
5 The debt crisis and neoliberalism (1980–?).

Colonialism to the debt crisis

As colonialism was established and consolidated through the sixteenth and seventeenth centuries, the Spanish and Portuguese focused on the extraction of a wide range of agricultural and mineral resources, paying no attention to the local environmental impacts of this or its long-run implications. Colonialism established a relationship of dependency with the imperial core, seeing nearly all surpluses expropriated outside of the region. Independence arrived beginning in the 1810s, but this did little to alter social conditions or the continent's role as a resource periphery. As long as the world economy grew, so would exports and progress, argued neoclassical thinkers. However, many of the activities were 'enclave' in nature, presenting few multiplier linkages into the socio-economy and concentrating surplus among a small landholding elite and foreign-oriented class.

In general, then, Latin America was a broadly outward-oriented region during colonial globalization and in the immediate postcolonial period. This changed somewhat following the First World War and the Great Depression, however, as the continent hit severe recession. Increasingly, doubt was cast upon the idea that simply exporting raw materials and unprocessed foodstuffs could sustain economic progress. The structuralist school of thought emerged out of this crisis (see discussion above). The recommended policy of ISI formed part of a wider move to inward-oriented policies combined with efforts towards regional integration. ISI dominated most economies until the 1980s, with some success in Brazil and Mexico. However, by as early as the late 1960s it had become obvious that structuralist solutions were not delivering broad economic development. Explosive rates of urbanization led to a crisis in the region's cities and the rural sector was left out of policy initiatives (Gilbert, 1996). Land reform progressed only slowly despite the insistence of structuralist academics (Kay, 2002). Dependency theory grew out of dissatisfaction with the performance of structuralist theories, suggesting that participation in the global economy was perpetuating underdevelopment in the continent. Based in part on Castro's 1959 revolution in Cuba, left-wing revolutionary movements were formed in many countries. In the context of the Cold War, socialist revolution was hotly contested by the USA through a mixture of economic, political and military interventions and revolutionary dependency was soon eclipsed. Non-revolutionary dependency has remained influential in regional academia, if not policy circles, until today however (Kay, 1989; Cupples, 2013).

The debt crisis and neoliberalism

The debt crisis was the major factor causing the shift to free-market neoliberal policies and a return to outward orientation in Latin America. The combination of debt and SAPs in the 1980s (see discussion above) led to the worst decade ever in terms of development indictors; income growth fell from 5 per cent between 1965 and 1980 to 1.4 per cent between 1980 and 1989, and was negative for Argentina and Bolivia over the latter period; unemployment and poverty rose – open unemployment stood at over 10 per cent in six countries in 1985; inflation skyrocketed, reaching an average of 1,502 per cent per annum in Bolivia between 1984 and 1993; and Western-backed dictatorships seized the reins of power across the continent (Silva, 2004). Once again the continent became articulated into the global economy as a resource periphery as SAPs insisted on export orientation based on comparative advantage. This saw a partial return to reliance on mineral exports and the creation of non-traditional agricultural export sectors serving affluent markets in the Northern Hemisphere (Murray and Silva, 2004).

By the end of the 1990s, democracy had returned to Latin American nation-states as Cold War geopolitics, which had led to the implicit and explicit support of neoliberal dictatorships by the USA, faded (Barton, 1997). The transition has seen a partial return to policies informed in part by structuralism, referred to as neostructuralism (see Box 8.3 above), involving greater concern for social equity and the environment as well as a revival of economic regionalism (Gwynne and Kay, 2000; Murray and Rabel, 2008). In reality, however, these policies represent an extension of the neoliberal orthodoxy. The legacy of neoliberalism in terms of poor socio-economic conditions is apparent. For example, in Chile income distribution was worse in the early 2000s than it was in 1966 (Barton and Murray, 2002; Gwynne et al., 2008). Despite some diversification at the continental level, most countries remain dangerously reliant on highly specialized primary product export sectors. In 2008, in 13 of the 18 largest countries, primary products accounted for over 50 per cent of total export earnings. In Bolivia, Paraguay, Ecuador and Panama this measure stood at 90 per cent. The leading export in 18 of the 20 major Latin American countries is a primary product, with only the largest countries, Brazil and Mexico, bucking this trend (see Table 8.5). Of most concern between 2000 and 2008, the reliance at the continental scale (heavily influenced by Mexico and Brazil which are relatively industrialized) had risen from just over 41 per cent to just

Table 8.5 *Export specialization in Latin America, 1970–2008 (proportional values)*

	1970	1980	1990	2000	2008
Bolivia	96.8	97.1	95.3	72.9	92.8
Paraguay	91.0	88.2	90.1	80.7	92.1
Ecuador	98.2	97.0	97.7	89.9	91.7
Panama	96.5	91.1	83.0	84.1	91.3
Nicaragua	82.2	81.9	91.8	92.5	89.9
Chile	95.2	88.0	89.1	84.0	88.0
Peru	98.2	83.1	81.6	83.1	86.6
Uruguay	82.4	61.8	61.5	58.5	71.3
Honduras	91.8	87.2	90.5	64.4	70.7
Argentina	86.1	76.9	70.9	67.9	69.2
Colombia	91.0	80.3	74.9	65.9	68.5
Guatemala	71.9	75.6	75.5	68.0	62.8
Brazil	86.6	62.9	48.1	42.0	55.4
El Salvador	71.3	64.6	64.5	51.6	45.3
Costa Rica	81.3	70.2	72.6	34.5	37.6
Mexico	66.7	87.9	56.7	16.5	27.1
Venezuela	99.0	98.5	89.1	90.9	–
Belize	–	82.4	84.6	–	–
Total	**89.2**	**82.6**	**67.2**	**41.7**	**52.9**

Source: Compiled by authors based on CEPAL (various years)

under 53 per cent. This was due largely to the demand for natural resources including minerals and metals as well as forestry and agricultural products in China. At the time of writing, China was negotiating a long-term concession with Nicaragua in order to construct a rival to the Panama Canal to further facilitate this trade. It was also extensively involved in purchasing land for agricultural production together with mines in products such as copper and silver, and building infrastructure across the region.

The central role which resource exports play suggests that the original concerns of the structuralists remain relevant today, and are likely to remain so as neoliberalism induces further specialization based on comparative advantage. Latin America's role as a resource periphery has arisen precisely because of its natural endowment in a wide range of items. In this sense resources can act as a curse and lead to what Karl (1997) calls a 'paradox of plenty' (see Box 8.6). With the rolling out of the Free Trade Area of the Americas (FTAA) and the TransPacific

Box 8.6

Resource peripheries and the paradox of plenty

Development history over the recent past challenges the orthodox concept of the relationship between resources and development, and calls into question policies which encourage poorer countries to specialize in primary product exports. This is based on the 'paradox of plenty' observed by Karl (1997). Two points are of special relevance in this context:

1 East Asian economies are relatively 'resource deficient' but they have achieved high rates of economic growth from the 1980s onwards.
2 Latin American economies are relatively 'resource rich' but have performed poorly from the 1980s onwards.

In resource peripheries such as Latin America, 'favourable' resource endowment can prevent progressive economic development – Auty (1995) refers to this as *resource curse* (see also Hayter *et al.*, 2003). 'Resource cure thesis' argues that there is a range of problems associated with placing heavy reliance on the primary product export sector (see also Gwynne, 2004; Barton *et al.*, 2008). Among others, the following trends may evolve:

- decline in the terms of trade;
- increased control by transnational corporations and expropriation of surplus;
- negative environmental impacts;
- the perpetuation of low value-added activity;
- increased vulnerability to fluctuating global commodity markets;
- problems associated with management of windfall profits in mineral sectors;
- the evolution of enclave economies;
- focus on the needs of external markets leading to problems of domestic food security.

Added to the above is the observation that, in mineral-rich countries, open conflict is often more prevalent and inequality higher. This has certainly been observed in African nations where commodities such as diamonds exist. Although the paradox of plenty is not necessarily inevitable and serves as a hypothesis rather than a strict rule, solving these issues requires regulation and intervention. Neoliberalism is not capable of facilitating such policies and serves only to fuel the paradox.

Partnership (Kelsey, 2011; Murray and Challies, 2011), it is possible that this will become entrenched further, although neoliberal thinkers see the FTAA as a potential saviour – illustrating the contested discourses which shape the geographies of the region. The rise of Asian economic power will place further pressure on the continent in this regard – China is already Latin America's largest trading partner and a range of multilateral and bilateral trade agreements are likely to see these flows burgeon. It is time to reconsider the principles of structuralism and dependency analysis in order to build policy which reforms the nature of globalization in the continent and makes development from within possible (Sunkel, 1993; Barton and Murray, 2009). As Karl argues with reference to the paradox of plenty (see Box 8.6): 'it is not inevitable. Paradoxes can be resolved and development trajectories altered' (Karl, 1997, p. 242).

Conclusion – new development geographies?

Historically, globalization has created and perpetuated inequalities in levels of well-being and development. During the first wave, this was reflected in terms of the colonial division of labour that evolved to subordinate the periphery. In the postcolonial wave, when the concept of development was invented, patterns of marginalization and deprivation have become more complex. Neoliberalism in particular is leading to new networks and flows that threaten to raise inequality to greater heights within, as well as between, nation-states. The old North/South classifications and core–periphery models are less relevant today as new networks of inclusion/exclusion are forged. The poorer world as a whole, however, remains saddled with poverty and debt and, despite economic advances in East Asia, there is little consensus on what might solve such problems.

There are three ways of conceptualizing the relationship between globalization and development – neoliberal (hyperglobalist), structuralist/ neostructuralist (transformationalist), and dependency/post-development (sceptical). The evidence presented in this chapter lends support to the second view. That is to say, globalization – when conceived of as stretched flows across space – is not *inherently* negative for the poorer world if it is regulated and managed effectively. However, history has taught us that this has rarely been the case, and globalization as currently practised is exacerbating global inequalities, failing to raise people out of relative deprivation and locking whole regions into an exploitative capitalist global economy.

In the past decade, new development geographies have begun to emerge, in part the result of a new neostructural aid regime – focused on poverty alleviation, building the capacity of states and further facilitating the expansion of the global economy – and, in part, the consequence of the changing global economy itself with the continuing emergence of East Asia and the relative decline following the GFC of Europe and North America. Economic growth in China in particular has helped achieve an aggregate decline in poverty and measurable progress towards achieving several other Millennium Development Goals. However, despite a raising of average global levels of income, attention is now turning to those regions and groups 'left behind'; those still suffering from absolute poverty and locked into networks of deprivation. When we begin to map these significant pockets of poverty we can appreciate how globalization is reconstructing patterns of inequality. Development geographers thus now need to look beyond aggregate and average patterns of development and highlight instead the extremes: both those (the extremely poor) who remain untouched – or further marginalized – by economic growth and those (the extremely wealthy) who have gained the most from the growth of the early 2000s and who have hitherto remained largely untouched by critical scrutiny of development. We reiterate Sachs' call for us to focus on 'wealth alleviation' as much as 'poverty alleviation' (2012, p.27).

Is there a way that globalization can be turned to the advantage of poorer nation-states and those which inhabit networks of deprivation? We are seeing the evolution of global movements that are resistant to the regressive impacts of globalization and development as currently practised. Whether such movements effectively capture the multiple voices that are vying to be heard is a matter for further geographical action-based research. Discourses of globalization and development are closely linked. The pressing task for development geographers is to map the concrete outcomes of these discourses and get inside new networks of power and wealth in order to understand how they restructure the multiple spaces of development. Only then can we hope to inform eclectic and holistic policy designed to face the immense global challenge presented by inequality.

Further reading

Cupples (2013): This is a very thorough and up-to-date analysis of development in the continent of Latin America – particularly strong on cultural geographies.

Gwynne and Kay (2004): This collection of work from an international panel of academics presents a systematic and broad-ranging assessment of the impacts of globalization and modernity in Latin America.

Kay (1989): In this book, the Latin American contributions to development theory are discussed in detail.

Murray and Overton (2013): This entry in the *Companion to Development Studies* analyses the links between globalization and development and offers signposts to further study.

Potter (2013): This collection of entries is the most comprehensive reference that exists in the discipline of development studies.

Potter *et al.* (2008): Now in its third edition, this is arguably the best text on geographies of development that exists.

Power (2012): This is an excellent chapter from Power, who writes radically on development geography. See also Power (2003) – a very readable critical analysis of development geographies.

Rappaport (2013): This is a wide-ranging collection of chapters on aspects of Pacific Island environment and society.

Roberts and Hite (2007): In this reader, the relationship between development and globalization is carefully dissected.

Overton and Scheyvens (1999): This is the most comprehensive collection of essays on issues of sustainability in the Pacific region.

Willis (2011): In this title, Katie Willis discusses some of the major theoretical innovations in development over the past seven decades and relates them to policy.

Willis (2009): This book provides a wide-ranging and highly readable coverage of geographies of development with examples from across the Global South.

Asia Pacific Viewpoint is a journal that has numerous articles on Pacific Islands society and economy and should be consulted by those researching Oceania. For Latin American studies *Bulletin of Latin American Research* is highly recommended.

9 Environment, sustainability and globalization

- Environmental degradation – definitions
- A brief history of global environmental degradation
- Global environmental problems?
- Population growth, globalization and resource consumption
- Globalization of environmental awareness
- Global environmental regulation from above and below
- Conclusion – transformed environments

The concepts of 'one world', 'interconnectedness' and 'globalism' have the environmental movement to thank for their broad diffusion around the globe and into the popular imagination. Since the first flight to the Moon in 1969 and the beaming back to Earth of pictures of the world as one fragile and floating globe, the movement has made increasing inroads into conventional politics, academia and the media. The idea that the global environment is at risk due to anthropogenic activity, and that something must be done quickly to abate this, is accepted by a growing majority of people across the world – especially in the relatively affluent West. The opinions of the International Panel on Climate Change with regard to global warming and other changes form a broad consensus. Environmentalist 'worldviews' are now more or less mainstream in the academic literature. Left-wing ecologists talk of the fragility of the Earth and the imbalance that humans have created, and relate it to concepts such as 'spaceship earth' and the *Gaia* thesis (O'Riordan, 1981). Almost all are agreed that processes which are global in scope and yet local in terms of their source threaten the Earth's environment. In the manifestos of virtually all major political parties across the West, the environment figures centrally. In some countries, such as Germany and New Zealand, Green parties have themselves carved out significant representation in Parliament. We have entered a period where those who deny that environmental problems exist, or that technology and human innovation in response to scarcity will solve all issues, seem like extremists. Just 30 years ago it was the other way around, and environmentalists were broadly seen as idealistic and esoteric. This in itself is great testimony to the globalization of concern for the environment and the building of unprecedented global networks that articulate and circulate these issues.

Many definitions of geography cite 'interactions between the environment and humans' as a core focus of the discipline. Human geography has moved through a number of phases in terms of its conceptualization of the human–environment relationship. In the early part of the twentieth century, the concept of *environmental determinism* was seen as *the* organizing principle of the discipline. In this argument, the environment set strict boundaries to human activity in any given place and conditioned resultant patterns. Although this concept has been widely discredited, some environmentalists still appeal to its logic when talking of 'limits to growth' at the planetary level, and current discussions concerning 'peak oil' reflect this. Experts are divided on when the peak will come, and price rises in the late 2000s seemed to suggest that it may be approaching. However subsequent continued exploitation and reduced prices seem to have averted a potential crisis – although it is likely to resurface over the next two decades. Human geography has moved through phases of *environmental possibilism* and *human determinism*. The former sees the environment presenting a range of opportunities for humans, and the latter argues that human ingenuity allows complete domination over the environment. Over the past 20 years there has been a steady rise in human geography's engagement with the environmental issues, from a possibilistic perspective in particular. This has involved both applied studies of human impacts, work on the politics of environmental change, and – more recently – critical geographies concerning the boundaries between and social construction of 'natural' and 'human' activity (Whatmore, 2009). Ideas from political ecology – the study of who controls the environment and for whom – have had an important influence on environmental geographers over the past 15 years. This field traces how patterns of power shape the interaction of humans and the environment, and how the outcomes of change are mediated and distributed (i.e. in whose interests is the environment used, and who bears the costs of change). Human geography, then, is a natural home for the study of interaction between ecological and social systems. What makes the discipline particularly useful in the current era is its attention to global–local relations, allowing an analysis of the implications of globalization for the environment at different scales.

What is the relationship between globalization and the environment? This question is explored in this chapter through three interlinked observations. First, it is widely argued that the *processes of globalization are part of the explanation for environmental degradation.* In the first part of this chapter, we discuss the concept of degradation and then seek to relate its history to the unfolding of the different waves of globalization. Second, *some environmental problems, including population growth, have become global-scale concerns.* In a subsequent section we look at what constitutes a 'global' environmental problem and provide a number of case studies to back up the arguments. We then move on to look at the relationship between demographic change and environmental issues. Third, *concern for the environment has become globalized and the search for sustainability requires global-scale action.* We conclude by considering the evolution of the global environmental movement and attempts to regulate the system from above and below.

Environmental degradation – definitions

In order to analyse the implications of the processes of globalization on the environment, it is necessary to have a working definition of environmental degradation. A practical definition is offered by Tyler-Millar (2002, G5):

> Depletion or destruction of a potentially renewable resource such as soil, grassland, forest or wildlife that is used faster than it is naturally replenished. If such use continues the resource can become non-renewable (on a human time scale) or non-existent (extinct).

Examples of such degradation could refer to the following processes: urbanization of productive land; waterlogging and salinization of soil; soil erosion; deforestation; groundwater depletion; overgrazing and desertification; biodiversity loss; air, water and soil pollution; ozone depletion; and anthropogenic climate change. Humans need not cause such degradation, although most environmentalists agree that human systems are depleting resources and degrading the environment at unprecedented rates. In reality, the definition of environmental degradation is complex and contested and turns on one's world view. A deep ecologist, for example, who views 'nature' as having equal existence rights as humans, would have a very different viewpoint from a cornucopian who believes that the environment's purpose is to serve humans. We distinguish here between two very broad models in order to establish a perspective for the remainder of the chapter:

1 *Anthropocentric* (human system approach) – Under this view degradation occurs when ecosystems, or parts of them, are transformed in such a way that the consequences have net negative impacts upon the lives and/or health of humans. This may also include aesthetic degradation, which will not necessarily have negative ecosystem effects. This definition is deceptively simple. For example, what do we mean by negative? To what extent are aesthetic values to be included? To what extent should the distribution of environmental degradation be considered?

2 *Ecocentric* (whole system approach) – In this definition environmental degradation takes place where ecosystems, or parts of them, are transformed in such a way that the consequences have net negative impacts on the ecosystem as a whole. Again, this definition is deceptively simple. For example, are all parts of the ecosystem to be valued equally? Science is not always clear on what the impacts of a given change will be in the short, medium and long term – how do we judge therefore? Finally, by measuring and possibly prioritizing certain parts of the ecosystem, we are implying that human judgement can and should be applied which leads towards a more anthropocentric perspective.

In this chapter we will be referring to debates and works that often adopt an anthropocentric perspective, as this is the approach that pervades most writing and thinking on the topic. Deciding on what broadly constitutes degradation is one thing; actually measuring it is another, and scientists are deeply divided, as Box 9.1 on global warming illustrates. The core issue in the context of this chapter, then, is whether we can reasonably argue that processes or agendas of globalization are causing such degradation.

A brief history of global environmental degradation

While scientific opinion is fairly unified on the general proposition that many aspects of the global environment are being degraded, and that humans are playing the major role in this process, the links between this and the processes of globalization are not immediately obvious. How has globalization related to environmental degradation across the various waves?

Environmental degradation of global significance during the premodern globalization period (see Chapter 3) included a range of global extinctions of mammals and birds from over-hunting (for example, the Moa in New Zealand hunted by Māori as well as the large-scale degradation of Rapa Nui [Easter Island]). The diffusion of microbes due to large-scale migratory movements led to epidemics and population collapse – such as the Black Death in Europe during the Middle Ages. Localized pollution began to accumulate during this period. The main driving forces of environmental degradation during this time included urbanization, large-scale migrations, overpopulation in local areas, warfare and poor agricultural practices. In general, the extensity, velocity and impact of such processes were relatively limited as both nomadic and settled peoples were incapable of transforming their environments to any great degree.

The colonial wave of globalization set in motion a range of significant changes. The diffusion of European populations, methods and principles to Latin America and parts of Asia led to unprecedented change in the former especially. The isolation between Europe and the Americas was broken during this period, which set a number of important linked changes in motion. The introduction of non-native species, intensive hunting and disease impacted on both animal and human populations greatly, leading to the near extinction of many indigenous groups, for example. The exploitation of both mineral and non-renewable resources for use in the colonies and beyond transformed landscapes. Within Europe, the agricultural revolution and its particular constellation of innovation, technological change, market incentives and higher investment led to a rapid expansion in cultivated areas and land degradation as populations grew. Forests were cut down, marshland drained and a number of species, including wolves and bears, were greatly diminished by the growing 'organic economies' of the time (Held *et al.*, 1999).

The industrial revolution in Northern Europe and its subsequent diffusion across that continent and beyond combined with the way it pulled 'peripheral' economies into the new globalizing industrial complex is *the* watershed in terms of the human impact on the environment. Within Europe new production techniques combined with intense urbanization raised the potential for local degradation and initiated systemic global environmental degradation through industrial pollution, principally via the use of coal. In the colonies, possibly the major impact of the industrialization phase was deforestation, as land was cleared for plantation agriculture to supply imperial heartlands and the colonies. Furthermore, the diffusion of exotic biota had a significant impact – often deleterious. The privatization of land and concept of private property also led to a widespread undermining of indigenous environmental management modes. At the same time, the idea of environmental conservation – such as in game reserves – was diffused. Southeast Asian tropical rainforest was greatly depleted beginning in this period. This phase of globalization involved a steep acceleration in the extensity of European impacts and an increase in the velocity of environmental change.

The postcolonial wave of globalization from the end of the Second World War led to increasingly profound environmental change. The rise of modernization as the organizing principle for the new concept of 'development', which resulted in the diffusion of industrialization and urbanization to the periphery, had major ramifications for environments in the newly independent 'developing' countries (see Chapter 7). In these places, new pressures on resources required for economic growth, combined with unprecedented population growth, placed extra strain on local environments. In countries that remained locked in as resource peripheries, such as many of those in Latin America and Africa, these effects were amplified. The rise of socialist industrialization in the USSR and China also increased both local and global degradation significantly, often in ways that ravaged ecosystems more than in the West. This was due to the fact that Marxist theory suggested that the environment existed to be exploited by humans. In the West the post-war boom fuelled rapid increases in per capita resource consumption, especially in the USA, where a new consumer culture grew and the motor car was widely adopted. All combined, these factors led to historically unprecedented increases in human impact on the environment. Many of the problems which remain salient today, including ozone depletion, the anthropogenic greenhouse effect, deforestation, desertification, marine and air pollution, acid rain, biodiversity decline and nuclear waste, were initiated during this period.

The neoliberal phase of the postcolonial wave has had the effect of magnifying the above problems through the shift to free-market economics, reduced state regulation, and the accelerated diffusion of Western industrial/urban/consumerist models of development to poorer countries. Free-trade reform under the WTO, for example, has led to the dismantling of environmental regulations under 'unfair' trading rules in some cases. The continuing debt crisis in poorer countries and the increasing global wealth gap divert resources away from environmental conservation and towards short-term solutions to poverty. Deprivation also fuels conflict, both overt and otherwise, which often has deleterious environmental impacts. The commodities boom of the 2000s which led to accelerated mineral exploitation in poorer economies in order to fuel the growth of China has had damaging environmental impacts both locally and in a global sense. Latin American countries have, in many cases, returned to heavy dependence on commodity exports with all of the environmental and associated socio-economic problems identified by the structuralist thinkers (see Chapter 8). Overall the diffusion of capitalism and its associated production and consumption patterns is greatly perpetuating global environmental degradation. A number of examples including the boxed case studies of global environmental change illustrate this point throughout the chapter.

Global environmental problems?

In this section, we look at the idea that environmental problems, or at least an increasing amount of them, have become global in their scope. There is nothing inherently good or bad about global environmental change. 'Problems' and 'issues' are generally defined only in terms of their relationship to humans – that is to say, through an anthropogenic model of degradation. In this sense, environmental 'problems' are socially constructed. So what do we mean by global environmental problems? This depends very much on the definition of 'global', and this echoes earlier debates concerning how we define different scales (see Chapter 2). On the one hand, no environmental change is global; nothing impacts on the Earth's surface and ecosystems in a uniform way. On the other hand, all environmental change is global in the sense that all transformations are inextricably linked. The global ecosystem is the sum total of intricately interlocking parts and there are very few, if any, closed ecological systems. (See also Pickering, 2012; Eden, 2014.)

Given the above, how can we conceptualize global environmental degradation in a practical way, and how can this be linked to globalization? It is useful to divide global degradation into global systemic degradation and global cumulative degradation. Examples of the former include climate change, sea-level rise and ozone depletion. Examples of global cumulative degradation include desertification, flooding, deforestation, air and water pollution, soil erosion, acid rain and biodiversity loss.

Global systemic degradation refers to change which affects systems that are inherently global in scope. These systems are sometimes also referred to as global 'commons', the main examples being the atmosphere and the oceans. Within such systems, borders are relatively meaningless (although the Law of the Sea has achieved this for the oceans to a certain extent). The essential point is that degradation may come from many sources, which may be quite specific in location, but the impacts affect the system as a whole, as illustrated through the case study of global warming in Box 9.1. This does not imply however that the impacts across the system are necessarily even, as the example of sea-level rise in Box 9.2 exemplifies.

Box 9.1

Systemic global degradation? Global warming and globalization

There is considerable scientific and political controversy over the issue of climate change and, in particular, 'global warming'. The process of global warming is based on a naturally occurring phenomenon known as the 'greenhouse effect'. Short-wave solar radiation passes through the Earth's atmosphere, is absorbed by the Earth's surface, and reradiated as longer-wave radiation back into the atmosphere. Here, it is absorbed by atmospheric elements, principally water vapour and carbon dioxide. Some of this radiation is then reradiated, heating the Earth's surface and creating a positive environmental feedback loop. Without this effect global temperature would average −19°C, compared to the current average of 15°C, making life on Earth impossible. There is overwhelming evidence to suggest that the CO_2 content of the atmosphere is rising markedly due to anthropogenic activity and that, through the feedback loop referred to above, this is increasing the average surface temperature (see Figures 9.1 and 9.2).

The International Panel on Climate Change (IPCC), a body formed by members of the World Meteorological Organization and the United Nations Environment Program (UNEP), draws together the state-of-the-art evidence on climate change based on the work of thousands of scientists and academics in more than 120 countries. It has produced four reports and is currently compiling a fifth. The IPCC estimated in its third report that the temperature has risen by 0.6°C over the past century (IPCC, 2001) and that the magnitude of atmospheric greenhouse gases has increased by around 35 per cent since 1750. Whether this is due to human activity or part of a longer-term climatic cycle is contested, although the IPCC claims evidence for the former is unequivocal. In the fourth report the IPPC (2007) concluded that temperature would rise between 1.1 and 6.4 degrees Centigrade, depending on the model of globalization pursued by the world – tellingly the homogenous globalization model with more economic focus would deliver the highest temperature change. The IPCC also concluded that the levels of greenhouse gases in the atmosphere released since 1750 were higher than they have been for over 650,000 years.

Box 9.1
continued

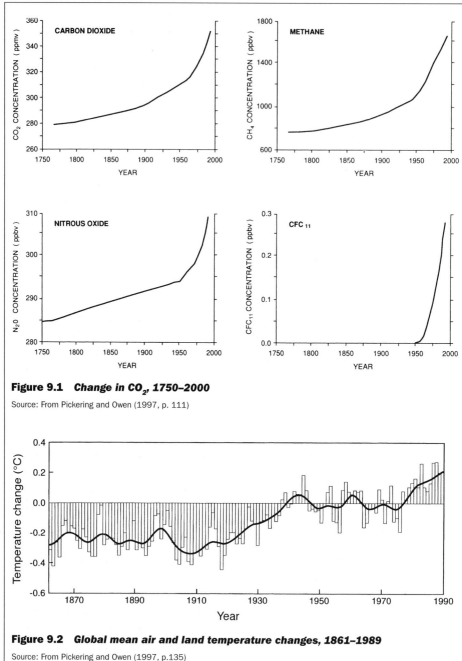

Figure 9.1 *Change in CO$_2$, 1750–2000*

Source: From Pickering and Owen (1997, p. 111)

Figure 9.2 *Global mean air and land temperature changes, 1861–1989*

Source: From Pickering and Owen (1997, p.135)

Box 9.1
continued

There were further devastating impacts predicted including the melting of the polar ice caps, and related flooding, alterations of the Gulf Stream, and major agro-system transformation (see also Figure 9.3). Sea levels were predicated to rise between 18 and 59 centimetres in the twenty-first century. The distributional impacts of climate change are uncertain, but it will cause substantial shifts in the climatic regimes across latitudes, leading to drought and potential famine in many mid-latitude locations, for example (IPCC, 2001, 2007). The IPCC estimates greater heatwaves, a more intense hydrological cycle and unprecedented sea-level rise. It is argued that the greatest proportion of negative impacts will fall on Third World countries, and poor persons within other countries. Other scientists, particularly from industrial lobby groups, have argued, however, that circulation models are not sophisticated or reliable enough and that rises of 0.5°C over the twenty-first century are more likely. Notwithstanding these arguments, there is a growing scientific consensus that global warming represents a serious and significant problem that must be addressed.

How does this relate to globalization? The rise in greenhouse gases is traceable to the industrial revolution. As industrialization has diffused across the world the effect has been magnified. Given discourses of modernization, many poorer countries aggressively pursue industrialization strategies and are adopting global consumer culture – including demand for petrol-driven cars and fossil fuel-powered appliances. Industrial expansion in heavily populated countries such as China and India has the potential to add greatly to greenhouse emissions. However, as we discuss

elsewhere in this chapter, it is per capita consumption in Western countries that is most responsible for the growth in greenhouse pollutants (see section below on consumption overpopulation). Some poorer countries have argued that the industrialized countries of today paid scant attention to the environment as they modernized, so why should they not be afforded the same luxury? A distributional dimension is thus inherent in this global problem and this is why the predications of the IPCC are differentiated according to the model of globalization the world is likely to pursue (see IPCC, 2007).

As such, the anthropogenic greenhouse effect has become the cause of major political controversy and has initiated globalized political reactions in the form of international cooperation, protocols and treaties which attempt to stabilise/reduce/eradicate polluting emissions (see discussion below on the Kyoto Protocol and what has come beyond it, for example). For a considerable amount of time, some national governments erred on the optimistic side and either failed to ratify or did not follow the guidelines contained in Kyoto Protocol. Across the industrialized world, and in the USA, there are a number of pro-industrial organizations that have sponsored media and research activities which play down the proposed seriousness of the global warming issue. The USA, under the administration of George W. Bush (2001–2009), was especially sceptical of global warming predictions. The US signed the Kyoto protocol but did not ratify it. Many commentators pointed to the Administration's links to the oil industry to explain the zeal with which the environmentalists have been fought. The Obama Administration has been more

Box 9.1
continued

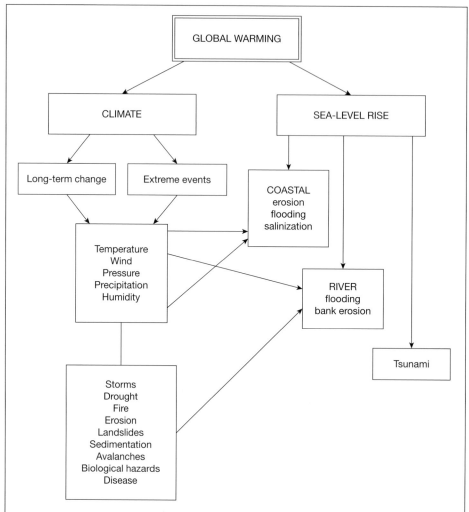

Figure 9.3 *Hazards due to global warming*

Source: Adapted from Pickering and Owen (1997, p. 149)

proactive in climate talks than its predecessor – establishing the Major Economies Forum on Climate, for example, and backing calls for emissions made through the G8; but critics are quick to point out that intended cuts in emissions remain proportionally small in the country that has contributed the most to the problem in a cumulative sense. In an absolute sense, in the late 2000s, China became the leading contributor to carbon emissions – and thus a solution incorporating China was increasingly urgently required. Overall, from a total ecosystem perspective, it is unclear whether global climate change of this nature can rightly be considered 'degradation'. From an anthropogenic perspective, however, it clearly is.

Box 9.2

Globalization, sea-level rise and the Pacific Islands

Significant scientific evidence exists which suggests that average sea level is rising across the planet. The IPCC claimed in 2001 that sea level had risen by 10–20cm over the past 100 years, and, in 2007, estimated that by 2100 it will have risen between 18cm and 59cm. The melting of the ice caps and the expansion of the surface of the ocean given raised temperatures across it bring this about. Island regions in general are especially vulnerable to the impacts of climate change and sea-level rise in particular (IPCC, 2001, 2007). In the Pacific Ocean there are a number of linked effects that arise from the combination of sea-level rise and warming (Nunn, 2003). Most crucially for the Pacific Island region such conditions are likely to increase the frequency, intensity and impact of cyclones. Such weather systems are devastating when they hit islands, as recent destruction in Niue in 2003/04, the Cook Islands in 2005, American Samoa in 2011, and Samoa and Fiji in 2012 exemplify. In the case of Niue, the country's infrastructure and agricultural production was almost completely destroyed by Cyclone Heta which hit during the New Year period 2003/04. Raised sea levels exacerbate the problems cyclones bring, increasing the intensity of storm surges and thus wave damage along coasts. A further problem is that protective coral reefs, which help dissipate the effects of wave damage, are under pressure, especially in tourist localities. This is a particular concern on the southern coast of Viti Levu, Fiji, for example, where resort tourism is concentrated.

Other effects of sea-level rise are troubling. Across the region, the vast majority of the population lives on the coast. As sea level rises it can salinate freshwater tables and lead to crop failure. These problems are magnified on atoll islands – ringed ancient reefs surrounding a submerged volcanic island – where limestone rock forms allow easier percolation. Processes such as this have been observed in Kiribati, Tokelau, the Federated States of Micronesia and the Marshall Islands. Although the press has tended to exaggerate claims that islands are likely to disappear and some politicians in the countries themselves have latched onto such claims for their own benefit (Barnett and Campbell, 2010; Connell, 2003b, 2009), contingency plans for the removal of people are being made in some cases. Ultimately the sovereignty of a number of smaller island nations such as Niue and Tuvalu could be under threat from processes whose source lies entirely outside of the Pacific Islands' territory and control. In some ways, Pacific Island territories have become canaries in the mine of global climate change. The distributional consequences of this are significant as the islands have contributed virtually nothing to the cumulative build-up of greenhouse gases (Farbotko, 2010).

Cumulative global environmental degradation occurs when local incidents of degradation become multiplied to such an extent that they begin to constitute a global-scale concern. As ever, what exactly is meant by 'global' here is subject to debate; however, where degradation accumulates to the extent that it impacts upon a large proportion of the Earth's population or has the ability to threaten global atmospheric, biological or hydrological systems, it can be deemed as such. An example of this type of problem is deforestation (see Box 9.3). Other examples include the increased incidence of flooding, desertification, air and water pollution, soil erosion and biodiversity loss. The nature of these problems is such that it is hard to regulate them within the context of existing state borders and there are thus overspill effects (*negative externalities*). Air pollution, for example, is subject to atmospheric system movements and may be carried across borders. This was the case in the 1980s during the height of the acid rain controversy in Europe, where pollution from the east caused tree loss in Scandinavia and other parts in the north-west of the continent. In another example, deforestation in the Indian Himalayas has caused extensive flooding of the Ganges downriver in Bangladesh. Although the negative externalities caused by cumulative degradation are generally not of the same magnitude as those caused by systemic degradation, it is clear that environmental governance to solve cumulative problems *requires* moving beyond nation-states. There are many overlaps between cumulative and systemic global environmental problems, and facing them requires interpreting them as part of one system.

Overall, there are a number of uncertainties in attempting to define and analyse global environmental degradation. Even if we were to arrive at a point of consensus with regard to what constitutes degradation, this would not imply agreement on its magnitude or what should be done about it. Science is highly imperfect, and our measurement and prediction of environmental change in general is laced with ambiguity. This uncertainty can lead to inaction and lack of resolve. Given this, many environmentalists favour a *precautionary principle*, arguing that we should restrict our consumption and restructure activities on the basis of worst-case scenarios. Other groups have argued that forgoing current consumption on the basis of unequivocal scientific evidence is foolhardy and possibly dangerous. In the case of the Third World, current needs are more pressing than elsewhere; hence the tension between resources and development is greatly amplified. However, irrespective of world view and debates over the details of environmental degradation, there can be

Box 9.3

Cumulative global degradation: deforestation

Deforestation refers to 'the removal of trees from a locality. This removal may be either temporary or permanent, leading to partial or complete eradication of the tree cover. It can be a gradual or rapid process, and may occur by means of natural or human agencies, or a combination of both' (Jones, 2000, p. 123). The major reason for deforestation at the global scale, however, is human activity. Among other things, deforestation may take place to create new land for urban or agricultural use; to provide wood for fuel or construction; to allow the exploitation of mineral deposits; or to create reservoirs or build highways. In adjacent areas the practice degrades the environment in a range of linked ways. Among other things it can lead to: a loss of nutrients; increased run-off and flooding; heightened erosion; increased landslides; and coral reef depletion due to sediment build-up. Deforestation has occurred since human populations roamed the Earth and was especially common in premodern slash-and-burn societies. In England it is believed that deforestation goes back at least 8,900 years. However, change at this time was relatively localized and minimal in terms of its scope. In Europe, from the eleventh century, extensive deforestation took place as land was cleared for use as fuel and settled agriculture. In 200 years the forests of Central Europe were almost completely depleted.

With the rise of colonial globalization, deforestation in the periphery began both to serve local needs and supply raw materials to the imperial heartlands (see Chapter 8). In North America, before the arrival of colonial peoples, the forests occupied 170 million hectares of land; currently only approximately 10 million hectares remain. European settler colonies established in the eighteenth and nineteenth centuries, including New Zealand, the USA, Canada and Australia, have exhibited very high rates of tree loss, clearing the equivalent of that felled in Europe in hundreds rather than thousands of years.

More recently, in Third World postcolonial states, exploitation for export has formed an important part of national development strategies. Deforestation rates are highest in the humid tropics, where great swathes of hardwood rainforest are felled daily in regions such as Southeast Asia, Central Africa and the Amazon (see Map 9.2 and Plate 9.1). Commercial pressures are such that this forest is unlikely to ever be regenerated, causing unquantifiable biodiversity losses and unknown climatic impacts. Rainforest covers approximately 7 per cent of the Earth's surface today; a few thousand years ago it covered 14 per cent, and the majority of this has been felled over the past two centuries. The rate of loss accelerated significantly up until the start of the 2000s according to the FAO (2011). At the turn of the millennium, it was estimated that rainforest was disappearing at the rate of one acre per second – that is to say an area equal per annum to the combined size

Box 9.3
continued

of England and Wales. There is great controversy regarding current rates of deforestation and some evidence that it has slowed marginally as easy forest resources become exhausted. However, experts are agreed that it remains a serious problem for all of humanity. The first region to cause global concern was the Amazon where between 1978 and 1988 total rainforest clearance had risen from 78,000 km^2 to 230,000 km^2, and by the middle of the 2000s this had risen to over half a million square kilometres. This fell significantly by the middle of the 2000s but the losses were still significant at a rate of approximately 20,000 km^2 per annum. The case of the Amazon clearly illustrates the dilemmas facing poorer countries in this position. Farmers expanding the agricultural frontier undertake some of the clear felling but the building of highways to support this endeavour has an especially devastating effect, as does the building of hydroelectric power dams. In the case of the Amazon approximately 90 per cent of forest cut down since 1970 is now pastoral land and recently soy farming has been a major contributor to forest losses in that country. In general, for poorer governments, a trade-off between environmental conservation and economic development often exists. These processes have led to very damaging losses in the 2000s in places as diverse as Madagascar, Nigeria, the Republic of Congo and, most significantly in a proportional sense, Central America. As a consequence, it is estimated by some that by 2030 only 10 per cent of the world's original rainforest will exist. International bodies have only just begun to

understand these linkages and to appreciate that in order to break out of the cycle of environmental degradation alternative livelihood strategies need to be generated in affected areas. The concept of reducing emissions from deforestation and degradation (REDD) was first mooted in the mid-1990s and entered the global agenda most notably at the 2009 UN Copenhagen Climate Change Conference. The scheme is effectively an offset scheme that allows pollutors to pay for the non-exploitation of forests in poorer countries. Supported widely as a solution, it is likely to figure in the Post-Kyoto agreements. However the distributional impacts of the REDD scheme as well as its environmental impact at the local scale have been questioned in the literature (McGregor, 2012)

Deforestation has become a global problem in three ways. First, it is affecting an increasing proportion of the Earth's surface, constituting an *extensive* problem. Second, it has an important impact on the global climatic system, reducing the carbon-absorbing capacity of the biosphere and adding to pollution through burning. Third, the driving force in deforestation is often not 'local'; pressure to deforest has come from processes associated with globalization as practised such as colonialism, modernization and neoliberalism (see Chapter 8). The demand for hardwoods in the West, for example, has a direct impact on poorer countries where forests exist. In many richer countries reforestation is taking place, but does not offset increasing rates of

Box 9.3
continued

Plate 9.1 *Deforestation in Southeast Asia*
Illegal loggers at work in Palau Pandang, Malaysia
Source: Potter and Cooke (2004, p. 340)

deforestation in the Third World. In sum then, deforestation is a global environmental problem because both the consequences and the causes of degradation are felt beyond national borders and at the scale of the whole world.

little doubt that the Earth's environment is being radically altered and that economic and demographic change will place extra pressure on finite resources, as we discuss in the subsequent section. Meyer and Turner (1995, p. 317) confirm this point, arguing that:

> Though the patterns, sources and impacts of environmental change are not uniform across the globe, humankind has so altered the physical conditions of the earth that we must recognise overall an 'earth transformed' and an earth certain to be transformed further. These changes will have different expression and consequences in different regions, but they will be significant almost everywhere. The question for humankind is whether it will be able to adapt to the foreseeable and unforeseeable consequences without destroying the only home it has.

Population growth, globalization and resource consumption

One of the most important factors in the environmental degradation debate is demographic change. Is systemic and cumulative global degradation due to population growth, as is commonly heard? Or, are other processes that have unleashed unprecedented levels of consumption, especially in the West, at work? The answer is a mixture of the two. Consumption in the West is far greater in magnitude than in the Third World, despite the fact that close to four-fifths of the planet live in the latter. However, as poorer regions' economic bases change and per-capita consumption rises, continued high population growth has the potential to shift the relative contributions to global environmental degradation and this certainly has the potential to exacerbate price hikes in key resources (see Box 9.4 on the oil and food crises of the late 2000s).

Population grows geometrically; that is to say, it increases by a proportion over a given time period, leading to exponential growth. In 10,000 BC the population of the Earth was approximately ten million. By the time of Christ it was around 200 million. By 1650 it reached 500 million and, by 1750, one billion. Eighty years later it equalled two billion, rising to three billion by 1960, four billion by 1976 and five billion by 1987 (UNFP, 2013). It now stands at substantially over seven billion, a milestone that was passed in 2012 (see Figure 9.4). According to the United Nations, current projections estimate a global population of 9.5 billion by 2050 (UNFP, 2013). Based on this explosive

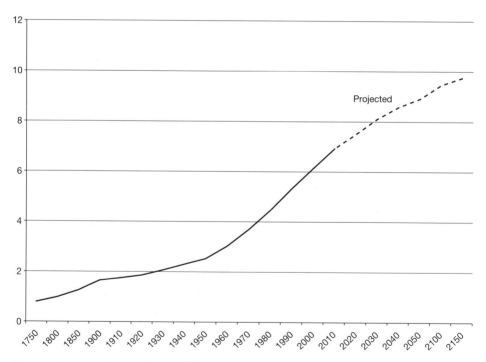

Figure 9.4 *Global population 1750–2150 (billions)*

Source: UNDP (http://esa.un.org/unpd/wpp/Excel-Data/population.htm accessed 2 September 2013)

Box 9.4

The link between food and oil crises and the concept of environmental limits

Towards the end of the 2000s, prices of both oil and food increased dramatically across the globe. This led some to argue that environmental limits were being illustrated quite clearly as overpopulation, particularly in the poorer world, and peak oil were being approached. In reality, the cause of the crisis in both cases was far more complex, representing the usefulness of a political ecology approach that looks at who controls environmental resources and for whom.

There have been two significant oil crises – one in 1973 and a second in 1979 – associated with political conflict in the Middle East and the policy of OPEC to restrict supply (see Chapter 8). Comparatively speaking, the oil price was low in the 1980s and 1990s. During the 2000s it began to rise and reached a peak in 2008 of US$147 per barrel, rising from around US$30 a barrel in 2003, which was a historic high. The reasons for the price rise have been debated, but experts agree that demand pressures – while significant given economic growth during the mid-2000s, particularly in East Asia – were only partially explanatory. Some have argued that geopolitical concerns were of equal importance including ongoing problems with supply from Iraq, problems with Iran and Venezuela and uncertainty concerning the Israel-Lebanon conflict, all in the context of the continued War on Terror. As prices rose, there were increased reports of the possibility of peak oil and this in turn is thought to have increased speculation of increased price rises, thus becoming a self-fulfilling prophecy. When the financial crisis hit the USA and the dollar fell sharply, this increased oil prices significantly also. By the end of 2009, given the reduced demand due to the GFC, the price had declined significantly, easing tensions with regard to the prospect of an oil crash. However, commentators have argued the GFC has perhaps only sheltered us temporarily from the long-term rise in oil prices – rising tensions in the Middle East at the time of writing were placing significant pressure on the oil price which had continued to rise to over US$100 a barrel again. This reminds us that continued reliance on oil as a energy source simply delays the inevitable crisis.

There were a number of consequences of the increased oil price – one of which was the contribution to the GFC in general. Perhaps the most profound was the impact on food prices, which in 2008 and 2009 increased dramatically. The FAO Food Price Index saw an increase from a figure of just under 100 in 2003 to double this in 2008 and to a peak of 227 in 2011. This led to a widespread food crisis especially in the poorer world where riots in a number of countries occurred. Some have credited this as being one of the sparks of the Arab Spring. Increased oil prices led to increased input, production and transport costs and

Box 9.4
continued

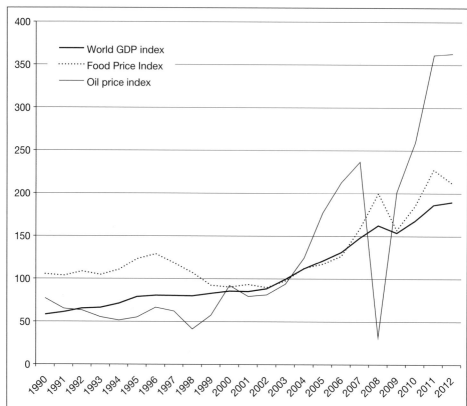

Figure 9.5 *Oil and food prices relative to GDP, 1990–2012*

Source: UNFAO (http://www.fao.org/worldfoodsituation/FoodPricesIndex/ accessed 2 September 2013)

Note: 2002–2004 = 100.

certainly contributed to the problem. In addition, population growth, shifts in diets in the evolving middle class in Asia, moves away from food exporting in poorer countries, and increased cultivation of biofuels on land used formerly for food production were cited as potential long-run factors. Partly as a consequence of this trend and the continued increase in global population, a number of countries, including the USA, UK, India and China, have been actively buying land in poorer countries –

especially in Latin America and Africa – in order to secure future supplies in a process that has been preferred to as 'land grabbing' (see Borras *et al.*, 2011, 2013). This factor has the impact of further increasing prices in those countries where land is purchased as well as a range of social and environmental impacts.

These two crises point to two important points. First, that so-called crises are interlinked and inherently political. Food and oil prices will continue to interact and it is

Box 9.4
continued

poorer countries that face the most serious consequences as they are already vulnerable to price increases and processes that seek to ameliorate such trends. The second point is that whilst the linked GFC may have dampened the trends that led to the crises of the late 2000s, these are likely to rise again, and, as such, these damaging outcomes have only been postponed in the context of the deregulated system that promotes such events.

growth, concern has been raised regarding the prospect of 'overpopulation' and its links to environmental degradation. This echoes the Malthusian view of the relationship between population and resources, which was accepted wisdom for much of the nineteenth and early twentieth centuries. It is wrong to think that the rate of population growth is increasing; it is actually falling at the scale of the globe as a whole. Between 1965 and 1970 total population grew on average at 2.04 per cent per annum, and between 1995 and 2000 this fell to 1.33 per cent per annum. By 2012, the UNFP estimated that annual population growth equalled just over one per cent per annum and that this would slow to zero by 2050 at the latest. There is the real prospect that by the mid-twenty-first century global population will actually begin to fall. Approximately 65 countries now have fertility rates below the replacement level.

Geographies of demographic change

Patterns of change and growth have varied widely across history and have been associated with important historical changes. After the industrial revolution in Europe there was a major population expansion as death rates declined. In the latter half of the last century, the major process was the decline in death rates in poorer countries brought about by the diffusion of Western medicine, which formed part of the modernization paradigm. It is this technological diffusion that explains the population 'explosion' in poorer countries. Despite these very broad generalizations, there are important spatial variations, and, as we progress down the geographic scale, generalizations become more difficult to make, which is one reason why abstracted models such as the 'demographic transition' are no longer useful. Today, rates of growth are generally lowest in the more affluent regions of the Earth – although there are significant variations, and a strong negative correlation between women's participation in the labour force and population growth has been observed (Buckingham-Hatfield, 2000). In the Third World, rates of growth are still relatively high, particularly in Sub-Saharan Africa, the Middle East and North Africa (see Table 9.1). There is undoubtedly a correlation between economic development and population growth. Based on differential growth rates at the regional level, the distribution of the global population is shifting markedly. The greatest challenge we face is that growth is projected to be greatest in regions and localities that are poorest. Migratory flows from rich to poorer regions will offset this trend only partially, not least as significant restrictions to migration have been erected in the West in the face of growing global disparities. People in high-growth regions will increasingly demand resource-intensive living standards as they emulate rich and 'globalized' lifestyles of the West.

For many years it was assumed, based on Malthusian ideas, that population growth was *causing* environmental degradation, and that this was leading to underdevelopment and poverty (see Ehrlich, 1971; Hardin, 1968). The causality between these factors has been questioned in more recent times, however. Today, *poverty is seen as a cause of population growth rather than a symptom.* High levels of poverty cause people to want more children in order to break the vicious circle of poverty. Children make economic sense and form a system of social security in many poorer countries. This view carried through at the 1994 Cairo UN Population Summit, for example, and was echoed in the Rio+20 summit of 2012 in Brazil. This implies that solving population growth is contingent upon reducing poverty, which in turn is likely to reduce local environmental degradation. In general, then, there has been a broadening in the debate on the relationship between population, development and the environment, which is now far more complex than was once thought.

Resource overpopulation and the spread of modernization

At the same time as the demographic debate has widened, we have seen the evolution of neo-Malthusian concepts (Pickering, 2012). These have more to do with concerns over the nature of consumption than population growth *per se*. There has been an increasing focus on the global environmental impacts of population growth combined with increased industrialization and consumption and the use of finite

Table 9.1 Current and projected population by country groupings, 1950–2050 (millions)

	1950	1960	1970	1980	1990	2000	2010	2020*	2050*
World	**2,525.8**	**3,026.0**	**3,691.2**	**4,449.0**	**5,320.8**	**6,127.7**	**6,916.2**	**7,716.7**	**9,550.9**
Africa	228.8	285.3	366.5	478.5	630.0	808.3	1,031.1	1,312.1	2,393.2
Asia	1,395.7	1,694.7	2,128.6	2,634.2	3,213.1	3,717.4	4,165.4	4,581.5	5,164.1
Europe	549.0	605.6	657.4	694.5	723.2	729.1	740.3	743.6	709.1
Latin America and Caribbean	167.9	220.4	287.6	364.2	445.2	526.3	596.2	661.7	781.6
North America	171.6	204.4	231.4	254.8	282.3	315.4	346.5	375.7	446.2
Oceania	12.7	15.8	19.7	23.0	27.0	31.2	36.7	42.1	56.9

Source: United Nations (http://esa.un.org/wpp/Excel-Data/population.htm accessed 12 July 2013)
Note: Medium fertility assumptions for projected populations; *projected

resources that this entails. Tyler-Miller (2002) refers to the current problem as one of 'consumption overpopulation' as opposed to 'people overpopulation'. In describing the environmental impact of human activity a simple identity may be used:

$I = P * A * T$ (where I = Impact; P = Population; A = Affluence; T = Technology)

The above implies that impact is equal to population multiplied by the amount of resources each person uses, multiplied by the environmental effects of the technologies used to provide and consume each unit of that resource. In poorer countries, population size and the resulting degradation of resources is the major issue. Overall, however, resource use per capita in such societies is relatively low. In richer countries high per-capita resource consumption and the resulting pollution and degradation are the major factors causing environmental problems. As Tyler-Millar notes (2002, p. 14):

> it is estimated that the average US citizen consumes 35 times as much as the average citizen of India and 100 times as much as the average person in the world's poorest countries. Thus poor parents in a developing country would need 70–100 children to have the same lifetime resource consumption as two children in a typical US family.

Such discrepancies give rise to major differences in terms of the relative contribution of different regions to global environmental problems. Table 9.2 shows the relative levels of resource consumption (per 100 people) in the 'industrialized' and 'developing' worlds as they are referred to there. Figure 9.6 shows contributions of different regions to annual greenhouse gas emissions. It is clear that it is the industrialized world which is adding significantly more to global environmental problems and, despite the relative economic rise of China, it still remains a long way off the per capita footprint of countries such as Australia, New Zealand and the United States for example. Notwithstanding, the economic growth of East Asia, and to a lesser extent South Asia does raise alarming possibilities with respect to the relationship between energy, resources and sustainable development in future years.

Global development, population and environment – links and challenges

It is too simplistic to blame population growth for environmental degradation and global environmental problems. Population growth

Table 9.2 *Relative levels of resource consumption, 2009*

Country group	CO_2 emissions per capita	Energy use (kg of oil equivalent per capita)	Motor vehicles per 1,000 people
Low income	0.29	356.8	nd
Lower middle income	1.58	648.4	26.6
Middle income	3.33	1,154.7	60.0
Upper middle income	5.13	1,668.7	89.2
High income	11.22	4,775.3	571.2
World	**4.71**	**1,787.9**	**176.0**

Source: World Bank (http://data.worldbank.org/ accessed 22 August 2013)

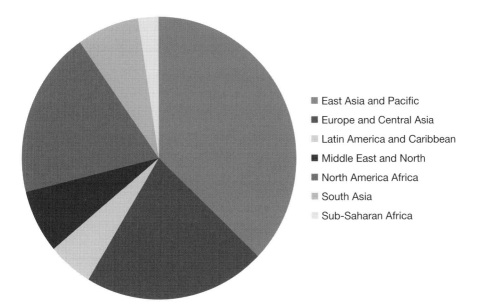

East Asia and Pacific

Europe and Central Asia

Latin America and Caribbean

Middle East and North

North America Africa

South Asia

Sub-Saharan Africa

Figure 9.6 *Contributions to greenhouse gas emissions, 2009 (% of total emissions)*
Source: World Bank (http://data.worldbank.org/ accessed 22 August 2013)

per se is not the cause of resource depletion – resource consumption is. At the present time this consumption is concentrated principally in the West – but this will not be the case for ever. We simply do not know if the world can accommodate a 'catch-up' by poorer nations if they pursue globalized Western economic models. Indeed, we believe that it is absolutely impossible without a revolutionary shift to a post-carbon energy regime combined with a more efficient use of the resources that currently exist. In this sense – as we argued in Chapter 8 – the notion of mainstream and conventional development being attainable by all has become a dangerous myth. Some argue that it has been in the interest of the West to curtail Third World populations rather than shoulder the burden of reducing the global impact of the industrialization that sustains their economies. Findlay (1995), for example, argues that the West has turned its attention to the 'overpopulation' explanation at the various summits and conferences which have taken place in order to avoid more costly solutions to the problem – such as emissions cutting. However, this is not to say that population growth is not a problem; accompanied by certain conditions in certain places it clearly is. In general, the approach required lies somewhere between the extreme views that exist. That is to say, we should accept that there are limits to growth but that appropriate technology can, in theory, reduce environmental pressures while at the same time increasing prosperity. However, alternative technologies require sustained investment and the world remains dangerously reliant on finite-resource-fuelled development at present. We need ways of increasing local well-being that do not degrade the local and global environment if we are to rise to the challenge of the 'population-consumption bomb'.

Globalization of environmental awareness

The modern environmental movement is often argued to have begun in the 1960s (Castree, 2009). Defined as the 'organised political expression of environmentalism' (ibid., p. 220), it was influenced by works including *Silent Spring* by Carson *et al.* (1962), and *The Limits to Growth* by Meadows (1972). It is associated with the post-Second World War generation and the 1960s liberal social revolution in the USA, the UK and the West more generally. As recently as the late 1970s, members of the environmental movement were seen as radical, but there has been a significant mainstreaming since then. Whether in the West this general concern has evolved out of genuine unease regarding the long-term evolution of the planet or out of individualistic and utilitarian notions of the loss of aesthetic and use value is a debatable point, however.

It is from within this movement that the maxim 'think global, act local' evolved, representing an early articulation of a globalization concept that has entered the popular imagination. According to Castree (2009), the environmental movement comprises six categories:

1 environmental NGOs (e.g. Friends of the Earth);
2 environmental new social movements (e.g. Namada Dam Movement);

3 Green parties;
4 governments with Green sensibilities;
5 businesses with Green sensibilities;
6 'Green' consumers.

To this list a seventh group may be added, namely *intergovernmental multilateral agencies* (such as the UNEP or the IPCC) that seek to coordinate intergovernmental regulation as well as to accommodate the views of pressure groups and other NGOs.

Environmentalism, defined as 'a concern that environment should be protected, particularly from the harmful effects of human activity' (Castree, 2000, p. 220), pre-dates the rise of this movement, however. The early Greeks expressed concern over human–environment relations, and religions such as Buddhism have profound environmentalist elements. In the USA, early environmentalism is associated particularly with the works of John Muir who, in the early twentieth century, helped establish the US National Parks system. Environmental concern and management is not just a concept pertaining to Western cultures. Many precolonial cultures in the South Pacific, Latin America and Africa had elaborate systems for the conservation of natural resources, and many evolved concepts which viewed humans as an inextricable part of the environment (e.g. Māori and Aboriginal culture in Oceania, *Vanua* in Fiji). We must be careful not to commit the error of 'Eurocentrism' when considering the evolution of environmentalism. In places across the Pacific region, for example, traditional and customary protection methods are combined with contemporary regulation to protect local marine stocks.

Notwithstanding the above, the genesis of the global environmental movement took place in Western society during the late 1960s, and this should be seen as a watershed period. It may be argued that the movement's rise at this time had three roots: (1) the evolution of the 'post-materialist' society which grew out of post-Second World War affluence and led to a changed value system which involved concern for the environment among other things (see Chapter 5); (2) space flight which introduced the concept of spaceship earth and brought the idea of limits to growth into explicit view; (3) advances in science and linked technology which allowed an understanding of the complexity of the environment that was hitherto impossible. The evolution of a globalized movement does not imply that there is agreement on the nature of the relationship between humans and the environment. On the contrary, the movement has become a very heterogeneous entity characterized by diverse groups of many political and philosophical shades. There are, according to O'Riordan (1981) (see also Eden, 2014), many different 'environmentalisms' flowing from different ideologies and world views, which result in very different regulation strategies (see Box 9.5).

Since the early 1970s, then, there has been an explosion of concern across the globe as well as resistance to, and regulation of, environmental degradation in a variety of forms (see Figure 9.6). Environmental concern may be said to have globalized in at least three, interlinked, ways, and we discuss points two and three in detail subsequently.

1 *A rapid increase in the physical and intellectual resources committed to researching the environment* and, in particular, the human impact upon it. In universities, environmental science and environmental studies have become mainstream undergraduate and postgraduate degrees. At the research level, dozens of new international journals

---

Box 9.5

World views and environmental policy

Increasing ecological awareness has seen the development of a number of world views concerning the relationship between humans and the environment. Many of these have their roots in older philosophies and development theories. At the extremes of a continuum there are two perspectives, which often clash in debates over appropriate policy:

- *Ecocentrism* – Science and technology is the problem. Humans must live *with* the Earth. There are inherent limits to growth. Ecocentrics are likely to be sceptics or radical transformationalists with respect to their views on globalization.
- *Technocentrism* – Science and technology is the solution. We will devise

ways of using the Earth more efficiently. There are no inherent limits to growth. Technocentrics are likely to be conservative transformationalists or hyperglobalists with respect to their views on globalization.

These world views have clearly different implications in terms of environmental policy. Essentially, the technocentric worldview dominates at the present time in the actions of governments and transnational agencies. The ecocentric view is institutionalized by some NGOs, new social movements and, significantly, Green parties, although these bodies occupy the fringes of policymaking in most democracies.

have evolved that are committed to environmental issues. Results are widely disseminated through such publications, international academic conferences and other network flows of academic activity. Unifying many of these approaches is the desire to define and implement sustainability (see Box 9.6).

2 *A flourishing of intergovernmental agencies which seek to manage and regulate the global environment from above,* giving rise to new international laws, protocols and conventions (see Figure 9.6).

Box 9.6

Sustainable development – organising concept or smokescreen?

Sustainability was *the* buzzword of the 1990s and replaced 'environmentalism' as shorthand for concerns over the environment and society (see Whitehead, 2013). It has a wide range of meanings, and they all have to do with continuity and are broadly positive. Longman's dictionary gives the following definitions of 'to sustain': to give support or relief to; to supply with sustenance; to cause to continue; to support the weight of; and to endure. In debate about environment and society the word 'sustainable' usually means to allow change while maintaining the best of what already exists. The most widely accepted definition of sustainable development, for example, comes from *Our Common Future* (Brundtland, 1987, p. 2): 'Development that meets the needs of the present without compromising the ability of future generations to meet their own needs'.

Sustainability has come to include a holistic concern for the nature of environmental change and incorporates the critical interlinkages between environment, economy, politics, society and culture. The concept has been derived from scientific theories (such as 'carrying capacity') and social scientific principles (such as 'maximum sustainable yield') – even if it does not always say much about them in action. It has become a term around which much work in both the physical and the natural sciences is now organized. The concept first emerged at the United Nations Conference on the Human Environment in Stockholm (1972) as a compromise intended to allow economic growth and environmental conservation at the same time. The meanings and usages of the term have proliferated since then and it has been hijacked as a concept to rationalize environmental exploitation taking place under politics of all shades. Some have argued that the use of the term by some, such as oil TNCs in their advertising and right-wing conservative

Box 9.6
continued

parties in their manifestos, amounts to the 'Green-washing' of society. Is the term *more* than just a buzzword? In many ways the term just re-labels the debate over human–environment interaction. As such, it hides many of the important intricacies and details that one must appreciate if meaningful policy is to be designed. To some, sustainable development is a 'theoretical black hole' (Adams, 2008; Whitehead, 2013). Overall, the concept *per se* is useful as an ideal construct to frame debates over the environmental consequences of the processes of globalization, even if it has been widely misappropriated and is now used largely to justify continued economic growth.

3 *A flourishing of political responses to environmental degradation from below*, including NGOs, new social movements, Green parties, environmentalist businesses and Green consumers.

Global environmental regulation from above and below

The environmental movement is characterized by the evolution of political engagement at all geographic levels. In this context, Held *et al.* claim that '[t]here has been an enormous growth in the number and scope of international institutions, laws and treaties that regulate the environment alongside the development of complex international alliances of environmental movements and organisations' (Held and McGrew, 2007, p. 376). In other words, there has been an evolution of attempts at regulation from above and below and activities at these different scales interact, as we explore below.

Regulation from above

Over a century ago, there was only nascent regulation from above. International agreements on issues such as the trade in exotic and rare species existed, but the magnitude of agreements and their scope pales in comparison to today. The proliferation of international agreements/ conventions/treaties outstrips globalizing trends in any other sphere (Calvert and Rengger, 2002). The first international convention took place in 1946 and sought to regulate international whaling. In the 1950s, there were various agreements on habitat protection schemes, the nuclear waste cycle and the passage of toxic waste.

It was the 1972 Stockholm Conference hosted by the UN which gave birth to the international regulatory bodies we see today. The agenda that was established there set the scene for subsequent decades, leading directly to the World Conservation Strategy of the 1980s. In the 1970s and 1980s there were a number of important developments. These included the regulation of international waters and marine pollution (e.g. the London Dumping Convention of 1972 and the UN Law of the Sea of 1982); and wildlife protection (e.g. agreements on Antarctica, wetlands, migratory birds, polar bears and seals). Through the 1980s, wide-ranging conventions were established on the passage of hazardous wastes (Basel Convention 1989), control of CFC emissions (Vienna and Montreal protocols of 1985 and 1987) and acid rain deposition in Europe. The EU has been especially active in terms of environmental regulation since the 1970s, passing directives on issues such as air pollution, water pollution, waste disposal and vehicle emissions and lobbying consistently for global agreements in these and other areas.

The *Bruntland Report* of 1987 theorized and discussed the possibility of 'Green' growth and opened the door to a global-scale debate about sustainability (see Box 9.6) that eventually resulted in the 1992 Rio Summit on Environment and Development. Organized under the auspices of UNEP, the summit attempted to evolve far-reaching policies to deal with a comprehensive set of environmental problems and was attended by nearly every national government and hundreds of NGOs. Agreements were forged on biodiversity, climate change, deforestation and desertification. The North/South division was highly visible in the agonizing preparatory and conference negotiations. Some poorer countries maintained that they should be allowed to use the natural resources within their boundaries as they saw fit, and industrialized countries should pay the price for environmental degradation in the past. Agenda 21 was the detailed policy-guidance document which resulted from Rio. This 600-page document was divided into four sections (environment, development, actors and policy) of ten chapters each and was intended to provide a blueprint for local sustainability within a global framework. Some critics have argued that the Agenda, while sound in its intentions, has allowed agencies and governments who are practising clearly non-sustainable policy to hide behind a whirlwind of buzzwords and spin. Many of the agreements put in place at Rio have not been implemented, and progress on others remains slow.

A significant development borne out of the Rio Summit was the Kyoto Protocol (1997) on greenhouse emissions and global warming which forms the most important example of an international agreement/

convention on climate among the many which have been put in place since the Second World War. The Protocol evolved out of the UN Framework Convention on Climate Change (UNFCCC) signed by 167 nations in Rio and enacted in 1994. The UNFCCC required parties to voluntarily commit to reducing greenhouse gas emissions to 1990 levels by the year 2000. In 1997 the signatories met at Kyoto, Japan and developed a document that would be legally binding if ratified. This flexible Protocol allowed countries to set varied targets, undertake to reach them on different dates (but before 2012) and to trade their emissions-reduction goals with other countries. The Protocol required an average reduction of 5.2 per cent in greenhouse gas emissions compared to 1990 levels. Required reductions were higher in some places than in others according to their proportional contribution to emissions. For example, Europe was required to reduce its emissions by 8 per cent, while a number of low-carbon-emitting countries in the Third World were actually permitted to increase their reductions. Full ratification required signatures from countries that together accounted for 55 per cent of total emissions in 1990. When Russia ratified the document in November 2004 this paved the way for the Protocol to come into force by February 2005, committing signatories to the appropriate reductions by 2012. The USA and Australia were the two largest nations to not ratify. Australia later ratified and adopted binding commitments for the period 2013–2020, but the USA still has not. In 2001, the USA pulled out of the negotiations entirely and was only coaxed back into talking through the watering-down of the agreement. At the Buenos Aires UN Climate Convention of late 2004, the US administration stated that it would never sign the Protocol, given what it saw as unfairly advantageous terms for developing countries. The USA also sought actively to block discussion of what should happen after the Protocol was completed. Partly as a consequence, the targets of the agreement have not been met to date. Between 1990 and 2009 per capita emissions of CO_2 increased globally from 4.2 to 4.7 metric tonnes. Per-capita emissions in high income countries rose from 12.2 to 12.5 metric tonnes between 1980 and 2000, but this had fallen slightly to 11.2 metric tonnes in 2009. By contrast, in 2009, low-income countries produced only 0.3 metric tonnes and middle income countries 3.3 metric tonnes of CO_2 per capita. Geographically, however, the most significant change in per capita emissions has been in the East Asia and Pacific region where per capita emissions grew from 1.8 to 4.6 metric tonnes between 1990 and 2009.

In 2012 the Protocol was extended to a second binding agreement covering the years from 2013 to 2020. At the time of writing, there were just under 40 countries that had signed up to binding targets within the overall objective of reducing emissions by half by 2050. The United States remained uncommitted to ratifying the treaty, Canada had pulled out in 2011 and New Zealand, Russia and Japan elected not to adopt binding targets in the second period. New Zealand wished to develop an agreement that more explicitly took the emissions of developing countries into account, which have remained unbinding. In Washington, in 2007, a number of major players signed to a non-binding agreement for a successor to Kyoto that would include a capping of developing and developed world emissions. A deadline has been set for 2015 to develop the successor to Kyoto. At the present, the countries that are bound to reduce their emissions by 2020 account for only approximately 15 per cent of total annual submissions. The major challenge will be to increase this proportion significantly in the agreement that supercedes the Kyoto Protocol. The nature of this agreement was debated at the 2009 United Nations Climate Change Summit in Copenhagen; however, apart from a number of non-binding undertakings, no agreement was reached. Instead a 'weak agreement', according to media reports, was reached that included some assistance for poorer countries to meet emissions targets and adapt to climate change, promoted the concept of the REDD programme to reduce deforestation and recognised the case for keeping temperature change below 2 degrees centigrade.

The UN World Summit on Sustainable Development (alternatively called Rio+10 or the Earth Summit) took place in September 2002 in Johannesburg, South Africa and was intended to build on the Rio Summit of ten years earlier. Hopes were high for the meeting as it brought together over 150 countries, a wide range of NGOs and representatives from new social movements. The Zimbabwe issue and the impending War on Iraq overshadowed the proceedings, however. US President Bush did not attend and many saw this as a significant snub in terms of the international consensus which it had been hoped was evolving. Despite achieving agreements in water and sanitation, global warming, energy, biodiversity, trade, human rights and health, these were significantly toned down from original goals. Perhaps the most notable outcome of Rio+10 was that the USA entered a phase of extreme aggression *vis-à-vis* the environmental agenda although this improved under the Obama Administration. In the long term, the Bush Administration's environmental stance is likely to be judged as the most backward-looking and self-interested in the history of the USA.

The Rio+20 meeting in Brazil was even more of a disappointment than its immediate predecessor for many critics. No new binding agreements were forged and the conference was overshadowed by political controversy – particularly including the attendance of the then Iranian President Ahmadinejad. It became increasingly clear from Rio+20 that a clear gulf had developed between the USA and EU in terms of environmental regulation; and this was clearly evident at the Summit, leaving the former isolated on many issues, not least its continued failure to ratify the Kyoto Protocol. Although progress has sometimes been disappointing, there can be little doubt that the evolution of global regulation in this area represents a significant and positive step. Environmental issues have now been placed on the agendas of orthodox institutions including the World Bank, the IMF, the G8 and the World Trade Organization. Much of the impetus for this evolution has come from below as the grass-roots movement has consistently lobbied national governments to act in a cooperative way with other states.

Regulation from below

There has been a widening in the range of political environmental actors that seek to regulate global degradation from below the scale of the state (Castree, 2009). The combined growth in the number of transnational environmental NGOs and environmental new social movements has outstripped the rise in intergovernmental agreements. McCormick (1989) argues that the grass-roots movement has built upon a commonality of environmental interests, providing fertile ground for a struggle against those in power and the technocentric approaches they employ. As already discussed, the 'grass-roots' or 'bottom-up' environmental movement began largely in richer countries and was institutionalized through the formation of NGOs such as Greenpeace and Friends of the Earth (both established in 1971). Often built by civil society, and usually on a voluntary basis, environmental NGOs played a major role in raising public and state awareness. New social-environmental movements have also played a crucial role in evolving grass-roots environmentalism. Such organizations, including, for example, the Namada Dam Movement and the Chipko Movement, are often ephemeral, local issues based, relatively informal and flexible. They have provided the NGO sector with salient and topical issues that are employed to illustrate their broader goals. Environmental NGOs and NSMs are increasingly intertwined at the global scale, and there exists a range of global networks representing these interests.

It has been argued that, in the West, environmentalism has been concerned primarily with issues of lifestyle and amenity (Escobar, 1995). There are some environmentalist NGOs based in poorer countries which share these concerns – often supported by the elite and urban middle classes. Many environmental NGOs in the poor world are branches of Northern NGOs or are set up by European expatriate or settler groups (as in Greenpeace Fiji). In general, however, these issues have not captured the general public's imagination as they have in Europe, North America or Australasia over the past 30 to 40 years. However, many new social movements have emerged among lower socio-economic groups directly affected by environmental degradation in poor countries (Routledge, 2002, 2013). In this way, Third World environmentalism is often concerned with livelihoods rather than with lifestyles.

Formal Green parties have played a major role in mobilizing grass-roots action on the environment. The German Green Party was established in the 1970s and was the first to win significant formal representation in a national parliament. A number of similar parties have evolved and have attempted to emulate this success in the UK, Australia, New Zealand and the USA among others. None has yet gained a controlling influence in any parliament and is unlikely ever to do so given the appropriation of ecological policies by mainstream parties. In New Zealand, representation is relatively strong and the party has benefited from the existence of a proportional representation system, polling at 11 per cent in the general election of 2011. In the USA a similar electoral system would deliver a considerable presence in the Congress for the Greens. The current system, however, sees this party locked out of mainstream politics. As it stands, Green parties tend to influence policy by lobbying or through issue-based alliances or coalitions with larger parties. A global Greens network, which draws together parties from across the world, represents an important platform for the shared vision of these groups.

A final important aspect of the grass-roots environmental movement is the rise of Green consumerism. A rising number of consumers has been voting with its wallet by actively seeking eco-friendly products and this includes the rise of the organic sector which has increased its market share notably in the West. Industry has responded to this and perpetuated this increasingly profitable niche. Taking both local and global environmental issues into account is more or less an industry standard in many richer countries and in markets for elites in poorer ones. Associated with this has been the rise of the food miles movement where consumers are encouraged to consume locally produced food in order to reduce the environmental footprint. The rise of so-called farmers' markets in towns and cities of the United Kingdom represent examples of this (Plate 9.2). Yet the food miles movement is a contentious issue as any reduction in trade from poorer countries may impact on the livelihoods of farmers there and this is yet to be taken into account effectively in this debate, as is an appreciation of the high-energy

Plate 9.2 *Ludlow farmers' market*

Box 9.7

An inconvenient truth or a convenient orthodoxy? Al Gore and environmental discourses

In 2006, the film *An Inconvenient Truth* was launched to critical acclaim. It was narrated by former US Vice President Al Gore and based on a slideshow he used as part of a campaign he had developed to educate the public on climate change issues. The film proved popular with the public, grossing over $50 million worldwide, and it received two Academy Awards. Al Gore and the film were at the forefront of raising public concerns about climate change in the mid-2000s. The film presented scientific evidence not only that the world's climate was warming, with subsequent severe effects, but also that these are significantly human-induced changes. It had the effect of reinvigorating the environmental movement globally and substantially spreading the environmental message.

The film's title spoke not only of the 'truth' of scientific evidence of rapid climate change but also of the 'inconvenience' of the measures that would have to be taken to change the course of climate change. Gore asked for individuals to change the way they consumed energy and material goods, the way they produced and processed waste, and to focus on reducing carbon emissions. It has proved to be a remarkably successful campaign and climate change is now part of the mainstream of much of the world's political discourse and public policy.

Yet one of the consequences of the film and the climate change campaign has been the way that other environmental discourses and campaigns have been relatively silenced. Climate change has become the orthodox centrepiece of environmentalism and one that has a strong global perspective: global in the scale of environmental changes and global in the response that is required. On the other hand, more local and short-term environmental concerns are less visible. For example, the global perception of environmental change on Pacific Island atolls is that they are disappearing under rising sea levels. That may very well be the case, but they are also highly susceptible to pollution of lagoon waters and marine resources from inadequate effluent disposal; they are at risk from salinization of the fresh water lens (often the main source of potable water) through over-extraction; and they suffer from continued loss of marine and terrestrial biodiversity. Similarly, the climate change campaign in the West arguably has diverted attention from pressing but less fashionable local environmental concerns, whether solid waste disposal or noise pollution, and from difficult political questions such as the future of nuclear energy.

In these ways, climate change orthodoxy – although demanding difficult and radical changes in human behaviour – may be seen to be convenient for some. It can take away the focus from the small-scale, local and immediate issues of dealing with

Box 9.7
continued

environmental change and instead place it on some distant, long-term and global stage. Effects and solutions become primarily the responsibility of others – greedy Western consumers or poor displaced Pacific Islanders – rather than an everyday responsibility of every person on the planet.

production methods of European food production systems that effectively offset gains from lower transport costs.

Conclusion – transformed environments

Environmental problems have become global in their scope. Systemic and cumulative degradation has occurred on an unprecedented scale, causing geographically differentiated outcomes that threaten the whole Earth system. This has been driven by the processes and agendas of globalization – particularly since the industrial revolution and colonialism and, more recently, through neoliberal economic policy. This has led to a global consciousness of unprecedented scope and scale concerning environmental issues. Many international agreements as well as less formal globalized networks have evolved in order to attempt to regulate this both from above and below. In the face of issues that threaten collective welfare in the global commons and create transboundary issues, nation-states are too small by themselves to supply the required regulation. Furthermore, they are often too large to deal with localized environmental concerns, which can accumulate into global problems. This is not to say that the nation-state does not have a role to play in regulating the global environment; it does, but it will only achieve success in conjunction with a wide range of interests operating at various geographic scales. The recent problems associated with increased food and oil prices will return as the dust of the GFC settles, and it is a fundamental tension between economic growth and the environment that has created these problems.

Despite the rising concern that is evident almost universally, commitment to action varies greatly across the world. Some of the largest states, in particular the USA, have been far too willing to pull out of or not ratify agreements which they perceive as damaging to their self-interest. The global environmental movement will apply increasing pressure on governments and international bodies over the coming years in debates on climate change, sea-level rise, air quality, water scarcity, food safety

Plate 9.3 *An eco-friendly product*

and genetic modification. The issue of degradation cannot be resolved, however, until global inequality is addressed in a coherent way. There are inherent distributional characteristics to the way the environmental issues have been dealt with at the global scale. The problem of climate change has come to dominate the environmental consciousness of the West, and other issues which often impact poorer nations such as resource depletion, soil fertility loss and deforestation have become of secondary importance in the context of the climate change lobby. Global awareness has been constructed upon a discourse that has a particular geography – of stranded polar bears, flooded atolls, disappearing rainforests and the rise of the climate 'bogeyman', China – all of which turn attention away from the simple fact that it is the consumption of the West and the model of globalization that it has developed that drives the environmental problem. In this sense, we require a reassertion of the local in environmental debates. The global environmental movement is divided, not least because the interests of the rich and poor world in the context of globalization processes are more divergent today than ever before. In this sense, environment and development are inextricably linked and represent a single global challenge of enormous magnitude.

Further reading

Adams (2008): This book seeks to link discourses of environmentalism with discourses of development and engages critically with the concept of sustainable development.

Bradshaw (2012): This chapter provides a useful overview of the energy crisis and some of the responses that have been put in place to address it.

Christoff and Eckersley (2013): This book explores the links between globalization and the environment from a range of perspectives with some excellent empirical analysis.

Held *et al.* (1999) ch. 8: This chapter traces out environmental degradation over history and seeks to link it explicitly to the evolution of globalization.

O'Riordan (1981): *Environmentalism* is the classic book on the rise of the environmentalist movement which traces its early nature and its philosophical roots.

Park *et al.* (2008): This is an excellent collection on global governance with respect to the environment.

Pickering (2012): This is an excellent chapter that introduces issues and concepts in environmental geography in a lively and engaging way.

Robbins (2012): See this excellent text for a critical introduction to political ecology.

Zedillo (2008): This discusses the prospect of global warming in a Post-Kyoto world.

⑩ Progressive globalization – long live geography

- Sixteen geographical questions about globalization
- Geographies of globalization – a schema
- An agenda for a more globalized and relevant human geography
- The gathering waves of globalization? Some final reflections

In this chapter, the major questions posed in the introduction to this book are returned to, the contribution of geographers to the study of globalization summarized, and the geographical elements of globalization theses drawn out. An agenda for human geographical research on globalization is presented, calling for a decentred critical regional geography that approaches big issues in a holistic way through the mapping of spatial differentiation on the ground. The central argument from a geographical point of view is *that space is relativized, place is hybridized and scale is collapsed* through the operation of globalization as a process and agenda. In general, it is argued that although globalization as currently practised is increasing global inequality and degrading the environment, it need not necessarily be a regressive process. Globalization, if reconceptualized, researched and regulated effectively, has the potential to improve global well-being, equity and sustainability. Cultivating a critical globalized imagination, which celebrates difference without idealizing or reifying it, is crucial in constructing an alternative and inclusive globalization process.

Sixteen geographical questions about globalization

By way of conclusion, we return to the questions posed in Chapter 1. Summary answers are offered below on the basis of the evidence and discussion presented throughout the book. The highly contested nature of globalization implies that each response is one of a number that could be made. In general, the arguments below adopt a transformationalist world view. It is this perspective that sits best with *most* theoretical and empirical work undertaken by geographers. The transformationalist 'thesis', as argued in Chapter 2, sits on a continuum, and the general tone of research on globalization in human geography might be more appropriately termed *radical transformationalist*. That is to say that while accepting that globalization is real, has concrete impacts and can be transformed by human action, geographical accounts have also emphasized the power of discourses of globalization and the driving role of capitalist/imperialist exploitation over space and time. The 'answers' presented here are expressed in relatively abstract form, and interpreting them requires revisiting the empirical evidence and theoretical arguments presented in relevant chapters.

How can globalization be defined?

A formal definition was offered in Chapter 1 and followed throughout the book. This represents only one of many possible definitions however. Most agree that globalization involves the *stretching of social relations across space* in all three spheres and that this results in *transnational flows which transcend borders*. This latter point gives rise to a *network society*, whose population is increasingly *conscious of the shrinking world*. Many definitions also refer to the *role of capitalism* as the driving force of the process. Of course, for sceptics who do not believe that globalization as a set of processes exists, it is seen as a discourse employed at various times in diverse ways to serve specific interests.

Is globalization new?

This book has argued consistently, and Chapter 3 has attempted to illustrate in detail, that globalization has *a long history*. When exactly it began is a matter of debate, however. Some argue that it started with the first intercontinental flows of people and ideas over 10,000 years ago. Others date it from the end of the Second World War. The periodization of globalization turns on its definition. If you believe, as this book has argued, that globalization is intimately tied to the expansion of capitalist circuits and the stretching of flows of accumulation across the world, then it *began in about 1500 with the rise of the Hispanic empires*. It then proceeded through *two waves*, colonial and postcolonial, each of which was further subdivided into two phases: mercantilist and industrialist in the case of the former, and modernization and neoliberal in the case of the latter. Each of these waves and phases was brought to an end by restructuring crises in global capitalism. Arguably we are currently in the midst of such a phase where the three crises – environmental, developmental and financial/economic – are unfolding and challenging capitalism as currently practised.

What drives globalization, now and in the past?

This is a source of great contention in globalization studies. Some see technology determining its evolution; others place culture or economics at the centre. Many accounts see capitalism as intimately involved, as part of a wider human imperative to shrink space. This book has argued that *capitalism is the central driving force now and in the past* and that time–space compression has resulted from the inherent need of capitalism to reduce the turnover time of capital and speed up its circuits, but it has also been constantly stressed that this is not an economic determinist position. *Capitalism is conceptualized as a broader culture* born of ideologies rooted in the Enlightenment period resulting in specific traits that drive its desire and ability to expand.

Is contemporary globalization different from the globalization of the past?

The current wave, especially the neoliberal phase, is characterized by increased extensity, intensity and velocity of global flows and the rise of dense networks that transmit them. According to the argument throughout this book, and which is not an uncommon one, *we are living in an era of vastly accelerated globalization*. This acceleration has been facilitated greatly by recent advances in social media and other technological change.

Is it homogenizing global society?

Permeating this book is the idea that *globalization exacerbates difference and fragments society*. There are powerful forces (e.g. cultural imperialism, superpower hegemony, development, modernization, Fordism) which, if unfolded on a homogeneous pre-existing landscape, have the potential to create a universalized society. However, these forces encounter locally specific histories and geographies and subsequent *interaction leads to hybridized or glocalized outcomes in culture, economy and politics.* Globalization works both ways at the same time along networks of local to local linkages that have become increasingly global in scope. This constantly reproduces scale and forging an ever-changing, heterogeneous, social landscape. Furthermore the pervasiveness of networks means that 'global' processes do not penetrate territory in a ubiquitous way and 'black holes' are left between the conduits of the net. As was argued in Chapter 2, the globalization of the capitalist system leads to a mosaicking of space at ever finer scales due to the processes that operate through new social, economic and political networks that link the local to the local at increasingly global reach.

Is globalization a process or an agenda?

Globalization comprises both processes and agendas that interact in complex ways. It has been argued in this book that agendas of globalization are tied up with imperatives of capitalism. These agendas unleash processes and flows that reconstitute agendas, which then go on to create another cycle of processes. Ideas from the regulation school are important here, since they provide arguments that support the idea of mutually constitutive agendas and processes.

How does globalization alter our concepts of space, place and scale?

Each of these core geographical elements is profoundly transformed by globalization, as was argued at length in Chapter 2. *Space is relativized, place is hybridized and scale is collapsed.* In general, deterritorialization and territorialization exist simultaneously and spaces of flows are disrupted and punctuated by spaces of places. In other words, networked flows emanate from particular localities, and some of these localities – like world cities – play a commanding role in directing such flows. A good way of summing this up is to talk of glocalization.

Is globalization the same thing as internationalization?

The sceptical argument posits that what we refer to as globalization is in fact just more intense internationalization. This book has argued that *internationalization is a component part of the broader process of globalization*, and contemporary globalization is qualitatively different to that of the past. All globalizing processes must take place in localities within nation-states, and networks link these processes to other localities in other nation-states. However, the networks that result are transnational, in that they exist above and beyond the state. New transnational networks do not necessarily use the nation-state system, which is a socially constructed entity, as a reference point. *The whole is greater than the sum of the parts* and the rise of a global consciousness exemplifies this.

Is globalization eroding the power of the nation-state?

Chapter 5 argued that *the state is being transformed by globalization as it seeks to regulate flows from above and below.* Some have argued that this implies a less important role for the state in a multilayered governance system. Most evidence points, however, to the continued salience of the nation-state as a container of economic activity, a reproducer of culture and a body that represents political will. Global geopolitics over the coming years will be defined by the relationship between the USA, China, the European Union and increasingly Brazil, India and Russia. In this sense, the nation-state will continue to be a highly influential unit in political and economic terms. Nation-states will proliferate but they may become more closely tied to actual nations – that is, ethnic and racial groupings – as they exert their identities beyond the confines of the state. The rise of diaspora is important in this regard and non-territorialized identities are increasingly important in all realms.

Is globalization bad for the environment?

As Chapter 9 illustrates, *the impact of globalization as currently practised has almost certainly degraded local and global environments seriously,* and this is associated with the spread of industrialization and rising consumption inherent in capitalism. Yet at the same time *globalization has created space for the expression of concern with this situation.* The global environmental movement has the potential to resolve the tensions if it can capture political will at the nation-state and global levels, although the continued reticence of the USA and other evolving superpowers to face global environmental issues does not bode well.

Is globalization bad for the 'poor world'?

As Chapter 8 argued, *globalization as currently practised has undoubtedly led to increased inequality within and between countries and regions,* and it has spread discourses that result in the subordination of millions of people. But the *new networks that are evolving from the bottom up seek to challenge the hegemony of neoliberalism and provide potential space for the practice of progressive development* which could yield equitable and sustainable improvements in well-being. Having said that, it is only the *potential* for progressive development which exists. The current power relations are such that achieving this kind of outcome will involve great struggle.

Does globalization create more winners than losers?

This is not the same question as the previous one because *winners and losers are created in both the Third World and the West.* Globalized society has become increasingly complex and, instead of seeing the existence of two or three 'worlds of development', or cores and peripheries, we are seeing the proliferation of networks of inclusion/ exclusion that are rewriting the old geographies of North/South and core/ periphery – although we argue that these still remain important and must not be forgotten. Some people fall through holes in the net in rich countries and the majority do so in poorer ones. There can be little doubt that, as currently practised, *globalization leads to greater concentration of power and affluence* and, relatively speaking, the majority of the world's population is left behind.

What has been the impact of the Global Financial Crisis on globalization?

The Global Financial Crisis which began in 2007 is related to globalization in intimate ways. We consider the crisis as merely the latest chapter in the unfolding of globalization as currently practised and the unfurling of non-sustainable and inequitable capitalism. However, we also believe that it is one of three interrelated crises in the environment, development and the economy. This crisis of crises suggests that any attempted solution needs to be broad and all-encompassing – no one crisis is solvable without solving the others. The impacts of the financial crisis were inherently uneven across the globe. Relatively speaking, those parts of the economy more heavily involved in financial speculation were more severely affected, and this had a major recessive impact on the middle class in Western countries in terms of access to credit. This rolled over into a sovereign debt crisis in a number of countries which still resonates today. The austerity measures associated with this have seen a return to neoliberal policies that reduce the role of the state. Having said that, the immediate reaction to the crisis was state intervention in the form bailout and financial stimulus. This has led to high levels of indebtedness between states – most particularly the USA which overall owes more than US$16 trillion with approximately 2 trillion dollars of that debt owned by China. This will have major geopolitical ramifications for many decades – though whether it will lead to greater stability or threaten peace is open to debate. As we discuss elsewhere, the intervention of the state does not imply a new wave of globalization, however; and to a large extent interventions were designed return to business as usual. There has been a partial downturn in global trade and rise in protectionism but coming agreements between the USA and European Union and other agreements such as the Trans Pacific Partnership seem likely to reverse this short-term trend. In the poorer world, the impacts of the GFC have been profound; but, in a relative sense, the middle class in Western countries has been more radically impacted. It may be that overall the GFC has had the effect at the macroeconomic level of reducing the GNP gap between Western and other states – Brazil and China, for example, grew through the crisis as the latter's demand for raw materials led to a new extractive phase in the economies of Latin America, Africa and Asia which, to an extent, sheltered them from the worst impacts of the crisis. At the same time, in an extraordinary way, the aid budgets of the West did not decrease through the crisis – though they pale in comparison to the bailouts. The nature of aid has shifted, however, as China and the West jockey for position through the provision of infrastructure. The West has also shifted many private sector interests to the aid budget. The regulatory response to the crisis will not solve it; if anything, it merely postpones an even greater downturn when the three crises unfold simultaneously.

Can globalization be reformed?

Globalization is not an inexorable force that exists outside of human control; it is socially constructed and perpetuated. People can reform the process through the construction of global regulation building on grass-roots concerns and actions. But this is obviously no easy task and will require a vigilant, well-informed and committed civil society. A great opportunity existed following the Global Financial Crisis which was not taken; and the interventions that did take place, while representing a divergence from pure neoliberalism, were intended to reproduce the hierarchies of power that had led to the problem in the first place. This approach will merely postpone an ever greater conflation of the three crises of the environmental, development and unstable capitalism that lie at the heart of the ongoing problems currently being experienced globally.

If so, how should it be reformed?

This is a crucial question that this book does not pretend to go near to answering. Some have argued for deglobalization and a return to localist or even anarchistic strategies. But delinking is probably not possible and arguably not desirable; it could lead to even greater relative deprivation and prevent the capturing of the opportunities presented by globalization. In approaching this question we need to move beyond simplistic concepts of globalization as good or bad; it is inherently neither. That is why the evolving *alter-globalization movement provides a more realistic option for meaningful reform* than the anti-globalization movement. Democratizing global institutions is a start, promoting local affirmation is important, and universal education is crucial to creating the kind of civil society that can face the challenges of globalization. Protests across the world at the time of writing suggest that there is a rising but uneven global consciousness of global civil society as symbolized, though not captured fully, in the Occupy movement. This movement is resisting the non-sustainable, inequitable and differentiated outcomes of the Global Financial Crisis – which should be seen as nothing more than the latest consequence of globalization as practised. The period of economic growth in the 2000s combined with the reassertion of the state – a neostructural period – raised the viability of the state *vis-à-vis* the market in terms of social and economic regulation. However, the use of the state to bail out private capital does not represent a new wave of globalization – rather it is a reform of the current wave designed to perpetuate the inequalities fostered by the neoliberal phase. By socialising the costs of the near-collapse of capitalism in the late 2000s, workers and individuals will pay to sustain private capital accumulation through monetary and fiscal reforms for many years to come. It is this fundamental inequity that the global civil movement has rallied against and lies at the centre of what must be addressed if a new progressive wave of globalization is to be forged.

What role can geography and geographers play in forging an alternative progressive globalization?

This is the subject of a subsequent section in this conclusion where an agenda for more progressive geographies of globalization is suggested.

Geographies of globalization – a schema

In Table 10.1, the geographical implications of the three theses of globalization in terms of the spheres and challenges considered in this book are summarized. Note that the hyperglobalist column refers to neoliberal, not radical, hyperglobalist theory. This book has argued consistently that it is the transformationalist viewpoint that most accurately captures the evolution of historical and contemporary globalization. The frustrating thing about this perspective, however, is that it is broad and offers a range of possibilities. If we adopt this view, we must therefore be careful to back up our conceptual claims with solid empirical work. We are compelled to make concrete suggestions in terms of how we can improve the world and, as addressed in the subsequent section, to do this we need to alter the way we approach geographies of globalization.

An agenda for a more globalized and relevant human geography

In the groundbreaking article 'Geographers and "globalization" (yet) another missed boat?' Dicken (2004) argues that geographers have

Table 10.1 *Geographies of globalization – a schema*

	Hyperglobalist	*Sceptical*	*Transformationalist*
Economic geographies	*Market is omnipresent and TNCs powerful.* Networks replace core–periphery division. Activity disembedded from nation-states. Deregulation and convergence.	*Trading blocs formed. Core–periphery division increased. Activity embedded in nation-states.* TNCs reflect national strategies. Regulation and divergence.	*Networks and structural patterns coexist.* Nation-states and TNCs govern markets. Re-regulation and convergence and divergence at the same time.
Political geographies	*Market-led global governance.* Nation-state replaced by 'natural' region states. Boundaries dissolve and sovereignty surrendered to global market.	*Core-led regionalism.* Sovereignty surrendered to regional groupings designed by powerful nation-states. Boundaries retrenched.	*Multilayered governance at three scales – global, national and local.* Nation-state remains central. Centralization and devolution at same time.
Cultural geographies	*New global civilization.* Homogeneous consumerist culture and global brands dominate. Universalization of cultural identities.	*Clashing cultures.* Civilizational blocs entrenched and cultural identities differentiated and relativized.	*New global and local hybrid cultures.* Possibility of progressive cultural change. But Westernization currently dominant.
Environment/ sustainability	*Sustainability solved by the market.* Market will price the environment efficiently and technology will solve scarcity.	*Sustainability threatened by global capitalism.* Spread of modernization and consumerism pushes environment to limits.	*Sustainability attainable through political action.* Environmental problems and concerns globalize. Regulation from above and below required.

Table 10.1 *continued*

	Hyperglobalist	Sceptical	Transformationalist
Development/ inequality	*Development attained by taking part in globalization.* Poverty solved by market. Integration unavoidable.	*Development threatened by global capitalist expansion.* Marginalization increased by spread of the market.	*Globalization offers threats and opportunities for development.* Progress contingent on careful management.
Moral message	*Pro-globalization.* Globalization is real, and is a moral good and force for progress.	*Anti-globalization in some readings.* Globalization is a discourse propagated by powerful interests.	*Alter-globalization.* Globalization is real and requires regulation to optimize it.

'developed a disturbing, even dysfunctional, habit of missing out on important intellectually and politically significant debates, even those in which geographers would seem to have a role to play' (p. 5). Dicken is particularly concerned that, as a group of scholars, geographers have had a limited impact on the evolution of the wider debate concerning globalization. This is despite the fact that geography is usually considered to be the 'world discipline' (p. 6). In a survey of some of the most influential books on the issue, he finds that less than one-third have references to geographers' work, and that of the total 14,000 references only 2 per cent are those of geographers. Indeed, often, even geographers do not reference other geographers' writing on geographies of globalization. Dicken (2004, p. 7) argues that:

> the evidence shows unequivocally that geographers are, indeed, marginal – at best – to the wider globalization debates. Geography is rather like the small child in the playground who always gets missed out when the big children are picking teams.

Although written over ten years ago – the arguments of this article remain as relevant today as they were then. The fact that nobody wants to play with us is troubling, first, because geographers cannot afford to miss out on the big issues of the day from a self-sustenance point of view and, second, because their insights are valuable for broader work in this field.

Plate 10.1 *Geographies of globalization in Dubai*

Source: Author's own

So, why are we left out? Geography is poorly understood by other academics and is misrepresented in society as a whole. In the case of the former, there is concern from some quarters, even within geography itself, that the discipline has become divorced from the 'big issues'. Some have blamed the 'cultural turn' and the influence of postmodernism on this (see Johnston and Sidaway, 2014). Hamnett (2003, p. 1), for example, argues:

> [The] rise of a 'post-modern' human geography, with its stress on textuality and texts, deconstruction, critique, 'reading' and interpretation has led human geography into a theoretical playground where its practitioners stimulate or entertain themselves and a handful of readers, but have in the process become increasingly detached from contemporary

social issues and concerns. The risk is that much of human geography will
cease to be taken seriously in the world beyond the narrow confines of
academe.

Under the influence of postmodernism geography has, according to some,
become increasingly parochial in its outlook and less willing to engage
with important shifts in actual places and regions across the world.
Instead, it has moved towards increasingly intricate studies of minute
events, processes and phenomena at a small scale. Arguably, geography
has lost its 'global studies' status in the desire to get 'local' and personal.
This in his 2004 article, and this echoes the criticisms of a number of
geographers over the past three decades including Stoddard (1987),
Johnston (1986) Potter (1993) (see Johnston and Sidaway, 2014).
Although the above concerns have some merit, they are probably
exaggerated. The 'cultural turn', while leading to esoteric geographies in
some cases, was founded on the belief that geography ought to be more
critical of hegemonic paradigms.

Many of the propositions of this 'turn' have provided richer and subtler
ways of interpreting the globalizing world (see Barnett, 2009; Crang,
2002, 2008; Imry, 2004). In this area, then, geographers need to stand
their ground. Despite this increased richness, in reality, the closure of
geography departments and the separation of human and physical
geography programmes in many places across the world including the
United Kingdom, New Zealand, Chile and beyond seems to suggest that
human geography is taken less seriously today than ever. This is
profoundly worrying as the three crises we face require the
interdisciplinary approach that very few disciplines other than geography
offer. Geography and geographers have not been strategic or relevant
enough and we all risk paying a price for this failure to engage with the
big issues of the day.

Outside of academia, however, the perception of geography is quite
different and even less favourable. The discipline is seen largely as a
descriptive compendium of facts – one of the shortcomings the cultural
turn sought to address – rather than a discipline that engages in the
analysis, interpretation and explanation of differentiation in global
society. The perception of geography as a glorified atlas may also be
discerned in other disciplines on occasion. Ironically, a number of
subjects, such as economics, have recently discovered a new spatiality.
However, the employment of a 'geographical perspective' elsewhere
often seems to misrepresent or under-acknowledge what has gone on in

geography. The type of geography which is being diffused across disciplinary boundaries is often reductionist and simplified. Given these multiple misconceptions, the discipline is often sidelined in important debates where it should, by rights, be leading the pack. Therefore, addressing questions of what geography is, and what it is relevant to focus upon, remain important tasks. Articulating clear and compelling responses to academia and the broader world represents an even more important challenge.

The central task for geographers, according to Dicken, is to pay more attention to the *outcomes* of globalization. In the past there has been a tendency for economic geographers in particular – although the same could be said of all human geographers – to focus on processes. Bridge says, '[t]he commitment by many economic geographers to a processual view of the world has the effect of collapsing conventional process–outcome distinctions: processes themselves become the object of explanation and are in themselves regarded as outcomes' (2002, p. 362).

Overall, while geographers have talked to each other fairly extensively about the outcomes of globalization, they have been largely ignored by the world outside of the discipline. To face big challenges such as globalization, we need to better define what we are and better represent this to the outside world. Massey, for example, argues that geography needs to be 'more confident of its own specificity' (2001, p. 5) and this, according to her, refers to the human–physical interface as well as our relatively sophisticated notions of space and place. Based on the above arguments, building on Dicken (2004) and other authors, we suggest a seven-point agenda for progressive geographies of globalization below.

1 *A critical regional geography* – We require what Dicken refers to as 'a revitalisation of "regional geography" within a relational framework' (2004, p. 19). This call is not a new one; Bradshaw (1990) made the case for a 'new regional geography'. Different regions exhibit distinct responses to globalization and the knowledge that critical area studies bring to research is crucial. This is not a call for a return to the regional studies of the 1950s and 1960s, which were descriptive and overgeneralized. Rather, it is a call for an analysis of the relationship between unique localities, places and regions and the wider global system. In this sense, we make a call for a reflexive and critical regional geography that takes places and regions seriously and tries to understand the mosaicked outcomes of the capitalist system made possible through networks of globalization. In this endeavour,

geography as a discipline needs to pay far more attention to research on regions outside of the Anglo-American zone (Murray, 2008). Non-Anglo-American geographers, located in the regions of concern, are undertaking a great deal of this research. However, their work is often marginalized.

2 *Empirically grounded research* – Much of the globalization literature is characterized by grand statements based on little empirical evidence. Studies that have sought to outline discourses of globalization, and the ways these are practised and performed, are important. However, such work should be complemented by empirical studies that seek to elucidate both the processes and outcomes of globalization. Again, geography is ideally placed. The subject has a long tradition of studies that seek to link change at lower scales to processes at higher scales and how restructuring at different levels interacts in mutually determining and complex ways. It is crucial, as noted previously, that we get below the level of the nation-state in what is researched. What geography needs are locality studies that have one eye firmly on the global.

3 *A more sophisticated geography* – Geographers need to develop more sophisticated frameworks of globalization upon which empirical work may be hung. Of crucial importance in this endeavour is the use of geographical concepts of space and place, often naively expressed in non-geographical globalization studies where space is seen as a fixed/territorial entity and places as concrete/bounded entities (see Chapter 2). A second crucial insight which geography has to offer is its relatively sophisticated notion of scale (see Chapter 2). Much of the non-geographical debate assumes a simplistic global/local dichotomy with the former determining the latter. As this book clearly argues, globalization is leading to a multi-scalar world. It is important to develop these debates in relation to research on concrete outcomes, however, as work of this nature can quickly become esoteric.

4 *A more democratic geography* – To understand globalization, and to provide suggestions for progressive reform, we need a truly global and democratic human geography. Contemporary geography is far from this; it is ironic that in a discipline where postcolonial perspectives have become so prominent the actual practice of research and publication is in many ways neocolonial. This is reflected by the fact that most geographical investigations of globalization study the West and are undertaken by Western academics. Furthermore, the geography journals where globalization research is published are located largely in the USA or the UK, dominated by a handful of

individual departments. Work by non-Anglo-American-based geographers on non-Anglo-American geographies should be encouraged more explicitly in the mainstream journals, and non-Western avenues of publication accorded greater status. We need a much less Eurocentric geography before we begin to describe, let alone explain, globalization in all its diversity. We believe that English-speaking geographers should have to learn a second language as part of their training, as this not only opens doors to a richer and more diverse literature it also exposes new world views that simply cannot be appreciated in monolingual academic practice.

5 *A more relevant geography* – There is an inherent tendency in academia to want to push back barriers and to shift to new frontiers of research. Many academics have a vocation towards this end, but the pressure to publish and the desire to differentiate work from that of others also drive the process. Globalization has been on the agenda in the social sciences, in a serious way at least, for only around 20 years. What must be avoided is the desire to add a 'post' prefix to globalization prematurely. We are just beginning to understand its importance and nature, and it is not yet time to move on. There is a habit in geography of shying away from 'big issues', focusing instead on smaller, less obvious concerns. These are easier to deal with and might get published more readily. This desire to constantly append a post- prefix is partly why geography missed the environment and development boats. Focusing on big issues does not mean that geographers should concentrate on the global at the expense of the local, however, since to do so would be to throw away one of the greatest strengths of the discipline.

6 *A more interdisciplinary and inclusive geography* – Geography is ideally placed to understand globalization given its inherently cross-disciplinary nature. Of particular importance are studies that seek to span the human/physical divide. As Chapter 9 illustrated, it is in the area of environmental globalization that some of the major challenges exist. In general, however, given the multidimensional and multi-thematic nature of globalization, we need greater interaction across the subdisciplines. There have been great strides taken towards this; new economic geography, for example, has built in many ideas from cultural geography. Naturally, more interaction is required with other disciplines but this will be fruitful only if we have a strong discipline characterized by cooperation across its component parts. The link between human and physical geography is crucial in this regard.

7 *A globalized imagination and moralistic geography* – There has been
something of an upsurge in moral geographies over recent years (see
Lee and Smith, 2004) building on the relevant debates of the 1970s.
Although we need strong empirical studies that flesh out critical
discourse-based work, we need to be less reticent in saying what we
feel the world should be like. Geography would do well to combine
empirical studies with globalized imaginations that suggest
alternatives to the status quo. This requires going beyond simplistic
anti- or pro-globalization arguments towards more sophisticated
analyses that acknowledge the potential opportunities and threats of
the process. This shouldn't be difficult – global civil society is already
doing it!

The gathering waves of globalization? Some final reflections

The rise of globalization adds a new urgency and dynamism to human
geography. The recognition that space, place and scale are important and
that a better appreciation of these, and other geographical concepts, might
help inform progressive policy implies a responsibility, however. We are
thus duty-bound to continue to interpret, model, test, measure, read, gaze
at, present and represent the unfolding of globalization and its
geographies. In terms of this book, what is perhaps frustrating is that it
has not provided *definitive* answers to the many questions it has posed.
Answers to such questions turn as much on one's own political and
philosophical make-up as anything else. What we hope the book has
provided, however, is a framework for evaluating contesting perspectives
on globalization. We also hope that the reader has been convinced that
the geographical take on the debates is distinct and valuable.

Our own 'thesis' is that globalization is real and we live in unprecedented
times. However, the links to the past are clear; the current wave of
globalization is reproducing and intensifying the unequal world. We live
in a neo-imperial age (Chomsky, 2012; Gregory, 2004; Harvey, 2003,
2005, 2011). The imperialism of today is insidious, it is rarely territorially
explicit, and the corporations, governments and elites that drive it are
non-accountable. The globalization agenda, and the processes which have
unfolded from it, are intimately linked to the expansion of monopoly
capitalism – although in rather Orwellian terms this is often referred to as
'*free*-market' economics. This imperative creates dependency and
vulnerability and is distinctly 'unfree' in many ways; it is detrimental for

the majority and it is built upon aggressive, sometimes military, penetration of marginalized economies and regions designed to maximize the wealth of the core. The only freedoms afforded through this process are fleeting, material benefits for those very few who are fortunately placed on networks of inclusion. But globalization is not like a lottery – some people will never win – there is too much at stake for it to be any other way. The capitalist elite has appropriated the process and made it its own – and is pursuing its own objectives at the expense of broader society, culture and the environment. As it is currently practised, globalization is perhaps the single greatest threat to human society. The Global Financial Crisis revealed this, and the resistance that has grown in terms of both the solutions applied and the factors that led to the crisis in the first place suggests that there is a growing global civil society consciousness. There was a chance for meaningful reform but vested interests meant that this was thrown away and the inexorable rise of globalization as currently practised continues into increasingly dangerous territory. Where we are headed towards is impossible to say, but we sense a geopolitical jockeying for position is taking place between the great powers of the EU, the USA and China that seeks to lay the ground for a post-carbon world order defined most probably by access to nuclear power. This jockeying for position is manifested in many ways including the current aid regime, military interventions in the Middle East and beyond, and policy that seeks to control the energy resource of the future. Political and economic geographies are often a function of access to resources and the next wave of globalization is likely to be defined by a new energy regime that is only just unfolding now. What is for sure is that the powers that have risen to prominence during the waves of globalization fuelled by carbon-producing energy will be extremely resistant to letting their power go.

However, there is another way, and globalization as currently practised is only one manifestation of a much broader potential. The cross-fertilization and hybridity which globalization creates could underpin a progressive society with the end-goals of welfare equality, global security and environmental sustainability. It is time to rewrite the agenda and to forge a new globalization from below that not only celebrates difference but also builds from it and perpetuates it. It is time to think local and act global, not shying away from big ideas for our small world whilst acknowledging and valuing diversity and difference as a force for good. This will entail the construction and application of new and alternative 'globalized imaginations'. If we believe that the global is constituted of the local and vice versa, then we can challenge globalization as currently practised through our own actions and carve a 'glocalized' future. Those of us who live on global networks of privilege have a duty to contribute to the construction of a better world for those who are not. The waves of globalization cannot be reversed, but they can be harnessed for good. The articulation of a more progressive discourse of globalization is the only way forward.

References

Aalbers, M. (2009) 'Geographies of the financial crisis', *Area*, 41(1), 34–42.

Adams, W. M. (1999) 'Sustainability', in Cloke, P. J., Crang, P. and Goodwin, M. (eds), *Introducing Human Geographies*, London, Arnold, 125–132.

Adams, W. M. (2001) *Green Development: Environment and Sustainability in the Third World* (2nd edn), London, Routledge.

Aguiton, C., Petrella, R. and Udry, C. (2001) 'A very different globalization: the globalization of resistence to the world economic system', in Hourtart, F. and Polet, F. (eds), *The Other Davos*, London, Zed Books, 63–68.

Albrow, M. (1996) *The Global Age: State and Society Beyond Modernity*, Cambridge, Polity Press.

Allen, J. and Hamnett, C. (1995) *A Shrinking World?: Global Unevenness and Inequality*, Oxford, Oxford University Press.

Allen, T. (2000) 'Taking culture seriously', in Allen, T. and Thomas, A. (eds), *Poverty and Development into the 21st Century*, Oxford, Oxford University Press, 443–467.

Allen, T. and Thomas, A. (2000) *Poverty and Development into the 21st Century* (revised edn), Oxford, Open University in association with Oxford University Press.

Amin, A. (ed.) (1994) *Post-Fordism: A Reader*, Oxford, Blackwell.

Amin, A. (1997) 'Placing globalization', *Theory, Culture and Society*, 14(2), 123–137.

Amin, A. (2002) 'Spatialities of globalization', *Environment and Planning A*, 34(3), 385–399.

Amin, A. (2004) 'Regulating economic globalization', *Transactions of the Institute of British Geographers*, 29(2), 217–233.

Amin, A. and Thrift, N. J. (1994) *Globalization, Institutions, and Regional Development in Europe*, Oxford, Oxford University Press.

Amin, A. and Thrift, N. J. (eds) (2004) *Cultural Economy: A Reader*, London, Sage.

Amin, S. (1997) *Capitalism in the Age of Globalization*, London and New York, Zed Books.

Anderson, J. (2010) *Understanding Cultural Geography*, London, Routledge.

Angel, D. (2002) 'Studying global economic change', *Economic Geography*, 78(3), 253–256.

Argent, N. (2002) 'From pillar to post? In search of the post-productivist countryside in Australia', *Australian Geographer*, 33(1), 97–114.

Armstrong, W. and McGee, T. G. (1985) *Theatres of Accumulation: Studies in Asian and Latin American Urbanization*, London, Methuen.

Aryeetey-Attoh, S. (2003) *Geography of Sub-Saharan Africa* (2nd edn), Upper Saddle River, NJ, Prentice Hall.

Auty, R. M. (1995) *Patterns of Development: Resources, Policy and Economic Growth*, London, Edward Arnold.

Banks, G., Murray, W. E., Scheyvens, R. A. and Overton, J. D. (2012) 'Paddling on one side of the canoe: the changing nature of New Zealand's development assistance programme', *Development Policy Review*, 30(2), 169–186.

Barcham, M., Scheyvens, R. and Overton, J. (2009) 'New Polynesian Triangle: rethinking Polynesian migration and development in the Pacific', *Asia Pacific Viewpoint*, 50(3), 322–337.

Barlow, M. and Clarke, T. (2001) *Global Showdown: How the New Activists are Fighting Global Corporate Rule*, Toronto, Stoddart.

Barnes, T. (2009) 'Economic geography', in Gregory, D., Johnston, R., Pratt, G., Watts, M. J. and Whatmore, S. (eds), *The Dictionary of Human Geography* (5th edn), Wiley-Blackwell, Oxford, 178–181.

Barnett, C. (2009) 'Cultural turn', in Gregory, D., Johnston, R., Pratt, G., Watts, M. J. and Whatmore, S. (eds), *The Dictionary of Human Geography* (5th edn), Wiley-Blackwell, Oxford, 134–135.

Barnett, J. and Campbell, J. (2010) *Climate Change and Small Island States: Power, Knowledge, and the South Pacific*, London, Earthscan.

Barrientos, S., Bee, A., Matear, A. and Vogel, I. (1999) *Women and Agribusiness: Working Miracles in the Chilean Fruit Export Sector*, London, Macmillan.

Barton, J. R. (1997) *A Political Geography of Latin America*, London, Routledge.

Barton, J. R. and Murray, W. E. (eds) (2002) *Chile: A Decade in Transition*, Special edition of the *Bulletin of Latin American Research*, 21(3).

Barton, J. R. and Murray, W. E. (2009) 'Grounding economic geographies of globalization: "Globalised spaces" in Chile's non-traditional export sector 1980–2005', *Tijdshrift of Social and Economic Geography*, 100(1), 81–100.

Barton, J. R., Gwynne, R. N. and Murray, W. E. (2008) 'Transformations in resource peripheries: an analysis of the Chilean experience', *Area*, 40(1), 24–33.

Batterbury, S. P. J. and Forsyth, T. J. (1997) 'Environmental transformations in developing countries', *Geographical Journal*, 163(2), 126–224.

Baudrillard, J. (1988) *Selected Writings*, Stanford, CA, Stanford University Press.

Beaumont, P. (2013) 'Global Protest grows as citizens lose faith in politics and the state', *Observer*, London, June 23.

Bebbington, A. (2004) 'Livelihood transitions, place transformations: grounding globalization and modernity', in Gwynne, R. N. and Kay, C. (eds) *Latin America Transformed: Globalization and Modernity* (2nd edn), London, Edward Arnold, 173–192.

Beck, U. (1992) *Risk Society: Towards a New Modernity*, London, Sage.

Beck, U. (2000) *What is Globalization?*, Cambridge, Polity Press.

Bell, D. (1974) *The Coming of Post-industrial Society: A Venture in Social Forecasting*, London, Heinemann Educational.

Bell, D. and Valentine, G. (1994) *Consuming Geographies: We Are What we Eat*, London, Routledge.

Bell, D. and Valentine, G. (1995) *Mapping Desire: Geographies of Sexualities*, London, Routledge.

Bell, D. and Binnie, J. (2000) *The Sexual Citizen: Queer Politics and Beyond*, London, Polity Press.

Bello, W. F. (2002) *Deglobalization: Ideas for a New World Economy*, London, Zed Books.

Bello, W. (2006) 'The capitalist conjuncture: over-accumulation, financial crises, and the retreat from globalisation', *Third World Quarterly*, 27(8), 1345–1367.

Bertram, G. (ed.) (2006) *MIRAB twenty years on*, special issue of *Asia Pacific Viewpoint*, 47(1).

Bertram, I. G. (1999) 'Economy', in Rapaport, M. (ed.) *Pacific Islands: Environment and Society*, Honolulu, University of Hawaii Press, 337–352.

Bertram, I. G. and Watters, R. F. (1985) 'The MIRAB economy in South Pacific microstates', *Pacific Viewpoint*, 23(3), 497–519.

Binns, T., Dixon, A. and Nel, E. (2012) *Africa – Diversity and Development*, Routledge, London.

Bisley, N. (2007) *Rethinking Globalization*, Basingstoke, Palgrave-Macmillan.

Blunt, A. and McEwan, C. (eds) (2003) *Postcolonial Geographies*, Continuum, London.

Borras Jr, S. M., Franco, J. C., Gomez, S., Kay, C. and Spoor, M. (2012) 'Land grabbing in Latin America and the Caribbean', *Journal of Peasant Studies*, 39(3–4), 845–872.

Boyer, R. and Drache, D. (1996) *States Against Markets: The Limits of Globalization*, London, Routledge.

Bradshaw, M. J. (1990) 'New regional geography, foreign area studies and perestroyka', *Area*, 22(4), 315–322.

Bradshaw, M. J. (2004a) *A New Economic Geography of Russia*, London, Routledge.

Bradshaw, M. J. (2004b) 'Resources and development', in Daniels, P. W., Bradshaw, M. J., Shaw, D. J. B. and Sidaway, J. D. (eds) *Human Geography: Issues for the 21st Century* (2nd edn), London, Longman, 113–144.

Bradshaw, M. J. (2012) 'Resources, energy, and development', in Daniels, P., Bradshaw, M. J., Sidaway, J. D. and Shaw, D. (eds) *Human Geography: Issues for the 21st Century* (4th edn), Edward Arnold, London, 111–139.

Brecher, J. and Costello, T. (1994) *Global Village or Global Pillage: Economic Reconstruction from the Bottom Up*, Boston, MA, South End Press.

Brecher, J., Childs, J. B. and Cutler, J. (1993) *Global Visions: Beyond the New World Order* (1st edn), Boston, MA, South End Press.

Brecher, J., Costello, T. and Smith, B. (2000) 'Globalization from below', *Nation*, 271(18), 19–22.

Brenner, N. and Theodore, N. (2003) *Spaces of Neoliberalism*, Oxford, Blackwell.

Bridge, G. (2002) 'Grounding globalization: the prospects and perils of linking economic processes of globalization to environmental outcomes', *Economic Geography*, 78(3), 361–386.

Britton, S. and Clarke, W. C. (1987) *Ambiguous Alternative: Tourism in Small Developing Countries*, Suva, University of the South Pacific.

Brohman, J. (1996) *Popular Development: Rethinking the Theory and Practice of Development*, Oxford, Blackwell.

Brookfield, H. C. (1975) *Interdependent Development*, London, Methuen.

Brundtland, G. H. (1987) *Our Common Future*, Oxford, World Commission on Environment and Development, Oxford University Press.

Bryceson, D., Kay, C. and Mooij, J. (eds) (2000) *Disappearing Peasantries? Rural Labour in Africa, Asia and Latin America*, London, Intermediate Technology Publications.

Bryson, J. R. (2012) 'Service economies, spatial divisions of expertise and the second global shift', in Daniels, P. W., Bradshaw, M. J., Shaw, D. J. B. and Sidaway, J. D. (eds) *Human Geography: Issues for the 21st Century* (4th edn), Harlow, Pearson, 359–378.

Bryson, J. and Henry, N. (2001) 'The global production system: from Fordism to Post-Fordism', in Daniels, P. W., Bradshaw, M. J., Shaw, D. J. B. and Sidaway, J. D. (eds) *Human Geography: Issues for the 21st Century* (1st edn), London, Prentice Hall.

Buchanan, K. M. (1963) 'The Third World: its emergence and contours', *New Left Review*, 18, 5–23.

Buchanan, K. M. (1964) 'Profiles of the Third World', *Pacific Viewpoint*, 5(2), 97–126.

Buckingham-Hatfield, S. (2000) *Gender and Environment*, London, Routledge.

Burton, J. W. (1972) *World Society*, Cambridge, Cambridge University Press.

Bygrave, M. (2002) 'Where have all the protestors gone', *Guardian Weekly*, 1–7 August.

Calhoun, C. (2007) *Nations Matter – Culture, History and the Cosmopolitan Dream*, London, Routledge.

Callinicos, A. (2001) 'Where now', in Bircham, E. and Charlton, J. (eds) *Anti-capitalism – A Guide to the Movement*, London, Bookmarks Publications.

Calvert, P. and Rengger, N. J. (2002) *Treaties and Alliances of the World* (7th edn), London, John Harper.

Carson, R., Darling, L. and Darling, L. (1962) *Silent Spring*, Boston, MA, Riverside.

Castells, M. (1996) *The Rise of the Network Society*, Malden, MA, Oxford, Blackwell.

Castles, S. and Miller, M. J. (1993) *The Age of Migration: International Population Movements in the Modern World*, New York, Guilford Press.

Castree, N. (2000) 'Environmental movement', in Johnston, R. J., Gregory, D., Pratt, G. and Watts, M. (eds) *The Dictionary of Human Geography*, London, Blackwell.

Castree, N. (2009) 'Environmental movement', in Kitchen, R. and Thrift, N. (eds) *International Encyclopedia of Human Geography*, London, Elsevier.

CEPAL (various) *Estadisticas annuarios*, Santiago, CEPAL.

Challies, E. R. T. and Murray, W. E. (2008) 'From neoliberalism to neostructuralism – the comparative politico-economic transition of Chile and New Zealand', *Asia Pacific Viewpoint*, 48(2), 123–145.

Challies, E. and Murray, W. E. (2010) 'The TPPA Agribusiness, and Rural Livelihoods', in Kelsey, J. (ed.) *No Ordinary Deal – Unmasking the Trans-Pacific Partnership Free Trade Agreement*, Allen and Unwin, New South Wales, 110–123.

Champion, T. (2001) 'Demographic transformations', in Daniels, P. W., Bradshaw, M. J., Shaw, D. J. B. and Sidaway, J. D. (eds) *Human Geography: Issues for the 21st Century*, Harlow, Prentice Hall.

Chandra, R. (1992) *Industrialization and Development in the Third World*, London, Routledge.

Chang, T. C. (ed.) (2005) 'Place, Memory and Identity in "New" Asia', special edition of *Asia Pacific Viewpoint*, 46(3), London and Melbourne, Blackwell.

Chant, S. (1999) 'Urban livelihoods, employment and gender', in Gwynne, R. N. and Kay, C. (eds) *Latin America Transformed: Globalization and Modernity*, London, Arnold.

Chant, S. (2004) 'Urban livelihoods, employment and gender', in Gwynne, R. N and Cristóbal, K. (eds) *Latin America Transformed: Globalization and Modernity.* Arnold Publication. Hodder Arnold, London, UK, 210–231.

Cho, G. (1995) *Trade, Aid, and Global Interdependence*, London, Routledge.

Chomsky, N. (2001) *9–11*, New York, Seven Stories.

Chomsky, N. (2004) *Hegemony or Survival? America's Quest for Global Dominance*, London, Penguin.

Chomsky, N. (2012) *Occupy*, London, Penguin Books.

Christoff, P. and Eckersley, R. (2013) *Globalization and the Environment*, Plymouth, Rowman & Littlefield Publishers.

Christopherson, S. (2002) 'Changing women's status in a global economy', in Johnston, R. J., Taylor, P. J. and Watts, M. (eds) *Geographies of Global Change: Remapping the World*, London, Blackwell, 236–247.

Cline-Cole, R. and Robson, E. (eds) (2003) *West African Worlds*, London, Pearson Education.

Cloke, P. J., Philo, C. and Sadler, D. (1991) *Approaching Human Geography: An Introduction to Contemporary Theoretical Debates*, London, Paul Chapman.

Cloke, P. J., Crang, P. and Goodwin, M. (1999) *Introducing Human Geographies*, London, Arnold.

Cloke, P., Crang, P. and Goodwin, M. (2005) *Introducing Human Geographies* (2nd edn), London, Routledge.

Cloke, P., Crang, P. and Goodwin, M. (2013) *Introducing Human Geographies* (3rd edn), London, Routledge.

Coe, N. M. (2012) 'The geographies of global production networks', in Daniels, P. W., Bradshaw, M. J., Shaw, D. J. B. and Sidaway, J. D. (eds) *Human Geography: Issues for the 21st Century* (4th edn), Harlow, Pearson, 334–358.

Coe, N., Kelly, P. F. and Olds, K. (2003) 'Globalization, transnationalism and the Asia Pacific', in Peck, J. A. and Yeung, H. W. C. (eds) *Remaking the Global Economy: Economic-Geographical Perspectives*, London, Sage, 45–60.

Coe, N. M., Kelly, P. F. and Yeung, H. W. C. (2007) *Economic Geography: A Contemporary Introduction*, Blackwell, Oxford.

Coe, N. M., Kelly, P. F. and Yeung, H. W. C. (2013) *Economic Geography: A Contemporary Introduction* (2nd edn), Blackwell, Oxford.

Cohen, R. (2008) *Global Diasporas – An Introduction*, London, Routledge.

Connell, J. (2003a) 'Regulation of space in the contemporary postcolonial Pacific city: Port Moresby and Suva', *Asia Pacific Viewpoint*, 44(3), 243–258.

Connell, J. (2003b) 'Losing ground? Tuvalu, the greenhouse effect and the garbage can', *Asia Pacific Viewpoint*, 44(2), 89–108.

Connell, J. (2008) 'Niue: embracing a culture of migration', *Journal of Ethnic and Migration Studies*, 34(6), 1021–1040.

Connell, J. and Gibson, C. (2003) *Sound Tracks: Popular Music, Identity and Place*, London, Routledge.

Cook, I. and Crang, P. (2012) 'Consumption and its geographies', in Daniels, P. W., Bradshaw, M. J., Shaw, D. J. B. and Sidaway, J. D. (eds) *Human Geography: Issues for the 21st Century* (3rd edn), Harlow, Pearson, 396–418.

Corbridge, S. (1993) *Debt and Development*, Oxford, Blackwell.

Corbridge, S. (2002) 'Third World debt', in Desai, V. and Potter, R. B. (eds) *The Companion to Development Studies*, London, Arnold, 477–479.

Corbridge, S., Thrift, N. and Martin, R. (eds) (1994) *Money, Power and Space*, Oxford, Blackwell.

Cosgrove, D. (2009) 'Cultural landscape', in Gregory, D., Johnston, R., Pratt, G., Watts, M. J. and Whatmore, S. (eds) *The Dictionary of Human Geography* (5th edn), Wiley-Blackwell, Oxford, 133–134.

Cowen, M. P. and Shenton, R. W. (1996) *Doctrines of Development*, London, Routledge.

Cox, K. R. (ed.) (1997) *Spaces of Globalization: Reasserting the Power of the Local*, New York, London, Guilford Press.

Crang, M. (1998) *Cultural Geography*, London, Routledge.

Crang, M. (2002) 'Qualitative methods: the new orthodoxy?', *Progress in Human Geography*, 26(5), 647–655.

Crang, M. (2008) *Cultural Geography*, London, Routledge.

Crang, M. (2009) 'Cultural geography', in Gregory, D., Johnston, R., Pratt, G., Watts, M. J. and Whatmore, S. (eds) *The Dictionary of Human Geography* (5th edn), Wiley-Blackwell, Oxford, 129–133.

Crang, M. and Thrift, N. J. (eds) (2000) *Thinking Space*, London, Routledge.

Crang, P. (1999) 'Local–global', in Cloke, P. J., Crang, P. and Goodwin, M. (eds) *Introducing Human Geographies*, London, Arnold, 24–34.

Crang, P. (2005) 'Local–global', in Cloke, P., Crang, P. and Goodwin, M. (eds) *Introducing Human Geography* (2nd edn), London, Routledge, 34–50.

Crang, P. (2013) 'Local–global', in Cloke, P. Crang, P. and Goodwin, M (eds) *Introducing Human Geography* (3rd edn), London, Routledge, 7–22.

Creswell, T (2013) 'Place', in Cloke, P., Crang, P. and Goodwin, M. (2013) *Introducing Human Geography* (3rd edn), London, Routledge.

Crush, J. (ed.) (1995) *Power of Development*, London, Routledge.

Crystal, D. (1997) *Cambridge Encyclopaedia of Language*, Cambridge, Cambridge University Press.

Cumbers, A. (2009) 'Regional integration', in Kitchen, R. and Thrift, N. (eds) *International Encyclopedia of Human Geography*, London, Elsevier.

Cumbers, A., Nativel, C. and Routledge, P. (2008) 'The entangled geographies of global justice networks', *Progress in Human Geography*, 32(2), 183–201.

Daniels, P. W. (2001) 'The geography of the economy', in Daniels, P. W., Bradshaw, M. J., Shaw, D. J. B. and Sidaway, J. D. (eds) *Human Geography: Issues for the 21st Century*, Harlow, Prentice Hall.

Daniels, P. W. and Lever, W. (eds) (1996) *The Global Economy in Transition*, Harlow, Longman.

Daniels, P. and Jones, A. (2012) 'Geographies of the economy', in Daniels, P. W., Bradshaw, M. J., Shaw, D. J. B. and Sidaway, J. D. (eds) *Human Geography: Issues for the 21st Century* (4th edn), Harlow, Pearson, 291–313.

Daniels, P. W., Bradshaw, M. J., Shaw, D. J. B. and Sidaway, J. D. (eds) (2001) *Human Geography: Issues for the 21st Century*, Harlow, Prentice Hall.

Daniels, P. W., Bradshaw, M. J., Shaw, D. J. B. and Sidaway, J. D. (eds) (2012) *Human Geography: Issues for the 21st Century* (4th edn), Harlow, Pearson.

Davids, T. and Van Driel, F. (2009) 'The unhappy marriage between gender and globalization', *Third World Quarterly*, 30(5), 905–920.

Desai, V. and Potter, R. B. (2013) *The Companion to Development Studies*, London, Routledge.

Dicken, P. (1998) *Global Shift: Transforming the World Economy* (3rd edn), London, Paul Chapman.

Dicken, P. (2000) 'Localization', in Johnston, R. J., Gregory, D., Taylor, P. J. and Watts, M. (eds) *The Dictionary of Human Geography*, London, Blackwell.

Dicken, P. (2003) *Global Shift: Reshaping the Global Economic Map in the 21st Century* (4th edn), London, Sage.

Dicken, P. (2004) 'Geographers and "globalization": (yet) another missed boat?', *Transactions of the Institute of British Geographers*, 29(1), 5–26.

Dicken, P. (2009) 'Globalization and transnational corporations', in Kitchen, R. and Thrift, N. (eds) *International Encyclopedia of Human Geography*, London, Elsevier.

Dicken, P. (2011) *Global Shift* (6th edn), Guildford Press, Guildford.

Dicken, P., Kelly, P. F., Olds, K. and Yeung, H. W. C. (2001) 'Chains and networks, territories and scales: towards a relational framework for analysing the global economy', *Global Networks*, 1(2), 89–112.

Dickenson, J. P., Gould, B., Clarke, C., Mather, S., Prothero, M., Siddle, D., Smith, C. and Thomas-Hope, E. (1996) *A Geography of the Third World* (2nd edn), London, Routledge.

Diener, A. C. and Hagen, J. (2009) 'Theorizing borders in a "borderless world": globalization, territory and identity', *Geography Compass*, 3(3), 1196–1216.

Dunning, J. H. (1988) *Multinationals, Technology and Competitiveness*, London, Unwin Hyman.

Dwyer, C. (1999) 'Migrations and diaspora', in Cloke, P. J., Crang, P. and Goodwin, M. (eds) *Introducing Human Geographies*, London, Arnold, 287–295.

Dwyer, C. (2013) 'Diasporas', in Cloke, P., Crang, P. and Goodwin, M. (eds) *Introducing Human Geography* (3rd edn), London, Routledge, 669–686.

Dymski, G. A. (2009) 'The global financial customer and the spatiality of exclusion after the "end of geography"', *Cambridge Journal of Regions Economy and Society*, 2(2): 267–285.

Economist, The (2003) 'The revenge of geography', *The Economist*, 15 March, 19–23.

Economist, The (2013) 'The march of protest: a global wave of anger is sweeping the cities of the world', Leader, *The Economist*, 29 July.

Eden, S. (2013) 'Global and local environmental problems', in Cloke, P., Crang, P. and Goodwin, M. (eds) *Introducing Human Geography* (3rd edn), London, Routledge, 431–447.

Ehrlich, P. R. (1971) *The Population Bomb*, Cutchogue, NY, Buccaneer Books.

Escobar, A. (1995) *Encountering Development: The Making and Unmaking of the Third World*, Princeton, NJ, Princeton University Press.

Esteva, G. (1992) 'Development', in Sachs, W. (ed.) *The Development Dictionary: A Guide to Knowledge as Power*, London, Zed Books.

Esteva, G. and Prakash, M. S. (1998) *Grassroots Post-modernism: Remaking the Soil of Cultures*, London, Zed Books.

Farbotko, C. (2010) 'Wishful sinking: disappearing islands, climate refugees and cosmopolitan experimentation', *Asia Pacific Viewpoint*, 51(1), 47–60.

Federal Reserve Bank of the USA (2011) *Financial Crisis Inquiry Commission – Final Report*, Washington, FRB.

Findlay, A. (1995) 'Population crises: the Malthusian specter?', in Johnston, R. J., Taylor, P. J. and Watts, M. (eds) (2002) *Geographies of Global Change: Remapping the World*, Oxford, Blackwell.

Firth, S. (2000) 'The Pacific Islands and the globalization agenda', *The Contemporary Pacific*, 12(1), 178–192.

Food and Agriculture Organization of the United Nations (2013) 'Trends in fruit' www.fao.org/docrep/005/y4358e/y4358e04.htm, accessed 13 September.

Frank, A. G. (1969) *Latin America: Essays on the Development of Underdevelopment*, New York, Monthly Review Press.

Frèobel, F., Heinrichs, J. and Kreye, O. (1980) *The New International Division of Labour: Structural Unemployment in Industrialised Countries and Industrialisation in Developing Countries*, Cambridge, Paris, Cambridge University Press, Maison des sciences de l'homme.

Friedmann, J. (1986) 'The world city hypothesis', *Development and Change*, 17(1), 69–83.

Friedmann, J. (1996) 'Where we stand: a decade of World City research', in Knox, P. L. and Taylor, P. J. (eds) *World Cities in a World System*, Cambridge, Cambridge University Press, 3–20.

Friedman, T. (2007) *The World is Flat 3.0 – A Brief History of the 21st Century*, New York, Picador.

Fukuyama, F. (1992) *The End of History and the Last Man*, London, Hamish Hamilton.

Gamlen, A. (2012) 'Creating and destroying diaspora strategies', *Transactions of the Institute of British Geographers*, 38(2), 238–253.

Gamlen, A and Marsh, K. (eds) (2011) *Global Migration and Governance*, Oxford, Edward Elgar.

Geddes, P. (1915) *Cities in Evolution, an Introduction to the Town Planning Movement and to the Study of Civics*, London, Benn.

George, S. (2001) 'Corporate globalization', in Bircham, E. and Charlton, J. (eds) *Anti-Capitalism: A Guide to the Movement*, London, Bookmark.

Gereffi, G. (1994) 'The organization of buyer-driven global commodity chains: how US retailers shape overseas production networks', in Gereffi, G. and Korzeniewicz, M. (eds) *Commodity Chains and Global Capitalism*, Westport, CT and London, Praeger.

Gereffi, G. (1996) 'Global commodity chains: new forms of coordination and control among nations and firms in international industries', *Competition & Change*, 1(4), 427–439.

Gibson, C. and Waitt, G. (2009) 'Cultural geography', in Kitchen, R. and Thrift, N. (eds) *International Encyclopedia of Human Geography*, London, Elsevier.

Gibson, L. (2004) 'Empty shells: demographic decline and opportunity in Niue', in Terry, J. P. and Murray, W. E. (eds) *Niue Island: Geographical Perspectives on the Rock of Polynesia*, Paris, INSULA, UNESCO, 203–216.

Gibson-Graham, J. K. (2002) 'Beyond global vs. local: economic politics outside the binary frame', in Herod, A. and Wright, M. W. (eds) *Geographies of Power: Placing Scale*, Oxford, Blackwell, 25–60.

Giddens, A. (1985) *A Contemporary Critique of Historical Materialism*, Cambridge, Polity Press.

Giddens, A. (1990) *The Consequences of Modernity*, Cambridge, Polity Press.

Giddens, A. (1991) *Modernity and Self-Identity: Self and Society in the Late Modern Age*, Cambridge, Polity Press.

Giddens, A. (1999) *Runaway World: How Globalization is Reshaping our Lives*, London, Profile.

Giddens, A. (2002) *Runaway World* (2nd edn), London, Profile Books.

Gilbert, A. (1996) *The Mega-City in Latin America*, Tokyo, United Nations University Press.

Gill, S. (1995) 'Globalization, market civilization, and disciplinary neoliberalism', *Millenium*, 24(3), 399–425.

Gilpin, R. (1987) *The Political Economy of International Relations*, Princeton, NJ, Princeton University Press.

Gilpin, R. (2001) *Global Political Economy: Understanding the International Economic Order*, Princeton, NJ, Princeton University Press.

Glasius, M., Kaldor, M. and Anheier, H. (eds) (2002) *Global Civil Society*, Oxford, Oxford University Press.

Goldsmith, E. and Mander, J. (eds) (2001) *The Case Against the Global Economy*, London, Earthscan.

Goodman, D. and Watts, M. (eds) (1997) *Globalizing Food*, London, Routledge.

Goodwin, M. (1999) 'Citizenship and governance', in Cloke, P.J., Crang, P. and Goodwin. M. (eds) *Introducing Human Geographies*, London, Arnold.

Gordon, D. M. (1988) 'The global economy: new edifice or crumbling foundations?', *New Left Review*, 168, 24–64.

Grassman, S. (1980) 'Long term trends in openness of international economies', *Oxford Economic Papers*, 32(1), 122–133.

Gray, J. (1998) *False Dawn: The Delusions of Global Capitalism*, London, Granta Books.

Gregory, D. (1994) *Geographical Imaginations*, Blackwell, Oxford.

Gregory, D. (2004) *The Colonial Present: Afghanistan, Palestine, Iraq*, Oxford, Blackwell.

Gregory, D., Johnston, R., Pratt, G., Watts, M. J. and Whatmore, S. (eds) (2009) *The Dictionary of Human Geography* (5th edn), Wiley-Blackwell, Oxford.

Greider, W. (1997) *One World, Ready or Not: The Manic Logic of Global Capitalism*, London, Allen Lane.

Gruffudd, P. (2013) 'Nationalism', in Cloke, P. J., Crang, P. and Goodwin, M. (eds) *Introducing Human Geographies* (2nd edn), London, Arnold, 378–389.

Guardian, The (2012) 'Developing world debt', www.theguardian.com/global-development/poverty-matters/2012/may/15/developing-world-of-debt, accessed 4 September, 2013.

Guardian, The (2013) 'Global development blog', www.theguardian.com/global-development, accessed 2September, 2013.

Guthman, J. (2009) 'Commodity chains', in Gregory, D., Johnston, R., Pratt, G., Watts, M. J. and Whatmore, S. (eds) *The Dictionary of Human Geography*, Wiley-Blackwell, Oxford, 101–102.

Gwynne, R. N. (1985) *Industrialization and Urbanization in Latin America*, London, Croom Helm.

Gwynne, R. N. (1990) *New Horizons? Third World Industrialization in an International Framework*, Harlow, Essex, Longman.

Gwynne, R. N. (2004) 'Political economy, resource use, and Latin American environments', *Singapore Journal of Tropical Geography*, 24(3), 247–260.

Gwynne, R. N. and Kay, C. (2000) 'Views from the periphery: future of neoliberalism in Latin America', *Third World Quarterly*, 21(1), 121–156.

Gwynne, R. N. and Kay, C. (2004) *Latin America Transformed: Globalization and Modernity* (2nd edn), London, Arnold.

Gwynne, R. N., Klak, T. and Shaw, D. J. B. (2003) *Alternative Capitalisms: Geographies of Emerging Regions*, London, Arnold.

Hägerstrand, T. (1968) *Innovation Diffusion as a Spatial Process*, Chicago, IL, University of Chicago Press.

Hägerstrand, T. (1975) 'Space, time and human conditions', in Karlquist, A. and Lundquist, F. (eds) *Dynamic Location of Urban Space*, Farnborough, Saxon House.

Hall, P. G. (1984) *The World Cities* (3rd edn), London, Weidenfeld & Nicolson.

Hamnett, C. (2003) 'Contemporary human geography: fiddling while Rome burns', *Geoforum*, 34(2), 1–4.

Hardin, G. J. (1968) 'The tragedy of the commons', *Science*, 162(3859), 1243–1248.

Hardt, M. and Negri, A. (2000) *Empire*, Cambridge, MA and London, Harvard University Press.

Hardt, M. and Negri, A. (2001) 'What the protestors in Genoa want', in Negri, A. (ed.) *On Fire: The Battle of Genoa and the Anti-capitalist Movement*, London, One-Off Press.

Hart, G. (2001) 'Development critiques in the 1990s: culs de sac and promising paths', *Progress in Human Geography*, 25(4), 649–658.

Harvey, D. (1989) *The Condition of Postmodernity: An Enquiry into the Origins of Cultural Change*, Oxford, Blackwell.

Harvey, D. (1995) 'Globalization in question', *Rethinking Marxism*, 8(4), 1–17.

Harvey, D. (1999) *The Limits to Capital* (new edn), London, Verso.

Harvey, D. (2000) *Spaces of Hope*, Edinburgh, Edinburgh University Press.

Harvey, D. (2001) *Spaces of Capital: Towards a Critical Geography*, Edinburgh, Edinburgh University Press.

Harvey, D. (2003) *The New Imperialism*, Oxford, Oxford University Press.

Harvey, D. (2005) *A Brief History of Neoliberalism*, Oxford, Oxford University Press.

Harvey, D. (2007) *The Limits to Capital*, London and New York, Verso.

Harvey, D. (2011) *The Enigma of Capital*, London, Polity Press.

Hau'ofa, E. (1993) 'Rediscovering our sea of islands' in Hau'ofa, E., Naidu, V. and Waddell, E. (eds) *A New Oceania: Rediscovering our Sea of Islands*, Institute of Pacific Studies, Suva.

Hay, C. and Marsh, D. (eds) (2000) *Demystfying Globalization*, London, Macmillan.

Hayter, R., Barnes, T. J. and Bradshaw, M. J. (2003) 'Relocating resource peripheries to the core of economic geography's theorizing: rationale and agenda', *Area*, 35(1), 15–23.

Held, D. (1995) *Democracy and the Global Order*, Cambridge, Polity Press.

Held, D. (ed.) (1991) *Political Theory Today*, Cambridge, Polity Press.

Held, D. (ed.) (2005) *Debating Globalization*, Cambridge, Polity Press.

Held, D. and McGrew, A. (2002) *Gloabalization/Anti-globalization*, Cambridge, Polity Press.

Held, D. and McGrew, A. (2007a) *Globalization/Antiglobalization*, Cambridge, Polity Press.

Held, D. and McGrew, A. (eds) (2007b) *Globalization Theory – Approaches and Controversies*, Cambridge, Polity.

Held, D., McGrew, A. G., Goldblatt, D. and Perraton, J. (1999) *Global Transformations: Politics, Economics and Culture*, Cambridge, Polity Press.

Henderson, J. (2003) 'The future of democracy in Melanesia: what role for outside powers?', *Asia Pacific Viewpoint*, 44(3), 225–242.

Herdt, G. H. (1997) *Same Sex, Different Cultures: Gays and Lesbians across Cultures*, Oxford, Westview Press.

Herman, E. S. and Chomsky, N. (1988) *Manufacturing Consent: The Political Economy of the Mass Media*, London, Vintage.

Herod, A. (2002) 'Global change in the world of organised labor', in Johnston, R. J., Taylor, P. J. and Watts, M. (eds) *Geographies of Global Change*, Oxford, Blackwell, 78–87.

Herod, A. (2003) 'Scale: the local and the global', in Holloway, S., Rice, S. and Valentine, G. (eds) *Key Concepts in Geography*, London, Sage.

Herod, A. (2009) 'Scale: the local and the global', in Clifford, N. J., Holloway, S. L., Rice, S. P. and Valentine, G. (eds) *Key Concepts in Geography* (2nd edn), London, Sage, 217–235.

Hertz, N. (2001) *The Silent Takeover: Global Capitalism and the Death of Democracy*, London, Heinemann.

Hill, R. C. (1989) 'Comparing transnational production systems: the automobile in the USA and Japan', *International Journal of Urban and Regional Research*, 13(3), 462–480.

Hillis, K. (1999) 'Cyberspace and cybercultures', in Cloke, P. J., Crang, P. and Goodwin, M. (eds) *Introducing Human Geographies*, London, Arnold, 324–331.

Hines, C. (2000) *Localization: A Global Manifesto*, London, Earthscan.

Hirst, P. Q. (1997) *From Statism to Pluralism: Democracy, Civil Society and Global Politics*, London, UCL Press.

Hirst, P. Q. and Thompson, G. (1999) *Globalization in Question: The International Economy and the Possibilities of Governance* (2nd edn), Cambridge, Polity Press.

Hirst, P., Thompson, G. and Bromley, S. (2009) *Globalization in Question* (3rd edn), Cambridge, Polity.

Houtart, F. (2001) 'Alternatives to the neoliberal model', in Hourtart, F. and Polet, F. (eds) *The Other Davos: The Globalization of Resistence to the World Economic System*, London, Zed Books, 47–59.

Hudson, R. (2006) 'Regions and place: music, identity and place', *Progress in Human Geography*, 30(5), 626–634.

Hughes, H. (2003) 'Aid has failed the Pacific', *Issue Analysis 33*, Sydney, The Centre for Independent Studies.

Huntington, S. P. (1991) *The Third Wave: Democratization in the Late Twentieth Century*, Norman, London, University of Oklahoma Press.

Huntington, S. P. (1996) *The Clash of Civilizations and the Remaking of the World Order*, New York, Simon & Schuster.

IFPI (International Federation of the Phonographic Industry) (2004) 'Global music retail sales', www.ifpi.org/site-content/publications/rin_order.html, accessed 12 November 2013.

IFPI (International Federation of the Phonographic Industry) (2013) 'The music industry in figures', www.ifpi.com, accessed 23 August 2013.

Ilbery, B. (2001) 'Changing geographies of global food production', in Daniels, P. W., Bradshaw, M. J., Shaw, D. J. B. and Sidaway, J. D. (eds) *Human Geography: Issues for the 21st Century*, Harlow, Pearson Education.

Imry, R. (2004) 'Urban geography, relevance, and resistance to the "policy turn"', *Urban Geography*, 25(8), 697–708.

Inglehart, R. (1990) *Culture Shift in Advanced Industrial Society*, Princeton, NJ, Princeton University Press.

IPCC (Intergovernmental Panel on Climate Change) (2001) *Climate Change 2001*, Cambridge, IPCC and Cambridge University Press.

Jackson, P. (1989) *Maps of Meaning*, London, Unwin Hyman.

Jackson, P. (2002) 'Consumption in a globalizing world', in Johnston, R. J., Taylor, P. J. and Watts, M. (eds) *Geographies of Global Change*, Oxford, Blackwell, 283–295.

Janelle, D. G. (1968) 'Central place development in a time-space framework', *Professional Geographer*, 20(1), 5–10.

Janelle, D. G. (1969) 'Spatial reorganization: a model and concept', *Annals of the Association of American Geographers*, 59(2), 348–364.

Janelle, D. G. (1973) 'Measuring human extensibility in a shrinking world', *Journal of Geography*, 72(5), 8–15.

Jenkins, R. O. (1987) *Transnational Corporations and Uneven Development: The Internationalization of Capital and the Third World*, London, Methuen.

Jens-Uwe, W. and Warrier, M. (2009) *A Dictionary of Globalization*, London, Routledge.

Johnston, R. J. (1986) *On Human Geography*, Oxford, Blackwell.

Johnston, R. J. (1996) *Nature, State and Economy: The Political Economy of Environmental Problems*, Chichester, John Wiley.

Johnston, R. J. (2000) 'Relevance', in Johnston, R. J., Gregory, D., Pratt, G. and Watts, M. (eds) *The Dictionary of Human Geography*, Oxford, Blackwell.

Johnston, R. J. and Sidaway, J. D. (2004) *Geography and Geographers: Anglo-American Human Geography since 1945* (6th edn), London, Arnold.

Johnston, R. J., Gregory, D., Pratt, G. and Watts, M. (eds) (2000) *The Dictionary of Human Geography* (4th edn), Blackwell, Oxford.

Johnston, R. J., Taylor, P. J. and Watts, M. (eds) (2002) *Geographies of Global Change: Remapping the World*, Oxford, Blackwell.

Jones, A. (2010) *Globalization – Key Thinkers*, Cambridge, Polity.

Jones, R. L. (2000) 'Deforestation', in Thomas, D. S. G. and Goudie, A. (eds) *The Dictionary of Physical Geography*, Oxford, Blackwell.

Jubilee 2000 (2002) 'About Jubilee research', UK, www.jubilee2000uk.org, accessed 12 March 2009.

Kaplan, D. H. (2009) 'Nationalism', in Kitchen, R. and Thrift, N. (eds) *International Encyclopedia of Human Geography*, London, Elsevier.

Karl, T. L. (1997) *The Paradox of Plenty: Oil Booms and Petro-States*, Los Angeles, University of California Press.

Kay, C. (1989) *Latin American Theories of Development and Underdevelopment*, London, Routledge.

Kay, C. (2001) 'Reflections on rural violence in Latin America', *Third World Quarterly*, 22 (5), 741–775.

Kay, C. (2002) 'Why East Asia overtook Latin America: agrarian reform, industrialisation and development', *Third World Quarterly*, 23(6), 1072–1102.

Kelly, P. F. (1999) 'The geographies and politics of globalization', *Progress in Human Geography*, 23(3), 379–400.

Kelsey, J. (1995) *The New Zealand Experiment: A World Model for Structural Adjustment?*, Auckland, NZ, Auckland University Press.

Kelsey, J. (ed.) (2010) *No Ordinary Deal*, Auckland, Bridget Williams Books.

Kelsey, J. (ed.) (2011) *No Ordinary Deal – Unmasking the Trans-Pacific Partnership Free Trade Agreement*, Allen and Unwin, New South Wales.

Kennedy, C. M. (2013) *Understanding inequality in Chile: A revisited dependency analysis of education*, PhD Thesis, Victoria University of Wellington.

Kennedy, C. and Murray, W. E. (2012) 'Centrando los Margenes', in Cousino Donoso, F. and Foxley Rioseco, A. M. (eds) *Chile Rumbo al Desarrollo: Miradas Criticas*, UNESCO, Chile, Santiago, 113–146.

Kennedy, C.M. and Murray, W.E. (2012) 'Growing apart? The persistence of inequality in Chile, 1964–2010' *Urbani Izziv,* 23(2), 22–35.

Keohane, R. O. (1995) 'Hobbes' dilemma and institutional change in world politics: sovereignty in international society', in Holm, H. H. and Sorensen, G. (ed.) *Whose World Order?*, Boulder, CO, Westview Press.

Khan, L. A. (1996) *The Extinction of the Nation State: A World Without Borders*, The Hague, Kluwer Law International.

Khanna, P. (2008) *The Second World: Empires and Influence in the New World Order*, New York, Random House.

Kindleberger, C. (1967) *Europe's Post-war Growth*, Oxford and New York, Oxford University Press.

Kindon, S., Pain, R. and Kesby, M. (2007) *Participatory Action Research Approaches and Methods: Connecting People, Participation and Place*, London: Routledge.

Kitchin, R. and Dodge, M. (2002) 'The emerging geographies of cyberspace', in Johnston, R. J., Taylor, P. J. and Watts, M. (eds) *Geographies of Global Change: Remapping the World* (2nd edn), Oxford, Blackwell, 340–354.

Klak, T. (1998) *Globalization and Neoliberalism: The Caribbean Context*, Lanham, Oxford, Rowman & Littlefield.

Klein, N. (2001) *No Logo: No Space, No Choice, No Jobs*, London, Flamingo.

Knox, P. L. (2002) 'World cities and organization of global space', in Johnston, R.J., Taylor, P. J. and Watts, M. J. (eds) *Geographies of Global Change* (2nd edn), Oxford: Blackwell, 328–338.

Knox, P. L. (2011) 'Extraordinary cities: world cities in global networks', in Agnew, J. A. and Duncan, J. S. (eds) *Companion to Human Geography*, Oxford: Blackwell.

Knox, P. L. and Agnew, J. A. (1998) *The Geography of the World-economy* (3rd edn), London, Edward Arnold.

Knox, P. L. and Taylor, P. J. (eds) (1995) *World Cities in a World Economy*, Cambridge, Cambridge University Press.

Korten, D.C. (1995) *When Corporations Rule the World*, West Hartford, Kumarian Press.

Krugman, P. R. (1996) *Pop Internationalism*, Cambridge, MA and London, MIT Press.

Krugman, P. R. (1998) 'What's new about the "new" economic geography?', *Oxford Review of Economic Policy*, 14, 7–17.

Krugman (2010) 'Inequality and crises: coincidence or causation?', Address at the Alphonse Weicuer Foundation, Luxembarg.

Larner, W. and LeHeron, R. B. (2002) 'From economic globalisation to globalising economic processes: towards post-structural political economies', *Geoforum*, 33(4), 415–419.

Lash, S. and Urry, J. (1994) *Economies of Signs and Space*, London, Sage.

Lee, H. (2004) 'Second generation Tongan transnationalism: hope for the future?', *Asia Pacific Viewpoint*, 45(2), 235–254.

Lee, R. (2000a) 'Economic geography', in Johnston, R. J., Gregory, D., Pratt, G. and Watts, M. (eds) *The Dictionary of Human Geography* (4th edn), Oxford, Blackwell.

Lee, R. (2000b) 'New international division of labour (NIDL)', in Johnston, R. J., Gregory, D., Pratt, G. and Watts, M. (eds) *The Dictionary of Human Geography* (4th edn), Oxford, Blackwell.

Lee, R. (2000c) 'Transational corporation', in Johnston, R. J., Gregory, D., Pratt, G. and Watts, M. (eds) *The Dictionary of Human Geography* (4th edn), Oxford, Blackwell.

Lee, R. (2002) '"Nice maps, shame about the theory"? Thinking geographically about the economic', *Progress in Human Geography*, 26(3), 333–355.

Lee, R. and Smith, D. M. (2004) *Geographies and Moralities: International Perspectives on Development, Justice and Place*, Malden, MA, Blackwell.

LeHeron, R. B. (1993) *Globalised Agriculture: Political Choice*, Oxford, Pergamon Press.

LeHeron, R. (2007) 'Globalisation, governance and post-structural political economy: perspectives from Australasia', *Asia Pacific Viewpoint*, 48(1), 26–40.

Levitt, P. (2001) *The Transnational Villagers*, Berkeley, London, University of California Press.

Lewis, A. (1981) 'The rate of growth of world trade, 1830–1973', in Grassman, S. and Lundberg, E. (eds) *The World Economic Order: Past and Prospects*, Basingstoke, Macmillan.

Lewis, W. A. (1955) *The Theory of Economic Growth*, London, Allen & Unwin.

Leyshon, A. (1995) 'Annihilating space? the speed up of communications', in Allen, J. and Hamnett, C. (eds) *A Shrinking World?*, Oxford, Oxford University Press, 11–54.

Leyshon, A. (1996) 'Dissolving difference? Money, disembedding and the creation of global financial space', in Daniels, P. W. and Lever, W. (eds) *The Global Economy in Transition*, London, Longman.

Lin Sien, C. (ed.) (2003) *Southeast Asia Transformed: A Geography of Change*, Institute of Southeast Asian Studies, Singapore.

Lipeitz, A. (1987) *Mirages and Miracles: The Crises of Global Fordism*, London, Verso.

Little, J. (2013) 'Society–space', in Cloke, P., Crang, P. and Goodwin, M. (2013) *Introducing Human Geography* (3rd edn), London, Routledge, 23–36.

Love, J. F. (1995) *McDonalds: Behind the Arches*, New York, Bantam.

Lowe, P., Murdoch, J., Marsden, T., Munton, R. and Flynn, A. (1993) 'Regulating the new rural spaces: the uneven development of land', *Journal of Rural Studies*, 9(3), 205–222.

Luard, E. (1990) *The Globalization of Politics: The Changed Focus of Political Action in the Modern World*, London, Macmillan.

McCarthy, J. (2000) 'Social movements', in Johnston, R. J., Gregory, D., Pratt, G. and Watts, M. (eds) *The Dictionary of Human Geography* (4th edn), Oxford, Blackwell.

McCarthy, J. (2005) 'Rural Geographies: multifunctional rural geographies reactionary or radical?', *Progress in Human Geography*, 29, 773–82.

McCarthy, J. (2007) 'Rural geography: Globalising the countryside', *Progress in Human Geography*, 32(1), 129–137.

McCormick, J. (1989) *Reclaiming Paradise: The Global Environmental Movement*, Bloomington, Indiana University Press.

McDowell, L. (1997) *Undoing Place?: A Geographical Reader*, London, Arnold.

McDowell, L. (1999) *Gender, Identity and Place: Understanding Feminist Geographies*, Cambridge, Polity Press.

McEwan, C. (2001) 'Geography, culture, and global change', in Daniels, P. W., Bradshaw, M. J., Shaw, D. J. B. and Sidaway, J. D. (eds) *Human Geography: Issues for the 21st Century*, London, Pearson Education.

McEwan, C. and Daya, S. (2013) 'Geography, culture and global change', in Daniels, P. W., Bradshaw, M. J., Shaw, D. J. B. and Sidaway, J. D. (eds) *Human Geography: Issues for the 21st Century* (4th edn), Harlow, Pearson, 272–289.

McFarlane, C. (2009) 'Translocal assemblages: space, power and social movements', *Geoforum*, 40(4): 561–567.

McGregor, A. (2008) *Southeast Asian Development*, Routledge, London.

McGregor, A. (2010) 'Green and REDD? Towards a political eoclogy of deforestation in Aceh, Indonesia', *Human Geography*, 3(2), 21–34.

McGrew, A. G. (2000) 'Sustainable globalization?', in Allen, T. and Thomas, A. (eds) *Poverty and Development: Into the 21st Century*, Oxford, Oxford University Press, 345–364.

McGrew, A (2008) *Politics Beyond Borders – The Principles of Global Politics*, Cambridge, Polity.

McGrew, A.G. and Lewis, P.G. (1992) *Global Politics: Globalization and the Nation-State*, Cambridge, Polity Press.

McIntyre, M. and Soulsby, J. (2004) 'Land use and land degradation on Niue', in Terry, J. P. and Murray, W. E. (eds) *Niue Island: Geographical Perspectives on the Rock of Polynesia*, Paris, INSULA, UNESCO, 217–226.

McKenna, M. K. L. and Murray, W. E. (2002) 'Jungle law in the orchard: comparing globalization in the New Zealand and Chilean apple industries', *Economic Geography*, 78(4), 495–514.

McLuhan, M. (1962) *The Gutenberg Galaxy: The Making of Typographic Man*, Toronto, University of Toronto Press.

McLuhan, M. (1964) *Understanding Media: The Extensions of Man* (1st edn), New York, McGraw-Hill.

McMichael, P. (2004) *Development and Social Change: A Global Perspective* (3rd edn), Thousand Oaks, CA, and London, Pine Forge.

McMichael, P. (2009) 'A food regime analysis of the "world food crisis"', *Agriculture and Human Values*, 26(4), 281–295.

McMichael, P. (2012) *Development and Social Change: A Global Perspective* (5th ed), Thousand Oaks, CA, and London, Pine Forge.

Malecki, E. J. (1991) *Technology and Economic Development*, New York, Longman.

Mann, M. (1988) *States, War and Capitalism: Studies in Political Sociology*, Oxford, Blackwell.

Manson, S. M. (2008) 'Does scale exist? An epistemological scale continuum for complex human-environment systems', *Geoforum*, 39(2), 776–788.

Mansvelt, J. (2005) *Geographies of Consumption*, London, Sage.

Marston, S. (2009) 'Scale', in Gregory, D., Johnston, R., Pratt, G., Watts, M. J. and Whatmore, S. (eds) *The Dictionary of Human Geography*, Wiley-Blackwell, Oxford, 664–666.

Martin, R. (1994) 'Stateless monies, global financial integration and national economic autonomy: the end of geography', in Corbridge, S., Martin, R. L. and Thrift, N. (eds) *Money, Power and Space*, Oxford: Blackwell, 253–278.

Martin, R. (2004) 'Geography: making a difference in a globalising world', *Transactions of the Institute of British Geographers*, 29(2), 147–150.

Mason, P. (2013) *Why It's Kicking Off Everywhere – The New Global Revolutions*, London, Verso.

Massey, D. B. (1984) *Spatial Divisions of Labour: Social Structures and the Geography of Production* (1st edn), London, Macmillan.

Massey, D. B. (1991) 'A global sense of place', *Marxism Today*, June, 24–29.

Massey, D. B. (1994) *Space, Place and Gender*, Cambridge, Polity Press.

Massey, D. B. (1995) *Spatial Divisions of Labour: Social Structures and the Geography of Production* (2nd edn), London, Macmillan.

Massey, D.B. (2001) 'Geography on the agenda', *Progress in Human Geography*, 25, 5–18.

Maye, D. and Ilbery, B. (2012) 'Changing geographies of food production', in Daniels, P. W., Bradshaw, M. J., Shaw, D. J. B. and Sidaway, J. D. (eds) *Human Geography: Issues for the 21st Century* (4th edn), Harlow, Pearson, 314–333.

Meadows, D. H. (1972) *The Limits to Growth: A Report for the Club of Rome's Project on the Predicament of Mankind*, London, Earth Island.

Mercier, K. (2003) 'The anti-globalisation movement in London', *Institute of Geography*, Victoria University of Wellington, Masters of Development Studies.

Meyer, W. R. and Turner, B. L. (1995) 'The Earth transformed: trends, trajectories and patterns', in Johnston, R. J., Taylor, P. J. and Watts, M. (eds) *Geographies of Global Change*, Oxford, Blackwell, 364–376.

Miller, T. E. and Shahriari, A. (2013) *World Music: A Global Journey*, London, Routledge.

Mittelman, J. H. (2000) *The Globalization Syndrome: Transformation and Resistance*, Princeton, NJ, Princeton University Press.

Morgan, S. (1984) *Sisterhood is Global*, London, Feminist Press.

Munck, R. (2006) *Globalization and Contestation: The New Great Counter-Movement*, London, Routledge.

Munck, R. (2008) 'Globalisation, governance and migration: an introduction', *Third World Quarterly*, 29(7) 1227–1246.

Munck, R. (2013) *Rethinking Latin America: Development, Hegemony, and Social Transformation*, Oxford, Palgrave Macmillan.

Murray, W. E. (1998) 'The globalisation of fruit, neoliberalism and the question of sustainability – lessons from Chile', *European Journal of Development Studies*, 10(1), 201–227.

Murray, W. E. (1999) 'Local responses to global change in the Chilean fruit complex', *European Review of Latin American and Caribbean Studies*, 66, 19–38.

Murray, W. E. (2000) 'Neoliberal globalisation, "exotic" agro-exports and local change in the Pacific Islands: a study of the Fijian kava sector', *Singapore Journal of Tropical Geography*, 21(3), 355–375.

Murray, W. E. (2001) 'The second wave of globalisation and agrarian change in the Pacific Islands', *Journal of Rural Studies*, 17(2), 135–148.

Murray, W. E. (2002a) 'From dependency to reform and back again: the Chilean peasantry in the twentieth century', *Journal of Peasant Studies*, 29(3/4), 190–227.

Murray, W. E. (2002b) 'Sustaining agro-exports in Niue: the inevitable failure of free-market restructuring', *Journal of Pacific Studies*, 24(2), 211–228.

Murray, W. E. (2004a) 'Mercosur', in Forsyth, T. J. (ed.) *Encyclopedia of International Development*, London, Routledge.

Murray, W. E. (2004b) 'Neocolonialism', in Forsyth, T. J. (ed) *Encyclopedia of International Development*, London, Routledge.

Murray, W. E. (2004c) 'Latin American Free Trade Area', in Forsyth, T.J. (ed.) *Encyclopaedia of International Development*, Routledge, London, 403.

Murray, W. E. (2006) 'Neo-feudalism in Latin America? Globalisation, agribusiness, and land re-concentration in Chile', *The Journal of Peasant Studies*, 33(4), 646–677.

Murray, W. E. (2008a) 'Latin America – challenges and opportunities and the role of New Zealand', in Murray, W. E. and Rabel, R. (eds) *Latin America Today*, NZ Institute for International Affairs.

Murray, W. E. (2008b) 'Neoliberalism, rural geography and the Global South', *Human Geography*, 1(1), 33–38.

Murray, W. E. (2009a) 'Neoliberalism and development', in Kitchen, R. and Thrift, N. (eds) *International Encyclopedia of Human Geography*, London, Elsevier.

Murray, W. E. (2009b) 'Oceania', in Kitchen, R. and Thrift, N. (eds) *International Encyclopedia of Human Geography*, London, Elsevier.

Murray, W. E. (2012) 'Rural worlds', in Daniels, P., Bradshaw, M. J., Sidaway, J. D. and Shaw, D. (eds) *Human Geography – Issues for the 21st Century* (4th edn), Edward Arnold, London, 237–255.

Murray, W. E. and Storey, D. (2003) *Post-colonial Transformations and Political Conflict in Oceania*, Melbourne, Blackwell, Special edition of *Asia Pacific Viewpoint*, 44(3).

Murray, W. E. and Silva, E. (2004) 'The political economy of sustainable development', in Gwynne, R. N. and Kay, C. (eds) *Latin America Transformed: Globalization and Modernity*, London, Arnold, 117–138.

Murray, W. E. and Terry, J. P. (2004) 'Niue's place in the Pacific', in Terry, J. P. and Murray, W. E. (eds) *Niue Island: Geography on the Rock of Polynesia*, Paris, INSULA, UNESCO, 9–30.

Murray, W. E. and Rabel R. (eds) (2008) *Latin America Today: Challenges, Opportunities and Trans-Pacific Perspectives*, New Zealand Institute of International Affairs, Wellington, New Zealand.

Murray, W. E. and Overton, J. (2011a) 'Neoliberalism is dead, long live neoliberalism. Neostructuralism and the new international aid regime of the 2000s', *Progress in Development Studies*, 11(4), 307–319.

Murray, W. E. and Overton, J. (2011b) The inverse sovereignty effect: aid, scale and neostructuralism in Oceania. *Asia Pacific Viewpoint*, 52(3), 272–284.

NAFTA (2013) 'Facts about NAFTA', www.naftanow.org/facts/, accessed 25 September 2013.

Nagar, R., Lawson, V., McDowell, L. and Hanson, S. (2002) 'Locating globalization: feminist (re)readings of the subjects and spaces of globalization', *Economic Geography*, 78(3), 257–284.

Nederveen Pieterse, J. (2004) *Globalization and Culture: Global Mélange*, Lanham, MD, Rowman and Littlefield.

Nicole, R. (2000) *The Word, the Pen and the Pistol: Orientalism in the Pacific*, New York, State University of New York Press.

Nunn, P. D. (2003) 'Revising ideas about environmental determinism: human–environment relations in the Pacific Islands', *Asia Pacific Viewpoint*, 44(1), 63–72.

O'Brien, R. (1992) *Global Financial Integration: The End of Geography*, London, Royal Institute of International Affairs, Pinter Publishers.

Ohmae, K. (1990) *The Borderless World*, London, Collins.

Ohmae, K. (1995) *The End of the Nation State*, New York, Free Press.

O'Riordan, T. (1981) *Environmentalism* (2nd edn), London, Pion.

Osbourne, R. (2013) *Vinyl: A History of the Analogue Record*, Ashgate Press, Oxford.

Ó Tuathail, G. (1996) *Critical Geopolitics*, Minneapolis, Minneapolis University Press.

Ó Tuathail, G. (2002) 'Post Cold-War geopolitics: contrasting superpowers in a world of global dangers', in Johnston, R. J., Taylor, P. J. and Watts, M. (eds) *Geographies of Global Change*, Oxford, Blackwell, 174–190.

Ó Tuathail, G. (2006) 'Thinking critically about geopolitics', in Ó Tuathail, G., Dalby, S. and Routledge, P. (eds) *The Geopolitics Reader* (2nd edn), Abingdon, Routledge, 55–64.

Overton, J. D. (2000) 'Vakavanua, vakamatanitu: discourses of development in Fiji', *Asia Pacific Viewpoint*, 40(2), 121–134.

Overton, J. D. (2003) 'Understanding coups', *Asia Pacific Viewpoint*, 44(3), 351–356.

Overton, J. D. and Scheyvens, R. (eds) (1999) *Strategies for Sustainable Development: Experiences from the Pacific*, London, Zed Books.

Overton, J. and Murray, W. E. (2011) 'Playing the scales: regional transformations and the differentiation of rural space in the Chilean wine industry', *Journal of Rural Studies*, 27(1), 63–72.

Overton, J. and Murray, W. E. (2012) 'Class in a glass: capital, neoliberalism and social space in the global wine industry', *Antipode*, 45(3), 702–718.

Overton, J. and Murray W. E. (2013) 'Sovereignty for sale – marginality in Niue', Hruati Geograkski Glasnik (*Croatian Geographical Bulletin*) 76(1), 5–25.

Overton, J., Murray, W. E. and Banks, G. (2012) 'The race to the bottom of the glass? Wine, geography, and globalization', *Globalizations*, 9(2), 273–287.

Overton, J., Murray, W. E. and McGregor, A. (2013) 'Geographies of aid: a critical research agenda', *Geography Compass*, 7(2), 116–127.

Pacione, M. (1999) *Applied Geography: Principles and Practice*, London, Routledge.

Painter, J. (1995) *Politics, Geography and Political Geography: A Critical Perspective*, London, New York, Arnold, Wiley.

Painter, J. (2000) 'Localization', in Johnston, R. J., Gregory, D., Pratt, G. and Watts, M. (eds) *The Dictionary of Human Geography* (4th edn), Oxford, Blackwell.

Painter, J. (2009) 'Political geography', in Gregory, D., Johnston, R., Pratt, G., Watts, M. J. and Whatmore, S. (eds) (2009) *The Dictionary of Human Geography* (5th edn), Wiley-Blackwell, Oxford.

Painter, J. and Philo, C. (1995) 'Spaces of citizenship', *Political Geography*, 14(2), 107–120.

Painter, J. and Jeffrey, A. (2009) *Political Geography*, London, Sage.

Palomino-Schalscha, M. A. (2011) *Indigeneity, autonomy and new cultural spaces: the decolonisation of practices, being and place through tourism in Alto Bío-Bío, Chile*, PhD Thesis, University of Canterbury, New Zealand.

Park, J., Conca, K. and Finger, M. (eds) (2008) *The Crisis of Global Environmental Governance: Towards a New Political Economy of Sustainability*, Abingdon, Routledge.

Peck, J. (2001) 'Neoliberalizing states', *Progress in Human Geography*, 25(3), 445–455.

Peck, J. and Tickell, A. (1994) 'Jungle law breaks out: neoliberalism and global–local disorder', *Area*, 26(4), 317–326.

Peck, J. and Tickell, A. (2002) 'Neoliberalizing space', *Antipode*, 34(3), 380–404.

Peck, J. and Yeung, H. W. C. (eds) (2003) *Remaking the Global Economy: Economic Geographical Perspectives*, London, Sage.

Peet, R. (1991) *Global Capitalism: Theories of Societal Development*, London, Routledge.

Peet, R. and Watts, M. (1993) 'Development theory and environment in an age of market triumphalism', *Economic Geography*, 69(3), 227–253.

Penrose, J. (2009) 'Nation', in Kitchen, R. and Thrift, N. (eds) *International Encyclopedia of Human Geography*, London, Elsevier.

Perlmutter, H. V. (1991) 'On the rocky road to the first global civilisation', *Human Relations*, 44(9), 902–906.

Perraton, J. (2001) 'The global economy – myths and realities', *Cambridge Journal of Economics*, 25(5), 669–684.

Perrons, D. and Posocco, S. (2009) 'Globalising failures', *Geoforum*, 40(2), 131–135.

Petras, J. (1993) 'Cultural imperialism in the late 20th century', *Journal of Contemporary Asia*, 23(2), 139–148.

Petras, J. (1999) *The Left Strikes Back: Class Conflict in Latin America in the Age of Neoliberalism*, Boulder, CO, Westview Press.

Petras, J. and Vettmeyer, H. (2001) *Globalization Unmasked*, London, Zed Books.

Philo, C. (ed.) (1991) *New Words, New Worlds: Reconceptualising Social and Cultural Geography*, Aberystwyth, Cambrian Printers.

Pickering, J. (2012) 'The environment and environmentalism', in Daniels, P., Bradshaw, M. J., Sidaway, J. D. and Shaw, D. (eds) *Human Geography – Issues for the 21st Century* (4th edn), Edward Arnold, London, 140–161.

Pickering, K. T. and Owen, L. A. (1997) *An Introduction to Global Environmental Issues* (2nd edn), London, Routledge.

Pieterse, J. N. (1995) 'Globalization as hybridisation', in Featherston, S., Nash, S. and Robertson, R. (eds) *Global Modernities*, London, Sage.

Pieterse, J. N. (2001) 'Globalization and collective action', in Hamel, P., Lustiger-Thaler, H., Pietrse, J. N. and Roseneil, S. (eds) *Globalization and Social Movements*, Basingstoke, Palgrave.

Pieterse, J. N. and Parekh, B. (eds) (1995) *The Decolonisation of Imagination: Cultural Knowledge and Power*, London, Zed Books.

Pollard, J. (2001) 'The global financial system: worlds of monies', in Daniels, P. W., Bradshaw, M. J., Shaw, D. J. B. and Sidaway, J. D. (eds) *Human Geography: Issues for the 21st Century*, Harlow, Prentice Hall.

Pollard, J. (2012) 'The global financial system – worlds of monies', in Daniels, P. Bradshaw, M. J., Sidaway, J. D. and Shaw, D. (eds) *Human Geography – Issues for the 21st Century*, (4th edn), Edward Arnold, London, 379–395.

Porter, M. E. (1990) *The Competitive Advantage of Nations*, London, Macmillan.

Potter, D. (2000) 'Democratization, good governance, and development', in Allen, T. and Thomas, A. (eds) *Poverty and Development into the 21st Century*, Oxford, Oxford University Press, 365–382.

Potter, L. and Cooke, F. M. (eds) (2004) *Negotiating Modernity and Globalisation in Rural and Marine Environments*, Special edition of *Asia Pacific Viewpoint*, Oxford, Blackwell.

Potter, R. B. (1993) 'Little England and little geography: reflections on Third World teaching and research', *Area*, 25, 291–294.

Potter, R. B., Binns, T., Elliott, J. A. and Smith, D. (1999) *Geographies of Development* (1st edn), Harlow, Prentice Hall.

Potter, R. B., Binns, T., Elliot, J. A. and Smith, D. (2004) *Geographies of Development* (2nd edn), Harlow, Prentice Hall.

Potter, R. B., Binns, T., Elliot, J. A. and Smith, D. (2008) *Geographies of Development* (3rd edn) Harlow, Prentice Hall.

Power, M. (2003) *Rethinking Development Geographies*, London, Routledge.

Power, M. (2012) 'Worlds apart – global difference and inequality', in Daniels, P., Bradshaw, M. J., Sidaway, J. D. and Shaw, D. (eds) *Human Geography – Issues for the 21st Century*, (4th edn), Edward Arnold, London, 177–199.

Power, M. and Sidaway, J. D. (2004) 'The degeneration of tropical geography', *Annals of the Association of American Geographers*, 94(3), 585–601.

Pratt, G. (2009) 'Femininst geographies', in Gregory, D., Johnston, R., Pratt, G., Watts, M. J. and Whatmore, S. (eds) (2009) *The Dictionary of Human Geography* (5th edn), Wiley-Blackwell, Oxford.

Pratt, G. and Craft, M. (2013) 'Masculinity/femininity', in Cloke, P., Crang, P. and Goodwin, M. (2013) *Introducing Human Geography* (3rd edn), London, Routledge, 81–95.

Prebisch, R. (1950) *The Economic Development of Latin America and its Principal Problems, New York, United Nations.*

Prebisch, R. (1964) *Towards a New Trade Policy for Development*, New York, UNCTAD.

Preston, P. W. (1996) *Development Theory*, Oxford, Blackwell.

Radcliffe, S. A. (2004) 'Civil society, grassroots politics and livelihoods', in Gwynne, R. N. and Kay, C. (eds) *Latin America Transformed: Globalization and Modernity*, London, Arnold, 193–209.

Rapaport, M. (ed.) (1999) *Pacific Islands: Environment and Society*, Honolulu, University of Hawai'i Press.

Rappaport, M. (2013) *The Pacific Islands – Environment and Society*, (2nd edn), Manoa, University of Hawai'i.

Raynolds, L. T. (2009a) 'Fair Trade', in Gregory, D., Johnston, R., Pratt, G., Watts, M. J. and Whatmore, S. (eds) (2009) *The Dictionary of Human Geography* (5th edn), Wiley-Blackwell, Oxford.

Raynolds, R T (2009b) 'Fair Trade' in Kitchen, R, and Thrift, N (eds.) *Encyclopedia of Human Geography*, Elsevier, p. 8–12.

Raynolds, L. T. (2012) 'Fair trade: social regulation in global food markets', *Journal of Rural Studies*, 28(3), 276–287.

Raynolds, L. T., Murray, D. L. and Wilkinson, J. (eds) (2007) *Fair Trade: The Challenges of Transforming Globalization*, Abingdon, Routledge.

Rennie-Short, J. (1993) *An Introduction to Political Geography* (2nd edn), London, Routledge.

Rennie-Short, J. (2004) 'Black holes and loose connections in a global urban network', *The Professional Geographer*, 56 (2), 295–302.

RIAA (2004) 'Anti-piracy', www.riaa.com/issues/piracy/default.asp.

Rigg, J. (2002) 'Of miracles and crises: (re-)interpretations of growth and decline in East and Southeast Asia', *Asia Pacific Viewpoint*, 43(2), 137–156.

Rigg, J. (2003) *Southeast Asia: The Human Landscape of Modernization and Development*, London, Routledge.

Ritzer, G. (1993) *The McDonaldizaton of Society*, London, Pine Forge/Sage.

Ritzer, G. (2007) *The Blackwell Companion to Globalization*, London, Blackwell.

Ritzer, G. (2008) *The McDonaldization of Society* (5th edn), California, Pine Forge Press.

Ritzer, G. (2010) *Globalizaiton – A Basic Text*, London, Wiley-Blackwell.

Ritzer, G. (2011) *Globalization – The Essentials*, London, Wiley-Blackwell.

Robbins, P. (2012) *Political Ecology*, John Wiley, London.

Roberts, D. (2001) *Guinness World Records: British Hit Singles and Albums*, Enfield, Guinness.

Roberts, J. T. and Hite, A. (eds) (2000) *From Modernization to Globalization: Perspectives on Development and Social Change*, Malden, MA, and Oxford, Blackwell.

Roberts, S. M. (2002) 'Global regulation and trans-state organization', in Johnston, R. J., Taylor, P. J. and Watts, M. (eds) *Geographies of Global Change*, Oxford, Blackwell, 143–157.

Robertson, R. (1992) *Globalization: Social Theory and Global Culture*, London, Sage.

Robertson, R. (2003) *The Three Waves of Globalization*, London, Zed Books.

Rosenau, J. N. (1980) *The Study of Global Interdependence: Essays on the Transnationalization of World Affairs*, London, Pinter.

Rosenau, J. N. (1990) *Turbulence in World Politics: A Theory of Change and Continuity*, New York, London, Harvester Wheatsheaf.

Rosenberg, J. (2005) 'Globalization theory: A post mortem', *International Politics*, 42(1), 2–74.

Rostow, W. (1960) *The Stages of Economic Growth: A Non-Communist Manifesto*, Cambridge, Cambridge University Press.

Routledge, P. (1999) 'Survival and resistence', in Cloke, P.J., Crang, P. and Goodwin, M. (eds) *Introducing Human Geographies*, London, Arnold, 76–83.

Routledge, P. (2002) 'Resisting and reshaping destructive development: social movements and globalizing networks', in Johnston, R. J., Taylor, P. J. and Watts, M. (eds) *Geographies of Global Change: Remapping the World* (2nd edn), Oxford, Blackwell, 310–327.

Routledge, P. (2009) 'Activist geographies', in Kitchen, R. and Thrift, N. (eds) *International Encyclopedia of Human Geography*, London, Elsevier.

Routledge, P. (2013) 'Survival and resistance', in Cloke, P., Crang, P. and Goodwin, M. (eds) *Introducing Human Geography* (3rd edn), London, Routledge, 325–338.

Rugman, A. (2000) *The End of Globalization*, London, Random House.

Ruigrok, W. and van Tulder, R. (1995) *The Logic of International Restructuring*, London, Routledge.

Rupert, M. (2000) *Ideologies of Globalization*, London, Routledge.

Rycroft, S. (2009) 'Cultural politics', in Kitchen, R. and Thrift, N. (eds) *International Encyclopedia of Human Geography*, London, Elsevier.

Sachs, J. (2008) *Common Wealth: Economics for a Crowded Planet*, New York, Penguin.

Sachs, J. (2011 [2005]) *The End of Poverty* (2nd edn), New York, Penguin.

Sachs, J. (2012) *The Price of Civilization*, New York, Turtleback Books.

Sassen, S. (2000) *Cities in a World Economy* (2nd edn), Thousand Oaks, CA, and London, Pine Forge Press.

Sassen, S. (2001) *The Global City: London, New York, Tokyo*, Princeton, NJ, Princeton University Press.

Sassen, S (2007) *A Sociology of Globalization*, New York, Norton.

Sassen, S. (2011) *Cities in a World Economy*, London, Sage.

Sauer, C. (1925) 'The morphology of landscape', *University of California Publications in Geography*, 2, 19–54.

Scheyvens, R. (2002) *Tourism for Development: Empowering Communities*, Harlow, Prentice Hall.

Scheyvens, R. (2011) *Tourism and Poverty*, London, Routledge.

Schirato, T. and Webb, J. (2003) *Understanding Globalization*, London, Sage.

Schoenburger, E. (2009) 'Transnational corporations', in Gregory, D., Johnston, R., Pratt, G., Watts, M. J. and Whatmore, S. (eds) (2009) *The Dictionary of Human Geography*, (5th edn), Wiley-Blackwell, Oxford.

Scholte, J. A. (2000) *Globalization: A Critical Introduction*, Basingstoke, Palgrave.

Scholte, J. A. (2005) 'Premature obituaries: a response to Justin Rosenberg', *International Politics*, 42(3), 390–399.

Scholte, J. A. and Robertson, R. (eds) (2007) *Encyclopedia of Globalization*, London, Routledge.

Scott, A. J. (1988) *New Industrial Space*, London, Pion Press.

Shurmer-Smith, P. and Hannam, K. (1994) *Worlds of Desire, Realms of Power: A Cultural Geography*, London, Edward Arnold.

Sidaway, J. D. (2000) 'Postcolonial geographies: an exploratory essay', *Progress in Human Geography*, 24(4), 591–612.

Sidaway, J. D. (2001a) 'Geopolitical traditions', in Daniels, P. W., Bradshaw, M. J., Shaw, D. J. B. and Sidaway, J. D. (eds) *Human Geography: Issues for the 21st Century*, London, Longman.

Sidaway, J. D. (2001b) 'The place of the nation-state', in Daniels, P. W., Bradshaw, M. J., Shaw, D. J. B. and Sidaway, J. D. (eds) *Human Geography: Issues for the 21st Century*, London, Longman.

Sidaway, J. D. (2008) 'Subprime crisis: American crisis or human crisis?', *Environment and planning D, Society and Space*, 26(2), 195–198.

Sidaway, J. D. (2011) 'The return and eclipse of border studies? Charting agendas 1', *Geopolitics*, 16(4), 969–976.

Sidaway, J. D. (2013) 'Geography, globalization, and the problematic of area studies. *Annais of the Association of American Geographers*, 103(4), 984–1002.

Sidaway, J. D and Grundy Warr, C. (2012) 'The place of the nation state', in Daniels, P. Bradshaw, M. J., Sidaway, J. D. and Shaw, D. (eds) *Human Geography – Issues for the 21st Century* (4th edn), Edward Arnold, London, 466 484.

Sidaway, J. D. and Mamadouh, V. (2012) 'Geopolitical traditions', in Daniels, P., Bradshaw, M. J., Sidaway, J. D. and Shaw, D. (eds) *Human Geography – Issues for the 21st Century* (4th edn), Edward Arnold, London, 419–441.

Sidaway, J. D., Bunnell, T. and Yeoh, B. S. A. (2003) 'Editors' introduction: Geography and postcolonialism', *Singapore Journal of Tropical Geography*, 24(3), 269–272.

Silva, E. (2004) 'Authoritarianism, democracy and development', in Gwynne, R. N. and Kay, C. (eds) *Latin America Transformed: Globalization and Modernity*, London, Edward Arnold, 141–156.

Sklair, L. (2001) *The Transnational Capitalist Class*, Oxford, Blackwell.

Smith, D. M. (1994) *Geography and Social Justice*, Oxford, Blackwell.

Smith, D. M. (2000) *Moral Geographies: Ethics in a World of Difference*, Edinburgh, Edinburgh University Press.

Smith, G. (2000a) 'Nation-state', in Johnston, R. J., Gregory, D., Pratt, G. and Watts, M. (eds) *The Dictionary of Human Geography*, Oxford, Blackwell.

Smith, G. (2000b) 'Geopolitics', in Johnston, R. J., Gregory, D., Pratt, G. and Watts, M. (eds) *The Dictionary of Human Geography*, Oxford, Blackwell.

Smith, N. (1984) *Uneven Development*, Oxford, Blackwell.

Smith, N. (2000) 'Global Seattle', *Environment and Planning D*, 18(1), 1–5.

Smith, N. (2003) *American Empire: Roosevelt's Geographer and the Prelude to Globalization*, Berkeley, University of California Press.

Soja, E.W. (1996) *Thirdspace: Journeys to Los Angeles and Other Real-and-Imagined Places*, Cambridge, MA and Oxford, Blackwell.

Sparke, M. (2009a) 'Nation state', in Gregory, D., Johnston, R., Pratt, G., Watts, M. J. and Whatmore, S. (eds) (2009) *The Dictionary of Human Geography* (5th edn), Wiley-Blackwell, Oxford.

Sparke, M. (2009b) 'Glocalisation', in Gregory, D., Johnston, R., Pratt, G., Watts, M. J. and Whatmore, S. (eds) (2009) *The Dictionary of Human Geography* (5th edn), Wiley-Blackwell, Oxford.

Sparke, M (2013) *Introducing Globalization: Ties, Tensions and Uneven Integration*, London, Wiley-Blackwell.

Spybey, T. (1996) *Globalization and World Society*, Cambridge, Polity Press.

Sriskandarajah, D. (2003) 'Inequality and conflict in Fiji: from purgatory to hell?', *Asia Pacific Viewpoint*, 44(3), 305–324.

Stalder, F. (2006) *Manuel Castells*, Cambridge, Polity.

Starr, A. (2000) *Naming the Enemy: Anti-Corporate Movements Confront Globalization*, London, Zed Books.

Stearns, P. N. (2009) *Globalization in World History*, London, Routledge.

Steger, M. B. (2002) *Globalism*, Maryland, Rowman and Littlefield.

Stiglitz, J. (2006) *Making Globalization Work*, London, Allen Lane.

Stiglitz, J. (2002) *Globalization and its Discontents*, London, Penguin.

Stiglitz, J. (2010) *Freefall*, London, Allen Lane.

Stiglitz, J. and Charlton, A. E. (2006) *Fair Trade for All*, Oxford, Oxford University Press.

Stoddard, D. R. (1987) 'To claim the high ground: geography for the end of the century', *Transactions of the Institute of British Geographers*, NS, 12, 327–336.

Storey, D. and Murray, W. E. (2001) 'Dilemmas of development in Oceania: the political economy of the Tongan agro-export sector', *Geographical Journal*, 167(4), 291–304.

Storey, D., Bulloch, H. and Overton, J. D. (2005) 'The poverty consensus: some limitations of the "popular agenda"', *Progress in Development Studies*, 5(1), 30–44.

Strange, S. (1996) *The Retreat of the State: The Diffusion of Power in the World Economy*, Cambridge, Cambridge University Press.

Stutz, F. P. and De Souza, A. R. (1998) *The World Economy: Resources, Location, Trade, and Development* (3rd edn), Upper Saddle River, NJ, Prentice Hall.

Sunkel, O. (ed.) (1993) *Development from Within: Towards a Neo-structuralist Approach for Latin America*, Boulder, CO and London, Lynne Rienner.

Swyngedouw, E. (1997) 'Neither global or local: "glocalization" and the politics of scale', in Cox, K. R. (ed.) *Spaces of Globalization: Reasserting the Power of the Local*, New York, London, Guilford Press, 137–166.

Taylor, G. (2002) 'We haven't gone away', *Observer Worldview* extra online www.observer.co.uk, 21 July.

Taylor, J. B. (2009) *The Financial Crisis and the Policy Responses: An Empirical Analysis of What Went Wrong* (No. w14631), Cambridge, MA, National Bureau of Economic Research.

Taylor, L. S. (2009) 'Agrofood system', in Gregory, D., Johnston, R., Pratt, G., Watts, M. J. and Whatmore, S. (eds) (2009) *The Dictionary of Human Geography* (5th edn), Wiley-Blackwell, Oxford.

Taylor, P. J. and Flint, C. (1989) *Political Geography: World Economy, Nation-State and Locality* (2nd edn), Harlow, Longman Scientific and Technical.

Taylor, P. J. and Flint, C. (2000) *Political Geography: World-Economy, Nation-State and Locality* (4th edn), Harlow, Prentice Hall.

Taylor, P. J., Watts, M. and Johnston, R. J. (2002) 'Geography/globalization', in Johnston, R. J., Taylor, P. J. and Watts, M. (eds) *Geographies of Global Change*, Oxford, Blackwell, 1–18.

Terry, J. P. and Murray, W. E. (eds) *Niue Island: Geographical Perspectives on the Rock of Polynesia*, Paris, INSULA, UNESCO.

Thompson, C. J. and Tambyah, S. K. (1999) 'Trying to be cosmopolitan', *Journal of Consumer Research*, 26(3), 214–241.

Thompson, G. and Allen, J. (1997) 'Think global, then think again: economic globalization in context', *Area*, 29(3), 213–227.

Thrift, N. J. (2000) 'Local–global dialectic', in Johnston, R. J., Gregory, D., Pratt, G. and Watts, M. (eds) *The Dictionary of Human Geography*, Oxford, Blackwell.

Thrift, N. J. (2002) 'A hyperactive world', in Johnston, R. J., Taylor, P. J. and Watts, M. (eds) *Geographies of Global Change*, Oxford, Blackwell, 29–42.

Tickell, A. (2005) 'Money and finance', in Cloke, P., Crang, P. and Goodwin, M. (2013) *Introducing Human Geography* (2nd edn), London, Routledge, 244–252.

Todaro, M.P. (1997) *Economic Development* (6th edn), London, Longman.

Toffler, A. (1970) *Future Shock*, London, The Bodley Head.

Tomlinson, J. (1991) *Cultural Imperialism: A Critical Introduction*, London, Pinter.

Tomlinson, J. (1999) *Globalization and Culture*, Cambridge, Polity Press.

Tyler-Millar, G. (2002) *Living in the Environment* (12th edn), London and New York, Brookes-Cole.

UNCTAD (2002) *The Least Developed Countries Report 2002*, New York, United Nations.

UNCTAD (2004) *World Investment Report*, New York, United Nations.

UNDP (various years) *Human Development Report*, Oxford, Oxford University Press.

UNFP (2013) *UNPF Population Fund Statistics 2013*, New York, UNFP.

United Nations (2013) *A New Global Partnership: Eradicate Poverty and Transform Economies through Sustainable Development. The Report of the High Level Panel of Eminent Persons on the Post 2015 Development Agenda*, New York, United Nations.

Urry, J. (2003) *Global Complexity*, Cambridge, Polity Press.

Vernon, R. (1977) *Storm Over the Multinationals: The Real Issues*, London, Macmillan.

Wallerstein, I. M. (1980) *The Modern World-System II*, New York, Academic Press.

Waters, M. (2001) *Globalization* (2nd edn), London, Routledge.

Weber, M., Roth, G., Wittich, C. and Fischoff, E. (1978) *Economy and Society: An Outline of Interpretive Sociology*, Berkeley, London, University of California Press.

Weiss, L. (1998) *The Myth of the Powerless State: Governing the Economy in a Global Era*, Cambridge, Polity Press.

Whatmore, S. (2002) 'From farming to agribusiness: global agri-food networks', in Johnston, R. J., Taylor, P. J. and Watts, M. (eds) *Geographies of Global Change*, Oxford, Blackwell, 57–67.

Whatmore, S. (2009a) 'Agribusiness', in Gregory, D., Johnston, R., Pratt, G., Watts, M. J. and Whatmore, S. (eds) (2009) *The Dictionary of Human Geography* (5th edn), Wiley-Blackwell, Oxford.

Whatmore, S. (2009b) 'Agricultural geography', in Gregory, D., Johnston, R., Pratt, G., Watts, M. J. and Whatmore, S. (eds) (2009) *The Dictionary of Human Geography* (5th edn), Wiley-Blackwell, Oxford.

Whatmore, S. and Thorne, J. (1997) 'Nourishing networks: alternative geographies of food', in Goodman, D. and Watts, M. (eds) *Globalizing Food*, London, Routledge, 211–224.

Whitehead, M. (2013) 'Sustainability', in Cloke, P., Crang, P. and Goodwin, M. (2013) *Introducing Human Geography* (3rd edn), London, Routledge, 448–460.

Wieringa, S. (ed.) (1995) *Subversive Women: Women's Movements in Africa, Asia, Latin America and the Caribbean*, London, Zed Books.

Williams, G., Meth, P. and Willis, K. (2009) *The Geography of Developing Areas*, London, Routledge.

Williamson, J. and Milner, C. (1991) *The World Economy: A Textbook in International Economics*, New York, London, Harvester Wheatsheaf.

Willis, K. (2013) *Theories of Development*, London, Routledge.

Wilson, G. A. (2001) 'From productivism to post-productivism . . . and back again? Exploring the (un) changed natural and mental landscapes of European agriculture', *Transactions of the Institute of British Geographers*, 26(1), 77–102.

Wilson, G. A. and Rigg, J. (2003) '"Post-productivist" agricultural regimes and the South: discordant concepts?', *Progress in Human Geography*, 27(6), 681–707.

Wolf, M. (2004) *Why Globalization Works*, Yale University Press.

Woods, M. (2007) 'Engaging the global countryside: globalization, hybridity and the reconstitution of rural place', *Progress in Human Geography*, 31(4), 485–507.

World Bank (2012) 'Global Finance Report' http://data.worldbank.org/sites/default/files/gdf_2012.pdf, accessed 12 August 2013.

World Bank (2013) 'Annual Statistics' http://wdi.worldbank.org/table/6.8, accessed 23 August 2013.

World Bank (various years) *Development Indicators*, Washington DC, World Bank.

World Bank (various years) *World Development Report*, Washington DC, World Bank.

Worth, O. and Buckley, K. (2009) 'The World Social Forum: postmodern prince or court jester?', *Third World Quarterly*, 30(4), 649–661.

Yeung, H. W. C. (2002) 'The limits to globalization theory: a geographic perspective on global economic change', *Economic Geography*, 78(3), 285–306.

Yeung, H. W. C. (2009) 'Economic globalization', in Kitchen, R. and Thrift, N. (eds) *International Encyclopedia of Human Geography*, London, Elsevier.

Young, M., Zuelow, E. and Sturm, A. (2007) *Nationalism in a Global Era: Why Nations Persist*, London, Taylor and Francis.

Zedillo, E. (2008) *Global Warming – Looking Beyond Kyoto*, Yale University.

Index

References to figures, maps, plates and cartoons are in *italics*; references to tables and boxes are in **bold**.